Algal Biotechnology

CABI Biotechnology Series

Biotechnology, in particular the use of transgenic organisms, has a wide range of applications including agriculture, forestry, food and health. There is evidence that it could make a major impact in producing plants and animals that are able to resist stresses and diseases, thereby increasing food security. There is also potential to produce pharmaceuticals in plants through biotechnology, and provide foods that are nutritionally enhanced. Genetically modified organisms can also be used in cleaning up pollution and contamination. However, the application of biotechnology has raised concerns about biosafety, and it is vital to ensure that genetically modified organisms do not pose new risks to the environment or health. To understand the full potential of biotechnology and the issues that relate to it, scientists need access to information that not only provides an overview of and background to the field, but also keeps them up to date with the latest research findings.

This series, which extends the scope of CABI's successful "Biotechnology in Agriculture" series, addresses all topics relating to biotechnology including transgenic organisms, molecular analysis techniques, molecular pharming, in vitro culture, public opinion, economics, development and biosafety. Aimed at researchers, upper-level students and policy makers, titles in the series provide international coverage of topics related to biotechnology, including both a synthesis of facts and discussions of future research perspectives and possible solutions.

Titles Available

1. **Animal Nutrition with Transgenic Plants**
   Edited by G. Flachowsky
2. **Plant-derived Pharmaceuticals: Principles and Applications for Developing Countries**
   Edited by K.L. Hefferon
3. **Transgenic Insects: Techniques and Applications**
   Edited by M.Q. Benedict
4. **Bt Resistance: Characterization and Strategies for GM Crops Producing Bacillus thuringiensis Toxins**
   Edited by Mario Soberón, Yulin Gao and Alejandra Bravo
5. **Plant Gene Silencing: Mechanisms and Applications**
   Edited by Tamas Dalmay
6. **Ethical Tensions from New Technology: The Case of Agricultural Biotechnology**
   Edited by Harvey James
7. **GM Food Systems and their Economic Impact**
   Tatjana Brankov and Koviljko Lovre
8. **Endophyte Biotechnology: Potential for Agriculture and Pharmacology**
   Edited by Alexander Schouten
9. **Forest Genomics and Biotechnology**
   Richard Meilan and Matias Kirst
10. **Transgenic Insects: Techniques and Applications, 2nd Edition**
    Edited by M.Q. Benedict and Maxwell J. Scott
11. **Plant Omics: Advances in Big Data Biology**
    Edited by Hajime Ohyanagi, Eiji Yamamoto, Ai Kitazumi and Kentaro Yano
12. **Next-generation Sequencing and Agriculture**
    Edited by Phillip Bayer and Dave Edwards
13. **Aquaculture and Fisheries Biotechnology: Genetic Approaches**
    Rex A Dunham
14. **Algal Biotechnology**
    Edited by Qiang Wang

Algal Biotechnology

Edited by

Qiang Wang

CABI

CABI is a trading name of CAB International

CABI
Nosworthy Way
Wallingford
Oxfordshire OX10 8DE
UK

CABI
200 Portland Street
Boston
MA 02114
USA

Tel: +44 (0)1491 832111
E-mail: info@cabi.org
Website: www.cabi.org

Tel: +1 (617)682-9015
E-mail: cabi-nao@cabi.org

© CAB International 2023. All rights reserved. No part of this publication may be reproduced in any form or by any means, electronically, mechanically, by photocopying, recording or otherwise, without the prior permission of the copyright owners.

The views expressed in this publication are those of the author(s) and do not necessarily represent those of, and should not be attributed to, CAB International (CABI). Any images, figures and tables not otherwise attributed are the author(s)' own. References to internet websites (URLs) were accurate at the time of writing.

CAB International and, where different, the copyright owner shall not be liable for technical or other errors or omissions contained herein. The information is supplied without obligation and on the understanding that any person who acts upon it, or otherwise changes their position in reliance thereon, does so entirely at their own risk. Information supplied is neither intended nor implied to be a substitute for professional advice. The reader/user accepts all risks and responsibility for losses, damages, costs and other consequences resulting directly or indirectly from using this information.

CABI's Terms and Conditions, including its full disclaimer, may be found at https://www.cabi.org/terms-and-conditions/.

A catalogue record for this book is available from the British Library, London, UK.

ISBN-13: 9781800621930 (hardback)
9781800621947 (ePDF)
9781800621954 (ePub)

DOI: 10.1079/9781800621954.0000

Commissioning Editor: David Hemming
Editorial Assistant: Emma McCann
Production Editor: James Bishop

Typeset by Straive, Pondicherry, India
Printed and bound in the UK by CPI Group (UK) Ltd, Croydon, CR0 4YY.

Contents

Contributors vii

Part I: Synthetic Biology and Toolkits

1. **Engineering Microalgae: Transition from Empirical Design to Programmable Cells** 1
   Yandu Lu, Xu Zhang, Hanzhi Lin and Anastasios Melis

2. **Application and Mechanism Analysis of Photosynthetic Microbial Coculture Systems for Bioproduction** 32
   Jin Wang, Xinyu Song, Lei Chen and Weiwen Zhang

3. **Genome Editing in Diatoms: Current Progress and Challenges** 54
   Xiahui Hao, Fan Hu, Yufang Pan, Wenxiu Yin and Hanhua Hu

4. ***Euglena*-based Synthetic Biology and Cell Factory** 61
   Chao Li, Zhenfan Chen, Zixi Chen, Anping Lei, Qiong Liu and Jiangxin Wang

5. **Visualizing Native Supramolecular Architectures of Photosynthetic Membranes in Cyanobacteria and Red Algae Using Atomic Force Microscopy** 77
   Long-Sheng Zhao, Yu-Zhong Zhang and Lu-Ning Liu

Part II: Algae Culture and Natural Products

6. **Heterotrophic High-cell-density Cultivation for Effective Production of Microalgae Biomass and Metabolites of Commercial Interest** 90
   Hu Jin and Feng Ge

7. **Metabolites of Microalgae** 102
   Gao Chen

8. **Green Extraction of Bioactive Compounds from Microalgae by Ionic Liquids** 115
   Xiangxiang Zhang, Yali Zhu and Quanyu Zhao

9. **Metabolomics Analysis of Algae: Current Status and Future Perspectives** 130
   Yanhua Li and Feng Ge

10. **The Need for Taxonomic Revision of *Phormidium treleasei* Gomont and Its Potential Use in Biotechnology** 140
    Kaitlin Simmons and Qingfang He

Part III: Algae-based Bioremediation

11. **Algae-based Aquaculture Wastewater Treatment and Resource Utilization** 154
    Pengfei Cheng, Chun Wang, Yahui Bo, Jiameng Guo, Xiaotong Song, Shengzhou Shan, Chengxu Zhou, Xiaojun Yan and Roger Ruan

12. **The Role of Microalgae in the Mitigation of the Impact of Chemical Pollution in Freshwater Habitat** 171
    Adamu Yunusa Ugya and Qiang Wang

13. **Microalgae-based Bioremediation of Heavy Metals and Emerging Contaminants** 183
    Vishal Rajput, Vinod Kumar, Krishna Kumar Jaiswal, Sanjay Gupta, Anna I. Kurbatova and Mikhail S. Vlaskin

14. **Construction of Microalgal Chloroplast Organelle Factory for the Carbon-neutralized Future** 201
    Zhu Zhen, Tian Jing and Cao Xupeng

Index 209

Contributors

Yahui Bo, College of Food and Pharmaceutical Sciences, Ningbo University, Ningbo, Zhejiang, China

Cao Xupeng, State Key Laboratory of Catalysis of China, Dalian Institute of Chemical Physics, Chinese Academy of Sciences, Dalian National Laboratory for Clean Energy, Dalian, China

Gao Chen, Shandong Academy of Agricultural Sciences, Jinan, Shandong, China

Lei Chen, Laboratory of Synthetic Microbiology, School of Chemical Engineering & Technology; and Key Laboratory of Systems Bioengineering and Frontier Science Center of Synthetic Biology, Ministry of Education of China, Tianjin University, Tianjin, China

Zhenfan Chen, Shenzhen Key Laboratory of Marine Bioresource and Eco-Environmental Science, Shenzhen Engineering Laboratory for Marine Algal Biotechnology, Guangdong Provincial Key Laboratory for Plant Epigenetics, College of Life Sciences and Oceanography, Shenzhen University, Shenzhen, China

Zixi Chen, Shenzhen Key Laboratory of Marine Bioresource and Eco-Environmental Science, Shenzhen Engineering Laboratory for Marine Algal Biotechnology, Guangdong Provincial Key Laboratory for Plant Epigenetics, College of Life Sciences and Oceanography, Shenzhen University, Shenzhen, China

Pengfei Cheng, College of Food and Pharmaceutical Sciences, Ningbo University, Ningbo, Zhejiang, China; Center for Biorefining and Department of Bioproducts and Biosystems Engineering, University of Minnesota-Twin Cities, Saint Paul, Minnesota, USA

Feng Ge, State Key Laboratory of Freshwater Ecology and Biotechnology, Institute of Hydrobiology, Chinese Academy of Sciences; and Key Laboratory of Algal Biology, Institute of Hydrobiology, Chinese Academy of Sciences, Wuhan, China

Jiameng Guo, College of Food and Pharmaceutical Sciences, Ningbo University, Ningbo, Zhejiang, China

Sanjay Gupta, Department of Biosciences, Swami Rama Himalayan University, Doiwala, Uttarakhand, India

Xiahui Hao, Institute of Hydrobiology, Chinese Academy of Sciences, Wuhan, China

Qingfang He, Department of Biology, University of Arkansas at Little Rock, Little Rock, Arkansas, USA

Fan Hu, School of Foreign Languages, China University of Geosciences, Wuhan, China

Hanhua Hu, Institute of Hydrobiology, Chinese Academy of Sciences, Wuhan, China

Krishna Kumar Jaiswal, Department of Green Energy Technology, Pondicherry University, Pondicherry, India

Hu Jin, State Key Laboratory of Freshwater Ecology and Biotechnology, Institute of Hydrobiology, Chinese Academy of Sciences; and Key Laboratory of Algal Biology, Institute of Hydrobiology, Chinese Academy of Sciences, Wuhan, China

Vinod Kumar, Algal Research and Bioenergy Laboratory, Department of Life Sciences, Graphic Era, Dehradun, Uttarakhand, India; Department of Environmental Safety and Product Quality Management, Institute of Environmental Engineering, Peoples' Friendship University of Russia (RUDN University), Moscow, Russia

Anna I. Kurbatova, Department of Environmental Safety and Product Quality Management, Institute of Environmental Engineering, Peoples' Friendship University of Russia (RUDN University), Moscow, Russia

Anping Lei, Shenzhen Key Laboratory of Marine Bioresource and Eco-Environmental Science, Shenzhen Engineering Laboratory for Marine Algal Biotechnology, Guangdong Provincial Key Laboratory for Plant Epigenetics, College of Life Sciences and Oceanography, Shenzhen University, Shenzhen, China

Chao Li, College of Food Engineering and Biotechnology, Hanshan Normal University, Chaozhou, China; Shenzhen Key Laboratory of Marine Bioresource and Eco-Environmental Science, Shenzhen Engineering Laboratory for Marine Algal Biotechnology, Guangdong Provincial Key Laboratory for Plant Epigenetics, College of Life Sciences and Oceanography, Shenzhen University, Shenzhen, China; Key Laboratory of Optoelectronic Devices and Systems of Ministry of Education and Guangdong Province, College of Optoelectronic Engineering, Shenzhen University, Shenzhen, China.

Yanhua Li, State Key Laboratory of Freshwater Ecology and Biotechnology; and Key Laboratory of Algal Biology, Institute of Hydrobiology, Chinese Academy of Sciences, Wuhan, China

Hanzhi Lin, Institute of Marine & Environmental Technology, Center for Environmental Science, University of Maryland, College Park, Maryland, USA

Lu-Ning Liu, Frontiers Science Center for Deep Ocean Multispheres and Earth System & College of Marine Life Sciences, Ocean University of China, Qingdao, China; Institute of Systems, Molecular and Integrative Biology, University of Liverpool, Liverpool, United Kingdom

Qiong Liu, Shenzhen Key Laboratory of Marine Bioresource and Eco-Environmental Science, Shenzhen Engineering Laboratory for Marine Algal Biotechnology, Guangdong Provincial Key Laboratory for Plant Epigenetics, College of Life Sciences and Oceanography, Shenzhen University, Shenzhen, China

Yandu Lu, State Key Laboratory of Marine Resource Utilization in the South China Sea, Hainan University, Haikou, Hainan, China; Department of Plant and Microbial Biology, University of California, Berkeley, California, USA

Anastasios Melis, Department of Plant and Microbial Biology, University of California, Berkeley, California, USA

Yufang Pan, Institute of Hydrobiology, Chinese Academy of Sciences, Wuhan, China

Vishal Rajput, Department of Biosciences, Swami Rama Himalayan University, Doiwala, Uttarakhand, India

Roger Ruan, Center for Biorefining and Department of Bioproducts and Biosystems Engineering, University of Minnesota-Twin Cities, Saint Paul, Minnesota, USA

Shengzhou Shan, College of Food and Pharmaceutical Sciences, Ningbo University, Ningbo, Zhejiang, China

Kaitlin Simmons, Department of Biology, University of Arkansas at Little Rock, Little Rock, Arkansas, USA

Xiaotong Song, College of Food and Pharmaceutical Sciences, Ningbo University, Ningbo, Zhejiang, China

Xinyu Song, Laboratory of Synthetic Microbiology, School of Chemical Engineering & Technology; and Key Laboratory of Systems Bioengineering and Frontier Science Center of Synthetic Biology, the Ministry of Education of China; and Center for Biosafety Research and Strategy, Tianjin University, Tianjin, China

Tian Jing, School of Bioengineering, Dalian Polytechnic University, Dalian, China

Adamu Yunusa Ugya, State Key Laboratory of Crop Stress Adaptation and Improvement; and School of Life Sciences, Academy for Advanced Interdisciplinary Studies, Henan University, Kaifeng, China; Department of Environmental Management, Kaduna State University, Kaduna State, Nigeria

Mikhail S. Vlaskin, Joint Institute for High Temperatures of the Russian Academy of Sciences, Moscow, Russia

Chun Wang, College of Food and Pharmaceutical Sciences, Ningbo University, Ningbo, Zhejiang, China

Jiangxin Wang, Shenzhen Key Laboratory of Marine Bioresource and Eco-Environmental Science, Shenzhen Engineering Laboratory for Marine Algal Biotechnology, Guangdong Provincial Key Laboratory for Plant Epigenetics, College of Life Sciences and Oceanography, Shenzhen University, Shenzhen, China

Jin Wang, Laboratory of Synthetic Microbiology, School of Chemical Engineering & Technology; and Key Laboratory of Systems Bioengineering and Frontier Science Center of Synthetic Biology, the Ministry of Education of China, Tianjin University, Tianjin, China

Qiang Wang, State Key Laboratory of Crop Stress Adaptation and Improvement; and School of Life Sciences; and Academy for Advanced Interdisciplinary Studies, Henan University, Kaifeng, China

Xiaojun Yan, Key Laboratory of Marine Biotechnology of Zhejiang Province, Ningbo University, Ningbo, Zhejiang, China

Wenxiu Yin, Institute of Hydrobiology, Chinese Academy of Sciences, Wuhan, China

Weiwen Zhang, Laboratory of Synthetic Microbiology, School of Chemical Engineering & Technology; and Key Laboratory of Systems Bioengineering and Frontier Science Center of Synthetic Biology, the Ministry of Education of China; and Center for Biosafety Research and Strategy, Tianjin University, Tianjin, China

Xiangxiang Zhang, School of Pharmaceutical Science, Nanjing Tech University, Nanjing, China

Xu Zhang, State Key Laboratory of Marine Resource Utilization in the South China Sea, College of Oceanology, Hainan University, Haikou, Hainan, China

Yu-Zhong Zhang, Frontiers Science Center for Deep Ocean Multispheres and Earth System & College of Marine Life Sciences, Ocean University of China, Qingdao, China; and Laboratory for Marine Biology and Biotechnology, Pilot National Laboratory for Marine Science and Technology, Qingdao, China; and Marine Biotechnology Research Center, State Key Laboratory of Microbial Technology, Shandong University, Qingdao, China

Long-Sheng Zhao, State Key Laboratory of Microbial Technology, Shandong University, Qingdao, China; Frontiers Science Center for Deep Ocean Multispheres and Earth System & College of Marine Life Sciences, Ocean University of China, Qingdao, China; Laboratory for Marine Biology and Biotechnology, Pilot National Laboratory for Marine Science and Technology, Qingdao, China

Quanyu Zhao, School of Pharmaceutical Science, Nanjing Tech University, Nanjing, China

Chengxu Zhou, College of Food and Pharmaceutical Sciences, Ningbo University, Ningbo, Zhejiang, China.

Yali Zhu, School of Pharmaceutical Science, Nanjing Tech University, Nanjing, China

Zhu Zhen, State Key Laboratory of Catalysis of China, Dalian Institute of Chemical Physics, Chinese Academy of Sciences, Dalian, National Laboratory for Clean Energy, Dalian, China; and School of Bioengineering, Dalian Polytechnic University, Dalian, China.

1 Engineering Microalgae: Transition from Empirical Design to Programmable Cells

Yandu Lu[1,2,4,5]*, Xu Zhang[1], Hanzhi Lin[3] and Anastasios Melis[2]*

[1]*State Key Laboratory of Marine Resource Utilization in South China Sea, School of Marine Biology and Fisheries, Hainan University, Haikou, China;* [2]*Department of Plant and Microbial Biology, University of California, Berkeley, California, USA;* [3]*Institute of Marine & Environmental Technology, Center for Environmental Science, University of Maryland, College Park, Maryland, USA;* [4]*Hainan Engineering & Research Center of Marine Bioactives and Bioproducts, Hainan University, Haikou, China;* [5]*Haikou Innovation Center for Research & Utilization of Marine Algal Bioresources, Hainan University, Haikou, China*

Abstract

Domesticated microalgae hold great promise for the sustainable provision of various bioresources for human domestic and industrial consumption. Efforts to exploit their potential are far from being fully realized due to limitations in the know-how of microalgal engineering. The associated technologies are not as well developed as those for heterotrophic microbes, cyanobacteria and plants. However, recent studies on microalgal metabolic engineering, genome editing and synthetic biology have helped immensely to enhance transformation efficiencies and are bringing new insights into this field. Therefore, this article summarizes recent developments in microalgal biotechnology and examines the prospect of generating specialty and commodity products through the processes of metabolic engineering and synthetic biology. After a brief examination of empirical engineering methods and vector design, the article focuses on quantitative transformation cassette design, elaborates on target-editing methods and emerging digital design of algal cellular metabolism to arrive at high yields of valuable products. These advances have enabled a transition of manners in microalgal engineering from single-gene and enzyme-based metabolic engineering to systems-level precision engineering, from cells created with genetically modified (GM) tags to those without GM tags, and ultimately from proof of concept to tangible industrial application. Finally, future trends are proposed in microalgal engineering, aiming to establish individualized transformation systems in newly identified species for strain-specific specialty and commodity products, while developing sophisticated universal toolkits in model algal species.

Keywords: genome editing tools; metabolic engineering; photosynthetic cell factories; synthetic biology; microalgae

---

*Address for correspondence: ydlu@hainanu.edu.cn; melis@berkeley.edu
Note: This chapter is based on an article that comes from a Creative Commons licence that allows reproduction.

© CAB International 2023. *Algal Biotechnology* (Qiang Wang)
DOI: 10.1079/9781800621954.0001

1. Introduction

Microalgae, with an estimate of 72,500 species (Guiry, 2012), are of great ecological importance, as they contribute almost half of the global organic carbon fixation (Melillo et al., 1993). They provide a variety of natural products to support ecosystems (e.g. coral reef ecosystem (Lu et al., 2020)) by photosynthesis with efficiencies approximately three times greater than those achieved by land plants (Melis, 2009). The microalgal life cycle of fast cell division and continuous biomass accumulation highlights the advantage gained from their sustainable cultivation (Scaife et al., 2015). Accordingly, various attempts have been made to exploit microalgae for production of commodity and specialty chemicals to meet human domestic and industrial demands. To identify and improve species that naturally produce valuable compounds, mutation breeding and genetic modifications have been utilized for a long period, but are largely via empirical approaches (Nielsen et al., 2014).

Admittedly, extensive reviews cover the bioengineering of cyanobacteria (Sengupta et al., 2018), the genomic context underpinning microalgal diversity (Brodie et al., 2017), and the ecology, evolution and applications of microalgae (Wijffels et al., 2013; Renuka et al., 2018). Therefore, this article primarily focuses on eukaryotic microalgae with a brief comparable study between microalgae and cyanobacteria. Two primary problems are preventing microalgal biotechnology from further development and industrialization. One is that only a few model microalgal species can be routinely transformed and ready for downstream industrialization. Exploitation of all but a handful of algal species is thus severely impeded by a limitation in molecular tools for competent engineering. The other problem resides in the extremely low transformation efficiencies (even compared to those achieved with plants) that are prevailing when working with microalgae, clearly demonstrating a need for novel and improved transformation techniques.

Recent studies on microalgal metabolic engineering, genome editing and synthetic biology are facilitating the improvement of transformation efficiencies and are bringing new insights into cellular metabolic processes (Jagadevan et al., 2018). There exists a demand to jointly consider these developments through a synthetic perspective of microalgal breeding. Therefore, after briefly reviewing empirical methods for vector design and engineering, specifically, this article focuses on (i) quantitative transformation cassette designs; (ii) elaborate target-editing methods; (iii) digital design of algal cellular metabolism for high yield of valuable products; and (iv) problems and countermeasures of industrial application. Drawing lessons from the broader history of the field and emerging advances, it is anticipated that a new era of rational design of digital microalgal cells is coming. This era would substantially benefit human society on food supply, energy consumption and environmental sustainability.

2. Relevance of Microalgae as Photosynthetic Cell Factories

Broadly speaking, strategies for engineering microalgal metabolic pathways can be divided into two categories: those based on endogenous pathways, and those derived from non-native pathways by involving heterologous genes.

2.1 Natural compounds

Long-term adaptation to a wide range of ecotypes has engendered diverse phenotypes and genotypes of microalgae, as well as helped algae evolve a robust acclimation plasticity (Brodie et al., 2017), enabling them to adapt to various niches and produce a vast array of compounds. Microalgae are highly efficient at sequestrating CO_2, accumulating biomass and many secondary metabolites including pharmaceutically and nutritionally active compounds for humans (or precursors for such compounds) (Table 1.1). It has therefore been suggested that genetically-tailored microalgae could serve as 'platform strains' to convert CO_2 into diverse useful compounds from metabolic intermediates (Tran *et al.*, 2013). However, current microalgal product yields are generally too low to meet the cost of commercial exploitation. Thus gene engineering approaches have been adopted in an attempt to overcome this drawback by modulating the activity of endogenous rate-limiting enzymes.

2.2 Heterologous compounds

Beyond producing endogenous compounds, microalgae could be recruited as cell factories to produce many different non-native compounds, ranging from small organic molecules to large recombinant proteins. Examples of the small molecules that could be produced using microalgae are provided in the report entitled *Top Value Added Chemicals from Biomass* (Werpy and Petersen, 2004), which identified 12 platform chemicals, i.e. small organic compounds that can be produced from sugars by microorganisms and subsequently converted into industrially relevant molecules.

The attractiveness of using microalgae as hosts to produce larger biomacromolecules (e.g. recombinant proteins) can be understood by comparing the properties with that of alternative biological production systems, such as bacteria, yeasts, mammals, insects or plants (Table 1.2). These desirable qualities, together with the rising demand for recombinant proteins, have driven the pursuit to introduce druggability into transgenic microalgae (Tran *et al.*, 2013). Consequently, increasing numbers of recombinant proteins, including antibodies, immunotoxins, vaccine antigens, and mammary-associated serum amyloid, have been produced from metabolic intermediates in microalgae (Table 1.1). Yet, for all that, it still remains challenging to the sustainable production of foreign chemicals by introducing a *de novo* engineered pathway, necessitating the development of new approaches and advanced engineering strategies.

As putative cell factories, cyanobacteria offer distinct advantages but also, usually, have some drawbacks, when compared with microalgae. Among the advantages are ease of transformation and the absence of epigenetic regulation or suppressor mutations to counter the effects of transformation. As a result, cyanobacteria have been successfully engineered to make a variety of heterologous fuels and useful chemicals (Lindberg *et al.*, 2010; Ungerer *et al.*, 2012; Formighieri and Melis, 2017). A breakthrough was achieved with the design of oligonucleotide fusion constructs (target genes are fused to the highly expressed endogenous (Formighieri and Melis, 2015) or exogenous genes (Betterle and Melis, 2018)), as protein overexpression vectors that have been used in cyanobacteria to produce plant and human genes that are otherwise difficult to express (Formighieri and Melis, 2015; Chaves *et al.*, 2017; Betterle and Melis, 2018). The 'fusion constructs' could facilitate the heterologous proteins being accumulated as dominant cyanobacterial proteins, accounting for 20–25% of the total cell proteins (Formighieri and Melis, 2015). However, pertinent in this respect is that genes from eukaryotic organisms, e.g. plants, animals, yeasts and humans, are consistently expressed at low levels, in both microalgae and cyanobacteria, in spite of the use of strong promoters designed to confer 'overexpression' of transgenes (Formighieri and Melis, 2015). Compared to microalgae, another drawback of the cyanobacterial system is that despite intensive industrial cultivations of cyanobacteria *Spirulina* (*Arthrospira*) species, productivity of most cyanobacteria is lower under mass culture and bright sunlight conditions (Melis, 2009). There are reports of very-fast-growth unicellular cyanobacteria (Bernstein *et al.*, 2016; Ungerer *et al.*, 2018); however, they have not yet been tested in industrial scales.

3. Engineering Vectors: Transition from Empirical to Quantitative Designs

To deliver transgenes into microalgae, several methods have been developed, such as

Table 1.1. Targeted compounds produced in microalgal cell factories.

Target products	Human benefits	Species and contents
Natural compounds		
Long-chained omega-3 PUFAs	Cardiovascular health	EPA: *Attheya septentrionalis* (4.58% DW), *Nannochloropsis oceanica* (23.52% TFA), *Nanofrustulum shiloi* (1.3% DW), *Phaeodactylum tricornutum* (3.4% DW), *Thalassiosira hispida* (4.63% DW); DHA: *Nitzschia laevis* (0.59% DW), *T. hispida* (0.61% DW), *A. septentrionalis* (0.60% DW); GLA: *Chlorella pyrenoidosa* (6.43% TFA) (Melillo et al., 1993; Guiry, 2012; Lu et al., 2020)
β-carotene	Antioxidants, improve cognitive function and skin and lung health, reduce macular degeneration and cancer risks	*Dunaliella salina* (10% DW), *Tetraselmis suecica* (8.6 mg/g oil), *Rhodomonas salina* (5.7 mg/g oil), *Thalassiosira pseudonana* (4.3 mg/g oil), *Nannochloropsis gaditana* (3.5 mg/g oil) (Melis, 2009; Scaife et al., 2015)
Astaxanthin	Anti-inflammatory effects, antioxidants, anti-cancer, and cardiovascular health	*Haematococcus pluvialis* (5.79% DW), *Chromochloris zofingiensis* (0.22% DW) (Nielsen et al., 2014; Sengupta et al., 2018)
Fucoxanthin	Anti-obesity and anti-oxidant	*P. tricornutum* (38.3 mg/g oil), *T. pseudonana* (31.9 mg/g oil), *Isochrysis* (19.2 mg/g oil), *Pavlova lutheri* (14.8 mg/g oil), *Odontella aurita* (0.8% DW), *Cyclotella cryptic* (1.29% DW) (Melis, 2009; Brodie et al., 2017)
Lutein	Prevents cataract, age-related macular degeneration, and cardiovascular diseases, and anti-oxidant and anti-cancer	*Scenedesmus obliquus* (4 mg/g), *R. salina* (8.7 mg/g oil), *Tetraselmis suecica* (5.6 mg/g oil) (Melis, 2009; Wijffels et al., 2013; Renuka et al., 2018)
Violaxanthin	Prevents night blindness and anti-oxidant property	*Nannochloropsis gaditana* (14.3 mg/g), *N. oculata* (11.5 mg/g), *T. suecica* (7.8 mg/g) (oil:biomass) (Melis, 2009)
Zeaxanthin	Prevents night blindness, anti-oxidant property, and prevents liver fibrosis	*N. gaditana* (3.4 mg/g), *Porphyridium cruentum* (3.2 mg/g), *T. suecica* (2.0 mg/g) (oil:biomass) (Melis, 2009)
Phytosterols	Cholesterol reduction, anti-inflammatory activity, and even anti-cancer properties	*Isochrysis* (14.9 mg/g oil), *N. gaditana* (17.0 mg/g oil), *N. oculata* (6.1 mg/g oil), *P. lutheri* (97 mg/g oil), *P. tricornutum* (16.5 mg/g oil), *P. cruentum* (26.5 mg/g oil), *R. salina* (26 mg/g oil), *T. suecica* (10.9 mg/g oil), *T. pseudonana* (34 mg/g oil) (Melis, 2009; Jagadevan et al., 2018)
Essential amino acids	Reduce symptoms of depression, boost mood and improve sleep	*C. pyrenoidosa* (21.76% DW) (Lu et al., 2020)
Triacylglycerol (TAG)	Biofuels	*Chlamydomonas reinhardtii* (20.5% DW), *P. tricornutum* (57.8% DW), *N. gaditana* (40–55% lipids of DW) (Werpy and Petersen, 2004; Lindberg et al., 2010; Tran et al., 2013)
Foreign compounds		
Immuotoxins	Inhibit tumor growth	*C. reinhardtii* (0.2–0.4% TSPs) (Ungerer et al., 2012)
Vaccine antigens	Block transmission of infectious diseases such as malaria	*C. reinhardtii* (0.09% TSPs) (Formighieri and Melis, 2015, 2017)
Interferon	Cell defense-signaling proteins in response to viral infections	*Synechocystis* PCC 6803 [18]

DHA, docosahexaenoic acid; DW, dry cell weight; EPA, eicosapentaenoic acid; GLA, linolenic acid; TSPs, total soluble proteins; TFA, total fatty acids

Table 1.2. Characteristics and advantages of representative expression systems.

Expression systems	Cost	Growth rate	Post-translational modification (PTMs)	Cultivation systems
Bacteria	Moderate	Rapid	Occur in a relatively low number of bacterial proteins in comparison with eukaryotic proteins (Chaves et al., 2017) Incorrect folding and assembly (Ungerer et al., 2018)	Heterotrophic cultivation Enclosed bioreactors
Cyano-bacteria	Low	Rapid	No (Bernstein et al., 2016)	Phototrophic or heterotrophic cultivation Enclosed bioreactors or open race ponds
Yeast	Moderate	Rapid	Yes Unsuitable glycosylation (Qin et al., 2012)	Heterotrophic cultivation Enclosed bioreactors
Mammals or insects	High cost	Slow	Yes Human-like PTMs with some discrepancies (Ungerer et al., 2018)	Heterotrophic cultivation Complex nutrient requirements Difficult to scale up Easily contaminated by animal pathogens (Matsunaga and Takeyama, 1995)
Plants	Less expensive	Slow	Yes (Matsunaga et al., 1991) The glycosylation patterns often differ from those in mammals (Demirer et al., 2019)	Phototrophic cultivation Easy to release the genetically modified materials (Kwak et al., 2019)
Microalgae	Low	Rapid	Yes Species-specific PTMs The glycosylation patterns are more similar to humans than E. coli and yeasts (Qu et al., 2012; Im et al., 2015)	Phototrophic or heterotrophic cultivation Enclosed bioreactors or open race ponds

the glass-bead method, *Agrobacterium*-mediated transformation, electroporation, and particle bombardment. The comparison of these transformation methods and characteristics of the transformed microalgae have been listed in Table 1.3. They have been intensively reviewed elsewhere (Qin et al., 2012) and thus are not dicussed in detail in this review. It should be noted that despite the pros and cons, electroporation-based transformation outweighs other methods in terms of the wide applicable range in microalgal species. Another promising and potentially species-independent method is nanoparticle-mediated DNA delivery, yet to be applied in microalgae. It was first developed by using bacterial magnetic particles (50–100 nm in diameter) in the 1990s (Matsunaga and Takeyama, 1995), and restricted due to difficulties in particle preparation (Matsunaga et al., 1991), but recently refined in plants (Demirer et al., 2019; Kwak et al., 2019). Moreover, despite still with many challenges, particularly for microalgae-holding tough cell wall, when combined with a droplet microfluidics platform, electroporation (Qu et al., 2012; Im et al., 2015) and nanoparticles (Bae et al., 2015) are promising methods to convert current 'population transformation' of microalgae into high-throughput 'single-cell engineering' (Kim et al., 2018). One of the necessary jobs remaining for practical application is to decrease the cost and simplify the manipulation of microfluidics.

Notwithstanding the progress achieved in developing tools for delivering exogenous DNA, routine transformation is available for

Table 1.3. DNA delivery methods of modern breeding strategies for microalgae.

Methods	Illustration	Pros	Cons	Examples
The glass-bead method		More convenient and much less costly	Less efficient	*Chlamydomonas reinhardtii* (Bae et al., 2015); *Dunaliella salina* (Kim et al., 2018); *Platymonas subcordiformis* (Bae et al., 2015; Kim et al., 2018; Bock, 2015)
Agrobacterium-mediated transformation		No need to generate protoplasts. Holds great promise to transform large DNA segments (> 100 kb) into cells	Low and unpredictable variation of efficiencies	Freshwater *C. reinhardtii* (Yadav et al., 2012); *Haematococcus pluvialis* (Doron et al., 2016); marine *Symbiodinium* spp. (Jiang et al., 2014); *Isochrysis* spp. (Gimpel et al., 2015); *Schizochytrium* sp. (Wang et al., 2016); *Nannochloropsis* sp. (Schroda et al., 2000); *Dunaliella bardawil* (Davies et al., 1992; Radakovits et al., 2012)
Biolistic transformation		Eliminates the need for production of plant protoplasts or infection by *Agrobacterium*	A rigid requirement to the diameters of the target cells. A tendency to generate rearranged and broken transgene copies	The green microalga *H. pluvialis* (Wei et al., 2017); the diatom *P. tricornutum* (Vieler et al., 2012); *Thalassiosira pseudonana* (Belliveau et al., 2018); *Cyclotella cryptic* (Zhang and Hu, 2014); *Navicula saprophila* (Zhang and Hu, 2014); *Chaetoceros* sp. (Cui et al., 2018); *Cylindrotheca fusiformis* (Mini et al., 2018)

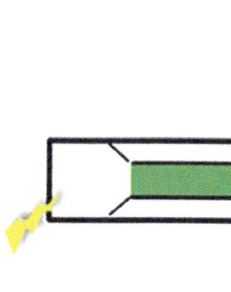

Electroporation	Simple, easily applicable, and very efficient	Needs a single-cell tissue homogenate	*C. reinhardtii* (Dunahay, 1993); *Scenedesmus obliquus* (Kindle, 1990); *Monoraphidium neglectum* (Boynton et al., 1988); *Nannochloropsis* (Mehrotra et al., 2017); *P. tricornutum* (Iddo et al., 2018); *Chlorella pyrenoidosa* (DeHoff and Soriaga, 2016); *Dunaliella* (Xue et al., 2015)
Nanoparticle-mediated DNA delivery	Species-independent. Protects polynucleotides against nuclease degradation	Has not yet been applied in microalgae	Both model and crop plants, including both dicots and monocots (Rosenwasser et al., 2014; Gimpel et al., 2015)

a restricted number of microalgae only (Bock, 2015). Moreover, even in established species, obstacles pertaining to gene delivery efficiency, transgene stability or heritability are preventing the transformation systems from practice. Therefore, it necessitates a methodological transition from tools used to manipulate metabolism, relying on experience-dependent strategies and constrained in particular species or strains (Yadav et al., 2012) to methods of quantitative and mathematic design.

3.1 Empirical designs

Critical to the creation of transgenic microalgae is the ability to transform cells with specific DNA sequences using vector constructs. To ensure the proper transcription, marker genes and/or genes of interest are typically expressed in individual cassette harboring a 5' promoter and a 3' terminator. Conventional protocols for selecting marker genes for microalgal transformation have been extensively reviewed (Doron et al., 2016). Promoters are critical components that work in concert with other genetic elements (enhancers, silencers, transcription factors and boundary elements/insulators) to direct the transcription of marker genes and other sequences. Promoter availability and selection thus profoundly influence the success of constructing robust genetic transformation systems. However, until now, few promoters are available for algal vectors, partialy due to the shortcomings of the respective genetic toolkits (particularly the limited number of known regulatory elements).

Most attempts to engineer microalgae have been conducted empirically by using a handful of repurposed tools. To achieve viable transformants, strong constitutive promoters from phylogenetically closely related algal species, viruses, or occasionally higher plants have been harnessed for transgene expression in microalgae. The viral promoters of CaMV35S (the cauliflower mosaic virus) and SV40 (Simian virus 40; an oncogenic simian polyomavirus) have been utilized for transient expression in some algal species, but heterologous promoter regions are usually inadequately recognized and regulated in microalgae (Jiang et al., 2014) (Table 1.4). In this respect, there are publications describing the successful transformation of microalgae with results that, unfortunately, could not be reproduced in other laboratories (Gimpel et al., 2015). For example, while the *Arabidopsis thaliana* U6 gene promoter was able to drive gene transcription in *Chlamydomonas reinhardtii* (Jiang et al., 2014), vectors featured in either *Arabidopsis* or *Chlamydomonas* promoters were not successfully recognized in *Nannochloropsis* sp. (Wang et al., 2016).

As alternatives to exogenous promoters, vectors can be constructed using an orthologous promoter related to one that has been previously characterized. For instance, heat shock protein (Schroda et al., 2000) and tubulin (Davies et al., 1992) are highly expressed proteins in microalgae such as *Chlamydomonas* (Table 1.4). Their promoters are therefore regularly used to drive constitutive nuclear expression of numerous genes in *Chlamydomonas*, and orthologous promoters have been used successfully in a number of other microalgae (Radakovits et al., 2012). However, microalgae tend to have numerous orthologs with quantitively unknown transcriptional and protein expression levels. In many cases, it is difficult to select suitable driving promoters for vector design. For example, *Nannochloropsis oceanica* strain IMET1 harbors nine orthologs for heat-shock proteins and eight orthologs for tubulin. Each of these promoter regions was used separately to construct vectors and drive gene expression (Wei et al., 2017). However, success was achieved with only 50% of the assembled vectors even though the same transformation protocol was used in all cases. Similarly, in *N. oceanica* strain 1779, the constructs harboring a *C. reinhardtii* α-tubulin promoter or a native lipid droplet surface protein (LDSP) promoter were employed to drive the expression of the *Streptomyces hygroscopicus aph7* gene, conferring resistance to Hygromycin B. The latter achieves a more than tenfold increase in

Table 1.4. Examples of genetic engineering in a variety of microalgae.

Species	Promoters	Target genes	Highlights	References
Nuclear transformation				
Pyrrophyta				
Symbiodinium spp.	CaMV35S or NOS promoter	GFP	The first report on stable nuclear transformation of dinoflagellate, *Symbiodinium* by *Agrobacterium*-mediated transformation	Jiang et al. (2014)
Chlorophyta				
Chlamydomonas reinhardtii	Endogenous promoter of *NIT*	*NIT*	The first report on stable nuclear transformation in microalgae; glass-bead-mediate transformation was used to recover a *NIT*-deficient strain with the wild type gene *NIT1*.	Bae et al. (2015)
C. reinhardtii	CaMV 35S promoter	β-glucuronidase (UIDA), green fluorescent protein (GFP) and hygromycin phosphotransferase (HPT) genes	The first *Agrobacterium*-mediated nuclear transformation in microalgae	Yadav et al. (2012)
C. reinhardtii	CaMV35S promoter, Arabidopsis U6 gene promoter, or endogenous promoter of the gene encoding Photosystem I reaction center subunit II (PSAD)	Cas9, guide RNA, mutant GFP (mGFP), or *Gaussia* luciferase (Gluc) genes	First attempt to express Cas9 and single guide RNA genes in microalgae	Salama et al. (2019)
C. reinhardtii	Endogenous HSP70A, RBCS2, or fused HSP70A–RBCS2 promoter	the eubacterial resistance gene AADA	Systematically studied the driving efficiency of endogenous algal promoters	Vila et al. (2012)
C. reinhardtii	β-TUB promoter	arylsulfatase gene	First evidence to show the driving efficiency of β-TUB promoter	Anderson et al. (2017)
C. reinhardtii	Endogenous PSAD promoter or a hybrid HSP70- RBCS2 promoter	APHVII" and yellow fluorescence protein (YFP) gene	Determined the relative contributions of GC content and codon usage to the efficiency of nuclear gene expression	Scranton et al. (2016)
C. reinhardtii	Endogenous PSAD or fused HSP70/RBCS2 promoter	Codon optimized Paromomycin (PARO) gene and *Renilla reniformis* luciferase (LUC) gene	Compared the efficiency, stability and insertion sites of *Agrobacterium*- versus electroporation-mediated transformation	Xin et al. (2017)

Continued

Table 1.4. Continued.

Species	Promoters	Target genes	Highlights	References
C. reinhardtii	Endogenous promoter of the gene encoding argininosuccinate lyase (ARG7)	ARG7	Employed a cell wall-less C. reinhardtii mutant devoid of ARG7 as host	Kilian et al. (2011)
C. reinhardtii	Endogenous promoter of ARG7 or the gene encoding nitrate reductase (NIT)	NIT1 and ARG7	Used silicon carbide whiskers to mediate the transformation of NIT or ARG7-deficient strain with the wild type gene NIT1 or ARG7	Eric et al. (2018)
C. reinhardtii	Endogenous RBCS2 promoter	BLE	Highlighted a potential role of introns as modulators of gene expression in microalgae	Adleragnon et al. (2017)
C. reinhardtii	Hybrid HSP70/RBCS2 promoter	BLE, mCerulean, or mCherry	Developed a set of genetic tools that enable proteins targeting to distinct subcellular locations	Neupert et al. (2009)
C. reinhardtii	The HSP70/RBCS2 promoter containing four copies of the first intron of RBCS2 between the HSP70A or RBCS2 promoter	BLE-2A-GFP and BLE-2A-XYLANASE 1 (XYN1)	Developed a nuclear multigene expression strategy using the foot-and-mouth-disease-virus 2A self-cleavage peptide (2A)	Jia et al. (2016)
C. reinhardtii	Endogenous PSAD promoter	Ferredoxin-hydrogenase (FD-HYD) gene	Showed that codon usage and mRNA folding energy in the vicinity of translation initiation affect foreign gene expression	Plucinak et al. (2015)
C. reinhardtii	Hybrid HSP70/RBCS2 promoter	Codon adapted LUC human erythropoietin (crEPO) gene	RBCS2 introns alone had a positive effect on expression; the secretion of the LUC protein into the medium was achieved by using the export sequence of the Chlamydomonas ARS2 gene	Muñoz et al. (2019)
Chlorella pyrenoidosa	Heterologous UBI promoter or endogenous promoters of heat shock protein70 (HSP70) or tubulin (TUB) gene	EGFP and NPTII	Electroporation-mediated transformation	Moog et al. (2019)
C. pyrenoidosa	Heterologous Ubiquitin (UBI) gene promoter	EGFP and the gene encoding neomycin phosphotransferase (NPT II)	Electroporation-mediated transformation	DeHoff and Soriaga (2016)
Dunaliella bardawil	CaMV 35S promoter	GFP, HPT and UIDA gene encoding β-glucuronidase (GUS)	Agrobacterium-mediated transformation	Davies et al. (1992)

Species	Promoters	Target genes	Highlights	References
Dunaliella salina	CaMV 35S or *C. reinhardtii* RBCS2 promoter	*BLE* and the gene encoding chloramphenicol acetyl transferase (*CAT*)	Electroporation-mediated transformation	Xue et al. (2015)
Haematococcus pluvialis	CaMV 35S promoter	*HPT*	*Agrobacterium*-mediated transformation	Doron et al. (2016)
H. pluvialis	Endogenous *PDS* promoter	phytoene desaturase (*PDS*) gene	Biolistic transformation	Wei et al. (2017)
Monoraphidium neglectum	Endogenous light-harvesting protein (*CAB2*) promoter	The genes encoding hygromycin B phosphotransferase (*APHVII*ʳ) and codon optimized mVenus variant (*mVenus*)	A pretreatment with lithium acetate and DTT increased electroporation efficiency	Boynton et al. (1988)
Neochloris oleoabundans	*C. reinhardtii* HSP70A-RBCS2 hybrid promoter	*GFP* and the hygromycin B-resistant (*HYG3*) gene	Electroporation-mediated transformation	Ramos-Martinez et al. (2017)
N. oleoabundans	CaMV35 promoter	genes encoding glycerol 3-phosphate acyltransferase, lysophosphatidic acid acyltransferase, and diacylglycerol acyltransferase	Manipulate lipid biosynthesis, use the self-cleaving peptide F2A to simultaneously co-express three genes	Parinov and Sundaresan (2000)
Scenedesmus obliquus	CaMV 35S promoter	*CAT* and *GFP*	Electroporation-mediated transformation	Dent et al. (2015)
Bacillariophyta				
Chaetoceros sp.	*T. pseudonana* FCP or NR gene promoters	nourseothricin resistance (*NAT*) gene and *GFP*	Biolistic transformation	Cui et al. (2018)
Cylindrotheca fusiformis	Endogenous promoter of fructanase gene	*ble* and *N. pelliculosa* frustulin (FRUE) genes	Biolistic transformation	Mini et al. (2018)
P. tricornutum	Endogenous fucoxanthin chlorophyll binding proteins (FCP) gene promoter	*Streptoaltoteichus hindustanus BLE* gene	Biolistic transformation	Vieler et al. (2012)
P. tricornutum	Endogenous *NR* or *FCPC* promoters	*CAT* and *EGFP*	Electroporation-mediated transformation; inducible promoter	Iddo et al. (2018)
P. tricornutum	Endogenous *FCPA* promoter	*BLE*, *UIDA*, *GFP*, *NPTII*, and genes encoding nourseothricin acetyl transferase (*NAT*) and streptothricin acetyl transferase (*SAT-1*)	Codon usage has a significant effect on the efficient expression of reporter genes in *P. tricornutum*.	Li et al. (2018)
P. tricornutum	Endogenous *FCPA* or *FCPB* promoters	*BLE*, *EGFP* and *UIDA*	First attempt to establish electroporation-based transformation for diatoms	Li et al. (2016)

Continued

Table 1.4. Continued.

Species	Promoters	Target genes	Highlights	References
P. tricornutum	Endogenous FCPA or FCPB promoters	a basta resistance (BAR) and the gene encoding a dual-function diacylglycerol acyltransferase (PtWS/DGAT)	revealed that PtWS/DGAT functions as either a wax ester synthase or a diacylglycerol acyltransferase, exhibiting a preference on saturated FA substrates	Bonente et al. (2011)
P. tricornutum	Endogenous FCPC promoter	Malic enzyme (ME) gene and CAT	An Omega leader sequence and 'ACC' nucleotide motif were added before PtME to enhance its translation; overexpression PtME boosted neutral lipid accumulation.	Tran et al. (2013)
P. tricornutum	Endogenous histone H4 promoter	GFP and BLE	Targeted GFP to specific organelles by using mitochondrial transit peptide, nuclear signal peptide or chloroplast transit peptide	Mackinder et al. (2017)
P. tricornutum	FCPA promoter or promoter of the gene encoding ammonium transporter, purine permease, or actin-like 2	EGFP and BLE	Characterized a series of promoters, either constitutive or inducible under nitrogen starvation	Freeman Rosenzweig et al. (2017)
Thalassiosira pseudonana	Endogenous FCP or nitrate reductase (NR) gene promoters	BLE and GFP	Biolistic transformation	Belliveau et al. (2018)
Chrysophyta				
Isochrysis spp.	CaMV35S promoter	PDS	Agrobacterium-mediated transformation	Gimpel et al. (2015)
Eustigmatophyta				
Nannochloropsis sp.	CaMV 35S promoter	UIDA	Agrobacterium-mediated transformation	Schroda et al. (2000)
Nannochloropsis sp.	Endogenous violaxanthin/chlorophyll a-binding protein (VCP2) gene promoter	BLE	Bidirectional promoter; Homologous recombination	Mehrotra et al. (2017)
Nannochloropsis sp.	Either endogenous or exogenous β-TUB promoter	BLE	Efficiency of PCR fragment-based transformation was higher than that based on plasmids.	Cerutti et al. (2011)
Nannochloropsis gaditana	Endogenous promoters of the genes encoding β-TUB, HSP70, and the ubiquitin extension protein (UEP)	BLE	Quantitative transforming cassette designs; the transformation efficiency of the chosen endogenous promoters approximately corresponds with their RNAseq quantification values; tested the P. tricornutum FCPB promoter, but without success.	Taylor et al. (2014)

Species	Promoters	Target genes	Highlights	References
N. gaditana	Endogenous UEP promoter	BLE	Targeted gene knockout via homologous recombination; a palmitic acid elongase affects eicosapentaenoic acid (EPA) and plastidial monogalactosyldiacylglycerol levels.	Zhao et al. (2007)
Nannochloropsis oceanica	C. reinhardtii α-tubulin promoter, 35S promoter, or endogenous LDSP promoter	Streptomyces hygroscopicus APH7 gene	Electroporation-mediated transformation; tested a serial of promoters	Tarver et al. (2015)
N. ocenica	Endogenous promoters of genes encoding VCP, V-type ATPase, HSP, and β-TUB	BLE, HYG, CAS9, guide RNA genes, hairpin structure DNAs, or a serial of DGATs	Select endogenous promoters based on the RNAseq quantification values	Huang et al. (2011); Gao et al. (2016)
N. ocenica	Endogenous ribosomal subunits (RIBI) or LDSP promoter	BLE, HYG, GFP, YFP, the cyan fluorescent protein (CFP), and Luciferase (LUX) gene	Ribi promoter is a bidirectional promoter; optimize 2A peptide ribosomal skipping efficiency; engineered EPA pathway for enhanced long-chain polyunsaturated fatty acid production	Geng et al. (2015)
N. oceanica	Endogenous DGTT5 or elongation factor (EF) promoter	GFP, type-2 DGAT-encoding genes (DGTT1–DGTT6), and Venus fluorescent protein (VFP) gene	Assessed six DGAT-encoding genes for TAG biosynthesis	Molnar et al. (2009)
N. oceanica	lipid droplet surface protein (LDSP) gene promoter	microsomal Δ12-desaturase gene and APHVII	Stress-inducible promoter; overexpression of endogenous Δ12-desaturase improved deposition of unsaturated fatty acids in TAG	De Riso et al. (2009)
N. oceanica	LDSP promoter	BLE, HYG, Aph7, VFP, GFP, lysophosphatidic acid acyltransferase (LPAT1–LPAT4) genes	Functional investigated on four LPATs for triacylglycerol biosynthesis	Hannon (2002)
Nannochloropsis salina	Endogenous UEP or TUB promoter	Endogenous basic helix-loop-helix (bHLH) transcription factors	Overexpression of bHLH2 led to increased growth rate, nutrient uptake, and productivity of biomass and fatty acids.	Fujiwara et al. (2013)
N. salina	N. oceanica VCP2 promoter or endogenous β-TUB promoter	GFP, BLE, and hairpin structure DNAs	Silencing of a pyruvate dehydrogenase kinase enhances TAG biosynthesis	Sodeinde and Kindle (1993)

Continued

Table 1.4. Continued.

Species	Promoters	Target genes	Highlights	References
Chloroplast transformation				
Chlorophyta				
C. reinhardtii	Endogenous promoter of chloroplast ATPB gene encoding the subunit of CF1 complex of the chloroplast adenosine triphosphate (ATP) synthase	A DNA fragment harboring the ATPB	The first report on microalgal stable chloroplast transformation; particle bombardment-mediated transformation Employed mutants defective in atpB and incapable of photosynthesis as hosts	Jeon et al. (2017)
C. reinhardtii	The 16S/ATPA promoter/UTR combination and endogenous 16S, RBCL, PSBD, ATPA and PSBA promoter	Codon-optimized luciferase gene LUXCT	Heterologous protein expression was improved in the chloroplast of through promoter and 5′ untranslated region optimization.	Greiner et al. (2017)
C. reinhardtii	Endogenous promoter of the gene encoding Core protein of photosystem I (PSAA)	PSAA or the gene encoding cytosine deaminase, endolysin, or hypothetical protein which is very toxic to E. coli	Biocontainment was built into the transgenes by replacing several tryptophan codons (UGG) with the UGA stop codon and using an orthogonal tryptophan tRNA to recognize these internal stop codons.	Daboussi et al. (2014)
C. reinhardtii	Variants of the library for the 5′-UTR of the genes encoding subunits of photosystem I (PSAA) and photosystem II (PSBD)	A codon-optimized luciferase reporter (LUXAB) gene	This study presents a synthetic biology approach to examine in vivo of designed variants of endogenous UTRs and quantitatively identify essential regions.	Gupta and Shukla (2017)
C. reinhardtii	—	—	This study represents the first step toward de novo creation of chloroplast genomes.	Serif et al. (2018)
Rhodophyta				
Porphyridium sp.	Native promoter of acetohydroxyacid synthase (AHAS) gene	A mutant AHAS gene	The first genetic transformation system for Rhodophytes; particle bombardment-mediated transformation	Meng et al. (2017)
Eustigmatophyta				
Nannochloropsis sp.	Endogenous promoter of large subunit of RuBisCO (RBCL) gene	GFP or BLE	Electroporation-based chloroplast transformation	Shin et al. (2016)
Diatom				
P. tricornutum	Endogenous promoter of RBCL	CAT and EGFP	The first plastid gene expression system for diatoms	Kao and Ng (2017)

transformation events compared with the former (Vieler et al., 2012). The limited transformation efficiency of microalgal expression systems using conventional strategies has resulted in poorly reproducible transformation protocols and constrained the scope of reverse-genetic tool development. Therefore, a more rational approach for promoter dissection and the design of biological engineering systems is needed.

3.2 Quantitative methods for transforming cassette design

The dissection of regulatory elements is essential for the design of engineering systems, which can in turn facilitate understanding of their natural counterparts (Belliveau et al., 2018) (Fig.1. 1). However, the regulatory mechanisms remain unenlightened in microalgae, even in the model species *C. reinhardtii*. Therefore, characterized endogenous promoters have been primary options when constructing customized vectors for specific microalgae. For example, a promoter for the gene encoding fucoxanthin-chlorophyll binding proteins (FCP) was isolated from diatoms, thoroughly tested and widely used for transformation vector design (Table 1.4). It is noticeable that vectors incorporating this promoter exhibited stable and relatively high transformation frequencies in both biolistic and electroporation-induced transformations (Zhang and Hu, 2014; Cui et al., 2018). Additionally, several endogenous regulatory elements were characterized and incorporated into vectors for *Chlamydomonas* transformations using *Agrobacterium* (Mini et al., 2018), electroporation (Mini et al., 2018), silicon carbide whiskers (Dunahay, 1993), glass beads (Kindle, 1990), and biolistic methods (Boynton et al., 1988), achieving comparable efficiencies. Thus, vector assembly is a key determinant of transformation efficiency. The frontiers of this field have been advanced by the development of novel promoter engineering strategies which are generally classified as: (a) random mutagenesis; (b) hybrid promoter design; (c) *de-novo* promoter synthesis (Mehrotra et al., 2017).

In addition to promoters, incorporating regulatory elements such as introns and featured transcript sequences were also applied to improve transformation efficiency and increase exogenous gene expression (Iddo et al., 2018). Synergistic effects were utilized to improve transgene expression by either incorporating different intron portfolios into vectors or including a consensus Kozak sequence in the 5' UTRs (DeHoff and Soriaga, 2016) (Table 1.4). To allow proteins to target to specific organelles, a leader-targeting sequence (Xue et al., 2015) comprising transit peptides (Rosenwasser et al., 2014) could be designed. These results highlight the key role of regulatory elements in design of engineering systems and the necessity of dissecting the mechanisms and functions of different regulatory elements. Unfortunately, the diversity and potential of these regulatory mechanisms (or elements) in microalgae are largely unknown (Gimpel et al., 2015) as a result of the low-throughput methods for characterizing the molecular mechanisms.

The ongoing expansion of sequenced genomes of microalgae (60 already completed or in the pipeline) and several cyanobacteria, and the associated omics information facilitates quantitative dissection of the mechanisms underpinning the functionality of promoters and other regulatory elements in a wide range of microalgae (Salama et al., 2019). Together with the development of sophisticated trapping systems (Vila et al., 2012) and the computational analysis of biological components, it is increasingly viable to rationally and systematically identify, characterize and standardize promoters, untranslated regions, terminators, enhancers, silencers, codon preferences, and other yet-to-be-discovered elements (Anderson et al., 2017). This knowledge-driven strategy offers several potential advantages over traditional methods for establishing microalgal transformation systems. For instance, *in silico* prediction and investigation of regulatory elements considerably increases the likelihood of discovering active regulatory components. Additionally, system-level investigation on gene structure can help identify species-specific regulatory mechanisms and biological components, which facilitates the customized design of synthetic promoters, biological bricks and circuits (Scranton et al., 2016). Finally, cross-species genome comparisons are helpful to unveil universal regulatory rules operating in different microalgae and thereby enable the design of universal (at the species or genus levels) transgene vectors.

We see notable progress in the dissection of algal genetic elements in such a knowledge-driven manner. A number of strong promoters have been discerned from different *Nannochloropsis* species (Wang et al., 2016; Xin et al., 2017) (Table 1.4). To enable multiple-gene expression, bidirectional promoters (Kilian et al., 2011; Eric et al., 2018) have also been isolated from this microalgal genus. Promoters, both constitutive and inducible under nitrogen starvation, were employed for customized transgene expression in the diatom *P. tricornutum* (Adleragnon et al., 2017). Quantitative profiling of such components can enable their use in BioBricks – DNA sequences conforming to a restriction enzyme assembly standard and encoding one or more functional units, which can be used as part of a scalable transgenic toolbox. The assembly and engineering of individual components with defined functions from libraries of such standardized interchangeable parts could enable the design and development of generalizable engineering tools.

4. Engineering Strategies: A Trend to Editing Methods

4.1 Gene overexpression

Deliberate overexpression of genes of interest (GOIs) is widely used for functional analysis (Fig. 1.2). Distinctive to the capacity for expressing multigenes in a single operon in prokaryotic or prokaryotic-derived genomes (e.g. microalgal chloroplasts), GOIs and marker

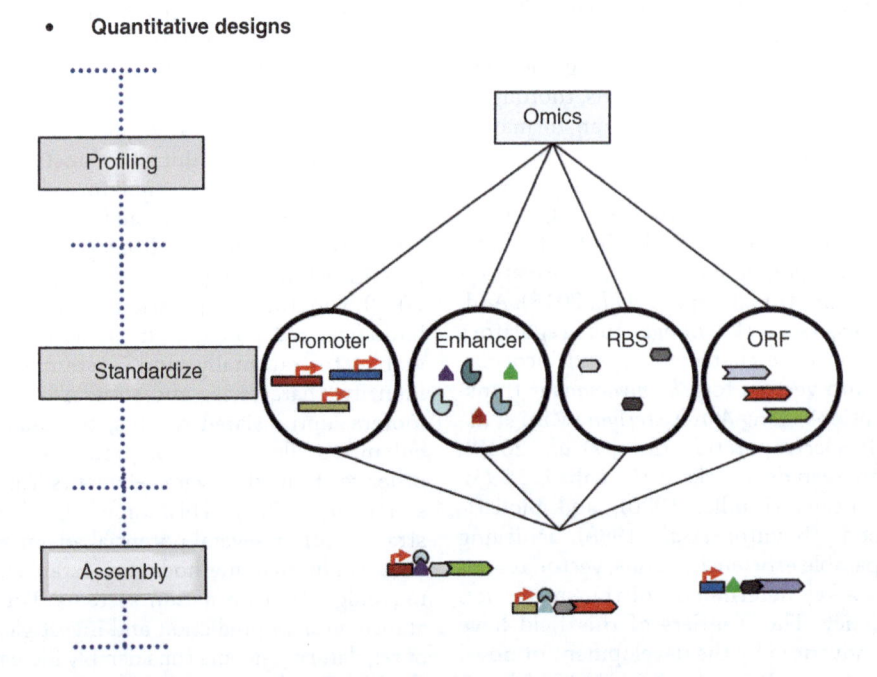

Fig. 1.1. Approaches of quantitative design for the engineering vectors. Abbreviations: RBS, ribosomal binding site; ORF, open reading frame.

genes should be located in different expression cassettes in microalgae. Overexpression of multiple genes is a distinct advantage and requirement for trait optimization. Various strategies have been designed for multigene expression in microalgae (Fig. 1.2a). For example, co-expressed genes can be cloned into separate vectors with different selective markers and then transformed into one strain sequentially (Neupert *et al.*, 2009). Alternatively, recently developed strategies for multigene engineering involve designing expression vectors using bidirectional promoters (Kilian *et al.*, 2011; Eric *et al.*, 2018), isocaudamers (Jia *et al.*, 2016), or using the self-cleaving peptide F2A (e.g. in *C. reinhardtii* (Plucinak *et al.*, 2015), *Nannochloropsis* (Eric *et al.*, 2018), or *Neochloris oleoabundans* (Muñoz *et al.*, 2019)). Gene-stacking overexpression systems have been established in some microalgae (e.g. *C. reiharditti*, *Nannochloropsis* sp., *Chlorella* sp. and *P. tricornutum*). This system is used to study specific metabolic pathways and to express intracellular (Moog *et al.*, 2019) or secreted (Ramos-Martinez *et al.*, 2017) recombinant proteins. These technologies simplify the transformation process, diminish the number of marker genes required for multi-step integration, and improve the dosage of target protein(s) after fusion with proper selectable markers.

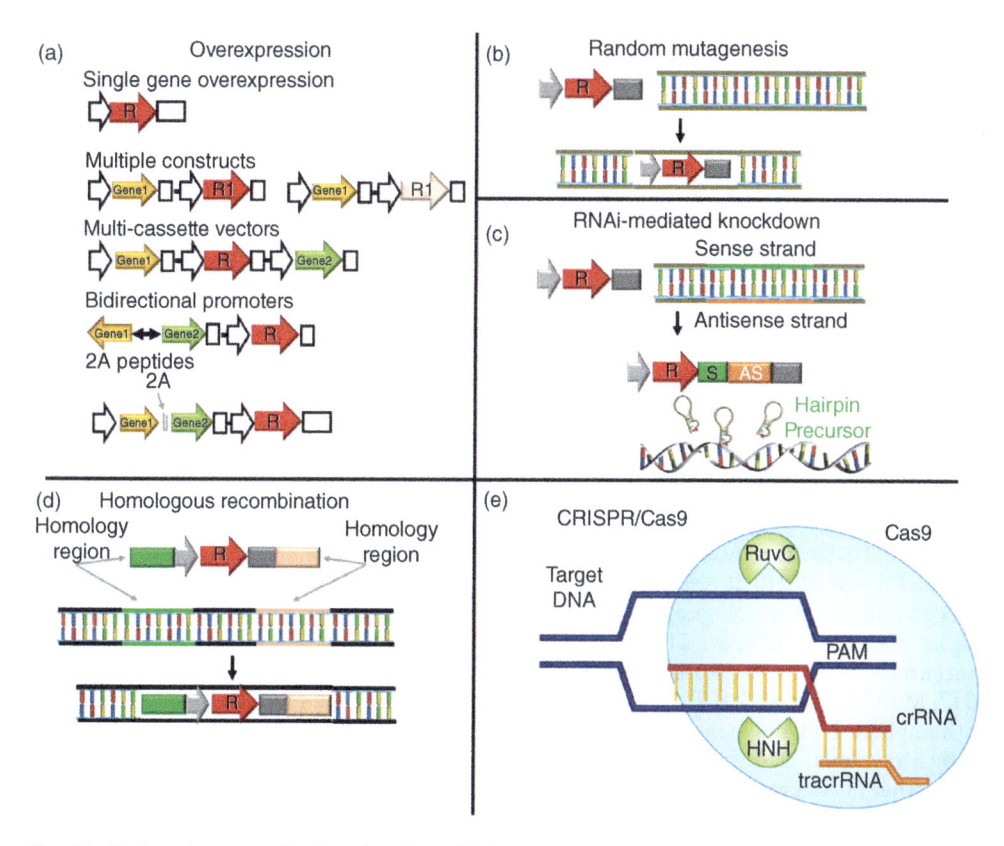

Fig. 1.2. Engineering strategies for microalgae: (a) Overexpression strategies for a single gene or multiple genes; (b) Random DNA insertional mutagenesis; (c) RNAi-mediated gene knockdown; (d) homologous recombination; (e) targeted genome editing by CRISPR/Cas9 technology. (R, resistance marker; Gene, genes(s) of interest; S, sense strand; AS, antisense strand; RuvC, RuvC nuclease domain; PAM, protospacer adjacent motif; HNH, HNH nuclease domain; crRNA, CRISPR RNA; tracrRNA, trans-activating crRNA.)

4.2 Random mutagenesis

Overexpression platforms serve as foundations for the establishment of a comprehensive engineering system incorporating technologies of insertional mutagenesis, RNAi-mediated gene knockdown, and genome editing. Identifying mutants with a phenotype of interest is a classical genetic approach for probing life processes (Fig. 1.2b). One way is to use transfer DNA (T-DNA), which is integrated into genomes randomly. It features tag-like sequences that facilitate identification of the affected gene(s) (Parinov and Sundaresan, 2000). Therefore, gene inactivation by T-DNA-mediated transposon insertion has been used in functional studies on diverse plant species that have in turn accelerated both basic research and applied plant biotechnology. However, T-DNA has not been demonstrated in algae thus far. Alternatively, by random insertion of transforming cassettes, whole-scale genome mutagenesis libraries have been generated only in a single microalga *C. reinhardtii* (Dent et al., 2015). The mutant population has proved to be effective to dissect the function of key genes in photosynthesis (Li et al., 2018) and TAG biosynthesis (Li et al., 2016). However, current methods for flanking sequence retrieval tend to be species-specific: the application of these methods is always constrained to a single species and can only determine insertion sites for that particular species, and, even worse, at a low frequency (Bonente et al., 2011). A tagged mutant library was recently created for *C. reinhardtii* in a similar fashion to T-DNA that was successful in *A. thaliana* (Li et al., 2016). It was utilized to dissect genes involved in biological process, such as photosynthesis (Li et al., 2018) and CO_2 concentration (Freeman Rosenzweig et al., 2017; Mackinder et al., 2017). The generality of this approach highlights its potential in investigating function of non-lethal genes in many other microalgae. However, a drawback is that loss-of-function mutations caused by random disruptions are unlikely to probe the function of lethal genes. In addition, the random mutagenesis approach is of limited utility in studies targeting specific genes and in hypothesis-driven tests.

4.3 RNA interference (RNAi)-mediated gene knockdown

RNAi-based gene silencing is a superior approach for oriented knockdown of arbitrarily chosen genes (Fig. 1.2c). Despite the apparent loss of core elements in many microalgal species (Cerutti et al., 2011), RNAi-mediated silencing appears to be widespread among algal lineages (Wei et al., 2017). Although its biological machinery remains elusive (Taylor et al., 2014), sophisticated systems for small RNAs have been discovered in green, red and brown algae, diatoms, and dinoflagellates (Zhao et al., 2007; Huang et al., 2011; Geng et al., 2015; Tarver et al., 2015; Gao et al., 2016). RNAi-mediated knockdown has been established for species including *C. reinhardtii* (Molnar et al., 2009), *N. oceanica* (Wei et al., 2017), and *P. tricornutum* (De Riso et al., 2009). Nevertheless, the RNAi method has several general shortcomings such as unstable phenotype heritability and unpredictable activities of target genes (Hannon, 2002).

4.4 Targeted genome editing

To avoid unwanted phenotypes brought by random transforming cassette insertion (when using overexpression) or off-target knockdown (when using RNAi), it is desirable to develop tools for precise modification of pre-selected sequences. Traditionally, homologous recombination is a routine practice for the genetic manipulation of prokaryotes. It was also developed in recombinogenic lower eukaryotes (Fujiwara et al., 2013) (Fig. 1.2d). However, this method delivers only limited success in most eukaryotic microalgae, where non-homologous recombination tends to occur in preference to the homologous one (Sodeinde and Kindle, 1993). These problems have been circumvented by developing strategies based on engineered nucleases, which substantially facilitates targeted gene disruption and is revolutionizing many areas of science (Jeon et al., 2017). Although the zinc-finger nuclease (ZFN) and transcription activator-like effectors (TALEN) are excellent tools for targeted gene knockout, they have only been proven to be useful in *C. reinhardtii* (Greiner et al., 2017) and *P. tricornutum* (Daboussi et al., 2014). Clustered regularly interspaced palindromic sequences (CRISPR/Cas9) technology is more efficient and much easier in design than ZFN and TALEN technologies (Gupta and Shukla, 2017). The straightforward design of CRISPR systems allows simultaneous disruption in multiple genes (Serif et al., 2018) and the creation of CRISPR mutant libraries (Meng et al., 2017) (Fig. 1.2e). The list of applications of CRISPR has expanded to include knock-in (Shin et al., 2016) and CRISPR interference (CRISPRi), which involves inducing sequence-specific interference of a target gene's transcription (Kao and Ng, 2017). By CRISPR/Cas9-derived activator systems, multiple endogenous (non)coding genes can be activated simultaneously (Chavez et al., 2015). The ability to upregulate any endogenous gene(s) provides unprecedented opportuninties to look into and reconstruct cellular behavior. Therefore, an increasing number of microalgal species are engineered by CRISPR/Cas9 systems, such as *C. reinhardtii* (Greiner et al., 2017), *Nannochloropsis* sp. (Ajjawi et al., 2017) and *P. tricornutum* (Nymark et al., 2016). However, the CRISPR/Cas9-based technology in microalgae is still in its infancy, as evidenced by low efficiencies and the need for more intensive mutant screening than in plants and animals. To ensure efficiency and accuracy, standardized targeted genome-editing protocols should be developed for microalgae on the basis of in-depth fundamental investigations, particularly on the effects of endogenous silence on introduced genetic materials (Jeong et al., 2002).

4.5 From nucleus to organelle engineering

Nuclei, chloroplasts and mitochondria are three membrane-bounded organelles in microalgal cells. Although most of preceding examples involve nuclear transformation, nuclear expression has the drawback of sensitivity to epigenetic effects and random insertion, which can cause variable transgene expression. In contrast, chloroplast engineering shows several key advantages, such as absence of epigenetic gene silencing, targeted localization of transgenes in the genomes, and relatively high expression levels (Daniell et al., 2016). Microalgal chloroplast genomes (e.g. in *Nannochloropsis* sp.) possess approximately 120 genes involved in photosynthesis and the gene-expression system (Wang et al., 2014). In principle, all of these genes could be precisely manipulated. Furthermore, a number of nuclear gene products (such as these involved in photosynthesis, oil metabolism or plant hormone synthesis) are targeted to plastids (Lu et al., 2014; Lu and Xu, 2015). For example, in *Arabidopsis*, more than 2000 nuclear-encoded proteins are predicted to localize in chloroplasts and contribute to cellular properties critical for growth or the production of essential cellular compounds (Leister, 2003). This further enlarges the category of biological traits which could be optimized by plastome engineering.

Although chloroplast transformation was initially developed in the green microalga *C. reinhardtii* (Boynton et al., 1988), plastome genetic engineering was more thoroughly developed in and applied to plant species. The use of chloroplast engineering to produce pharmaceuticals in crop plants presents several fundamental challenges (Zhang et al., 2017). Problems include the slow plant growth (Melis, 2009), especially in cases when a high yield and recombinant product amount are required. This problem is further accentuated by the seasonal growth of crops, as opposed to a year-round cultivation of microalgae. Moreover, transgene containment (Scotti et al., 2012) is always more difficult to cope with in plants than in photobioreactor-enclosed photosynthetic microalgae. Microalgal platforms avoid many undesirable issues, since these systems can be cultivated under tightly controlled and contained conditions in closed photobioreactors. In addition, the difference of codon usage in microalgae from that of plants helps avoid contamination of the human food chain with heterologous proteins due to cross-pollination (Wilson and Roberts, 2012). Biocontainment can be improved further by codon reassignment of the transgenes in the chloroplast (Young and Purton, 2016). Taken together, barriers encountered in the use of plant-based systems to produce recombinant protein or other heterologous products would make them less attractive than photosynthetic microorganisms, e.g. microalgae and cyanobacteria (Betterle et al., 2020). More than 100 foreign or native proteins have thus far been produced in algal chloroplasts, including 40 different therapeutic proteins produced in *C. reinhardtii* chloroplasts (Dyo and Purton, 2018). Table 1.4 shows some case studies of chloroplast transformation. More examples are given in recent reviews (Dyo and Purton, 2018).

However, despite some notable advances, microalgal plastid transformation is only available for a relatively small number of species (Bock, 2015). Chloroplast transformation has mainly relied on bombardment with gold microparticles laced with the transforming DNA, which is intractable for most microalgae. The smallest readily available gold particles have a diameter of around 0.6 μm, which exceeds the dimensions of most microalgal chloroplasts (Gan et al., 2018). An alternative method is via polyethylene glycol (PEG)-mediated protoplast transformation (Golds et al., 1993). However, an intrinsic drawback of this method is the requirement of cell wall removal prior to transformation. Protoplast preparation is technically demanding and effectively impossible for many microalgal species with complex cell walls (Maliga, 2004). A simple and straightforward electroporation-based method was recently devised for transforming *Nannochloropsis* chloroplasts (Gan et al., 2018). However, expression levels of transgenes in microalgal chloroplasts remain low compared to plant plastid and cyanobacterial expression systems. To bring plastome engineering into a full play, it is critical to establish protocols that enable efficient and reliable expression for individual transgenes or entire pathways. It is thus indispensable to identify several components, such as constitutive promoters for robust expression (Rasala et al., 2011), controllable expression systems for spatially or temporally inducible expression, and optimized genetic elements for multigene expression (Specht and Mayfield, 2013; Doron et al., 2016). Synthetic biology principles have been applied to plastome engineering, allowing *ex vivo* assembly, modification and duplication of the entire *Chlamydomonas* chloroplast genome (O'Neill et al., 2011). This proof-of-concept experiment demonstrates that *de novo* synthesis of algal chloroplast genomes is now possible.

4.6 From genetically modified (GM) to 'non-GM' engineering

GM crops have been used commercially for decades. Available impact studies of pest-resistant crops show that these technologies are beneficial to farmers and consumers, producing large aggregate welfare gains as well as positive effects on the environment and human health (Qaim, 2009; Li et al., 2014). Nonetheless, there has been substantial debate regarding the merits and ethics of GM and non-GM breeding technologies. Widespread public reservations have led to a complex system of regulations, and in most of the countries, the commercial use and cultivation of GM organisms (GMOs) is heavily regulated (Wasmer, 2019). For example, the EU, arguably, has the strictest regulations in the world for the use of GMOs in food and feed. Specifically, organisms obtained by techniques of genome editing (that cannot be distinguished from a conventionally bred variety or a naturally occurring variant; without GM tags) are regarded as GMOs and subject to the same restriction as other transgenic organisms (Wasmer, 2019). GM microalgae are created by inserting DNA-containing marker genes and additional GOIs into the host's genome. A critical concern regarding the GM approach is the risk of promoting antibiotic resistance. To address this concern, techniques for generating transgenic algae should not rely on antibiotic resistance markers. Thus, a successful *C. reinhardtii* chloroplast transformation and strain selection protocol with zero false-positives was achieved upon reconstitution of a functional *rbcL* gene and recovery of Rubisco catalytic activity, coupled with the heterologous expression of the *Saccharomyces cerevisiae* alcohol dehydrogenase *ADH1* gene in the algal chloroplast (Hsu-Ching and Anastasios, 2013). This recovery of function, however, requires that photosynthesis-lethal microalgal mutants, which act as the transgene recipient strain, can be propagated heterotrophically. Alternatively, a recombinase has been applied to remove marker genes in engineered cyanobacteria (Cheah *et al.*, 2013). Extrachromosomal vectors, or episomes, offer another tool for 'non-GM' breeding in microalgae (Karas *et al.*, 2015). Combined with genome-editing technique, engineered cells of diatoms (Slattery *et al.*, 2018) and *N. oceanica* (Poliner *et al.*, 2018) maintained the desired traits while episomes (harboring the marker gene) were lost. It highlights the potential to create a 'non-GM' microalga by the development of tools derived from episomes with specific and universal algal hosts (Nora *et al.*, 2019). This highly efficient genetic system will accelerate knowledge transfer from fundamental research to tangible applications.

5. History of Microalgal Engineering Approaches: A Starting Point for Future Development

Genetic manipulation has been accomplished in a wide spectrum of microalgae. In this section, we will revisit viable ways to generate useful products by engineered microalgae.

5.1 Conventional gene engineering – technologies based on single-enzyme encoding genes

Metabolic engineering is a purposeful modification of metabolic pathways for customized or improved production of useful compounds (Bailey, 1991) (Fig. 1.3). Historically, metabolic engineering efforts in microalgae have largely centered on modifying a single gene at a time, typically one encoding a rate-limiting enzyme. Biosynthetic pathways are retrofitted to preferentially flux endogenous carbon toward desirable-end chemicals (Xue *et al.*, 2015), but not to by-products (Trentacoste *et al.*, 2013). Strategies adopted for this purpose include yield improvement (Cui *et al.*, 2018; Han *et al.*,

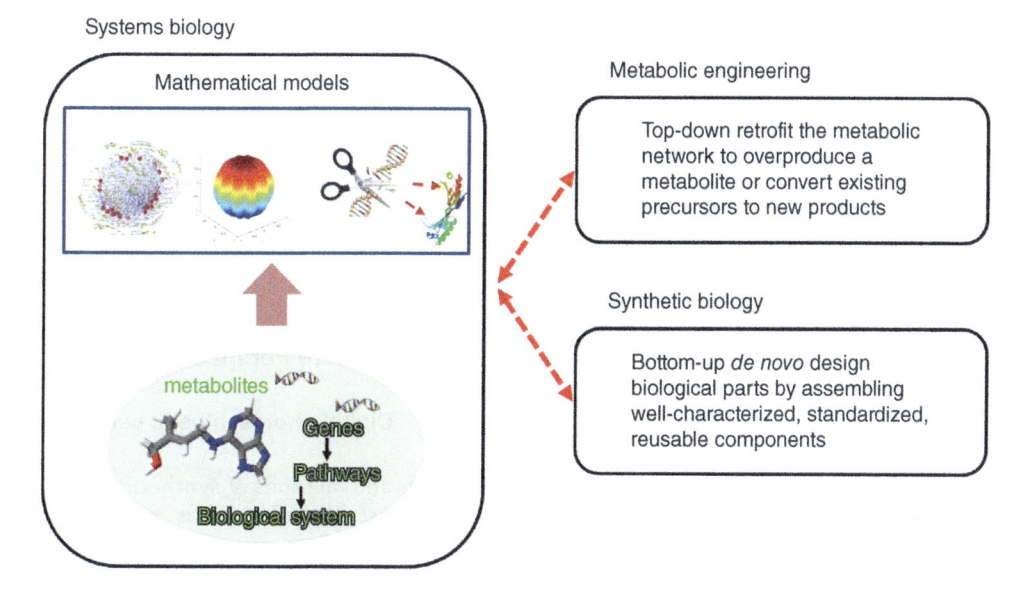

Fig. 1.3. Comparison between metabolic engineering and systems biology. (Adapted from Nielsen *et al.*, 2014)

2020), substrate range expansion (Xin et al., 2017), novel chemicals synthesis (Tran et al., 2013), and optimization of cellular robustness (Loera-Quezada et al., 2016; Lu et al., 2021). Subsequent developments in molecular and chemical biology led to design and reconstruction of complete metabolic networks and even whole organisms rather than just individual reactions. It is from these studies that the phrase 'cell factories' emerged. High-throughput omics techniques make it possible to rapidly link genotypes and phenotypes and to accelerate the engineering processes (Darby and Hall, 2008). These revolutionary techniques and mindsets have transformed metabolic engineering into a new field comprising genome-scale engineering, which incorporates computational biology, versatile approaches for interrogating and understanding cellular metabolism, and multiplex optimizations to reach diverse phenotypes (Ngan et al., 2015).

5.2 Transcriptional engineering – a control-knob gene-based approach

Mixed success has been achieved in microalgae using the aforementioned conventional genetic engineering strategies. In some cases, the abundance of the desired metabolite remained unaltered. Transcriptional engineering provides an alternative strategy where a control-knob gene can be modified to simultaneously modulate multiple steps of a metabolic pathway (Chen et al., 2018b). Initial attempts at trait improvement of microalgae by this strategy involved expressing plant transcription factors (TFs) and resulted in enhanced production of target chemicals (Zhang et al., 2014). This success prompted a pursuit of genome-wide identification of microalgal TFs (Thirietrupert et al., 2016), which were subsequently utilized to optimize the desired properties (Ngan et al., 2015; Ajjawi et al., 2017).

5.3 Synthetic biology – a systems-level precise engineering

The advance of metabolic engineering along with the increasing blending of DNA technology gave genesis to the field of synthetic biology (Fig. 1.3). A genetic toggle switch prototype was first devised in bacteria by metabolic engineers (Chen et al., 1993), elaborated by synthetic biologists (Gardner et al., 2000) and consummated in mice (Kemmer et al., 2010). Unlike metabolic engineering, synthetic biology may use unnatural molecules to create artificial biological systems (Nielsen et al., 2014). While this approach has been successfully implemented in bacteria, yeasts, higher plants and animals by characterizing and collecting reusable and standard biological parts (Ignea et al., 2016; Chen et al., 2018a), microalgal synthetic biology remains in its infancy. Attempts have been undertaken in only a handful of species. Their successes have been modest because of the underdevelopment of genetic tools and the limited knowledge of gene-regulation mechanisms in microalgae, as previously discussed (Hlavova et al., 2015). However, advances in several species (e.g. *Nannochloropsis* sp., *C. reinhardtii* and *P. tricornutum*) are now allowing an expansion to the realm of genome-wide reprogramming beyond conventional individual gene manipulation (Butler et al., 2020). Together with the ongoing expansion of microalgal omics datasets, it is possible to use these species as chassis for synthetic biology. A versatile and modular vector toolkit has been generated for *C. reinhardtii*, which features a standardized collection of reporters, selectable markers and targeting peptides that can be assembled and repurposed (Lauersen et al., 2015). Broadly speaking, this toolkit generation highlights the inaugural application of synthetic biology in microalgal engineering and enables the development of digital photosynthetic cells.

6. Future Perspectives

6.1 Digital photosynthetic cells

Principles and methods of synthetic biology from well-established systems are increasingly introduced to microalgae. However, the development of microalgae as true 'programmable' entities is hampered not only by the accessibility of well-characterized and repurposed genetic elements but also by a lack of high-throughput methodology to identify, standardize and rationalize basic parts and modules, not to mention systems-level circuitry. This is true even for *C. reinhardtii*, of which, among all microalgal species, the transcriptional regulation and cellular metabolism are best understood. The systematic knowledge of its regulatory elements and potential BioBricks remains elusive. Despite these challenges, the development of standardized and scalable methods for module assembly (e.g. Gibson assembly and Golden Gate cloning), engineering (i.e. transformation), colony handling (e.g. colony picking robots), and high-throughput strain screening (e.g. microfluidics) will give microalgal researchers unprecedented access to 'digital photosynthetic cells' in which each gene, pathway, and biological process can be quantitatively designed and modulated.

6.2 From proof of concept to tangible industrial application

Enhancement in the microalgal biofuel R&D occurred in the period between 2006 and 2011 because the global petroleum price skyrocketed (Chen *et al.*, 2015). However, with the 2014 collapse in the price of fossil fuels, the renewable biofuels field is at a crossroads and its further development becomes foggy. End-products with interest of the algal community now mainly focus on primary and secondary metabolites, such as proteins, essential oils, specialty chemicals, and biopharmaceuticals. These comprise sizable markets, albeit not as large as that offered by the transportation fuels field. Moreover, the diversity of natural secondary metabolite in microalgae is far from being fully exploited, suggesting the promising development of future R&D and commercialization. Meanwhile, it may be necessary to fill the gaps between laboratory R&D and tangible industrial production by engineering microalgae with crucial and economically relevant pathways of value-added compounds, such as wax (Cui *et al.*, 2018), astaxanthin (Roth *et al.*, 2017) and non-native isoprenoids (e.g. β-phellandrene; lupeol, bisabolene, patchoulol, geranyllinalool and 13R(+) manoyl oxide) (D'Adamo *et al.*, 2018; Wichmann *et al.*, 2018). Plant secondary metabolites are a valuable reservoir of drugs and numerous of these compounds, such as ginsenosides and artemisinin, are among the most promising and commercially important biopharmaceuticals that can be generated in microalgae and cyanobacteria. These pathways and the associated key genes could be characterized using microalgal systems which resemble those of plants, but possess unique advantages as mentioned above. Moreover, akin to yeast or bacterial systems, microalgae are promising hosts to produce these metabolites in scalable culturing systems. The development of market-ready drug-producing systems using industrial microalgal strains is just beginning (Lauersen *et al.*, 2016; D'Adamo *et al.*, 2018; Slattery *et al.*, 2018; Wichmann *et al.*, 2018).

To realize the economic viability of commercial microalgal products, there is a need to inform on several major areas, including: (i) for newly identified species with industrial potential, the challenge would be to deliver exogenous DNA into the cells; (ii) universally applied sophisticated engineering toolkits are essential in efforts to produce versatile biochemicals and biomaterials for agricultural, biomedical and industrial ends; (iii) the development of genetic circuits, gene switches, systems-level circuitry, synthetic devices, synthetic genomes, genome-editing technologies, and practices for microbiome engineering (Lawson *et al.*, 2019) will benefit fundamental studies in microalgal synthetic biology; (iv) genome-wide activation/knockdown/knockout of multiple endogenous genes and simultaneous overexpression of multiple exogenous genes involved in target pathways or networks would greatly facilitate the creation of 'super' microalgal strains as customized cell factories; (v) although the application of glycosylation mechanisms is largely untapped in microalgae (Baiet *et al.*, 2011; Schulze *et al.*, 2017), 'humanization' of microalgae glycans via glyco-engineering would greatly facilitate the development of active and safe biopharmaceuticals production with proper folding and addition of immunogenic glycans (Barolo *et al.*, 2020).

The sustainability of the target chemical productivity in engineered algal cells has been assessed in bench-scale (10 L) (Cui et al., 2018) or pilot-scale photobioreactors (PBRs; 550 L) (D'Adamo et al., 2018) in controllable laboratory conditions or in mimicked outdoor environments (Ajjawi et al., 2017). The scalability and stability of introduced traits in GM microalgae were also probed by culturing cells at different PBR configurations and volumes (Wei et al., 2019). Although engineered phenotypes appeared to be stable and robust across these tested scales (from 800 mL to 100 L) and PBR configurations (i.e. vertical column, vertical flat-plate and open raceway pond), reports on life-cycle assessments of bioenergy or other value-added chemicals using engineered microalgae are not yet available.

A barrier to the commercial-scale algal biofuel production is the relatively high capital and operating costs of PBRs, combined with the need to deliver fuel products at very low cost, typically less than $1 per kg. Coupling biofuel production with environmental applications, or the co-generation of high-value products, could achieve mutual benefits in renewable energy production and mitigation of greenhouse gas emission (Wu et al., 2017) or removal of pollutants from wastewater (Cheah et al., 2016). Moreover, extensive R&D is essential to study on lowering the cost of down-stream processing, such as industrial-scale cultivation, contamination control, cell harvesting and disruption, and extraction of target biomolecules. There is no doubt that engineering technology – in particular, simultaneously targeting/activation/overexpression/knockout of multiple genes – would play a critical role in achieving the objective. For example, truncated light-harvesting antennae of the photosystems can be generated to improve sunlight penetration and utilization, resulting in greater photosynthetic productivity of cultures with a high cell density under bright-sunlight conditions (Kirst et al., 2012). Versatile contaminant control strategies have also been developed for open-pond cultivation by modifying nutrition mode (Loera-Quezada et al., 2016), expressing herbicide resistance genes (Bruggeman et al., 2014) or genome-based chemical biology (Gan et al., 2017). Target biomolecules could be extracted using cost-effective cell-disruption methods by expressing algicides (Demuez et al., 2015) or upon bypassing the disruption step, employing an inducible green recovery strategy (Liu et al., 2011). Overall, sound opportunities exist for the development of market-ready multiple-product systems by employing microalgae as digital photosynthetic cells.

7. Conclusions

We have reviewed the trend of microalgal engineering in the era of synthetic biology. In this article, the bioengineering of cyanobacteria, genomes and evolution, and applications of microalgae were not in the center of our attention, but the bottleneck and the development of the engineering techniques of eukaryotic microalgae were considered. As mentioned, microalgal metabolic engineering, genome editing and synthetic biology are increasingly intervening in microalgal biotechnology due to their numerous advantages over mutagenesis strategies using chemical and physical mutagens. Microalgal engineering technologies rapidly transfer from empirical methods to quantitative design, as target genome-editing and synthetic biology play more and more important roles in breeding new algae and the digital design of algal cellular metabolism. The current trend explicitly demonstrates that alongside technological progress in transforming technique, algal engineering develops mostly towards customized transformation systems, systems-level precision engineering, and cells created without GM tags. All these aspects are going to be integrated with big-data analysis of the increasing omics dataset. Despite positive horizons, the costs, the development of scale-up configurations, as well as the public concerns on GMOs, are dominant challenges.

Declaration of competing interests

The authors report no declarations of interest.

Acknowledgements

We thank Dr Jian Xu (Qingdao Institute of BioEnergy and Bioprocess Technology, Chinese Academy of Sciences) for very helpful discussion. This manuscript was supported by the the National Key R&D Program of China (2021YFE0110100; 2021YFA0909602), the Key R&D Program of Hainan Province (ZDYF2022XDNY140), the National Natural Science Foundation of China (32060061 and 32370380), the Foreign Expert Foundation of Hainan Province (G20230607016E) and the Maryland Industrial Partnerships Program (07432143 and07431629).

References

Adleragnon, Z., Leu, S., Zarka, A., Boussiba, S. and Khozingoldberg, I. (2017) Novel promoters for constitutive and inducible expression of transgenes in the diatom *Phaeodactylum tricornutum* under varied nitrate availability. *Journal of Applied Phycology*, 1–10.

Ajjawi, I. *et al.* (2017) Lipid production in *Nannochloropsis gaditana* is doubled by decreasing expression of a single transcriptional regulator. *Nature Biotechnology* 35, 647.

Anderson, M., Muff, T.J., Georgianna, D.R. and Mayfield, S.P. (2017) Towards a synthetic nuclear transcription system in green algae: characterization of *Chlamydomonas reinhardtii* nuclear transcription factors and identification of targeted promoters. *Algal Research-Biomass Biofuels and Bioproducts* 22, 47–55.

Bae, S. *et al.* (2015) Exogenous gene integration for microalgal cell transformation using a nanowire-incorporated microdevice. *ACS Applied Materials & Interfaces* 7, 27554–27561.

Baiet, B. *et al.* (2011) N-glycans of *Phaeodactylum tricornutum* diatom and functional characterization of its N-acetylglucosaminyltransferase I enzyme. *Journal of Biological Chemistry* 286, 6152–6164.

Bailey, J.E. (1991) Toward a science of metabolic engineering. *Science* 252, 1668–1675.

Barolo, L. *et al.* (2020) Perspectives for glyco-engineering of recombinant biopharmaceuticals from microalgae. *Cells* 9, 633.

Belliveau, N.M. *et al.* (2018) Systematic approach for dissecting the molecular mechanisms of transcriptional regulation in bacteria. *Proceedings of the National Academy of Sciences*.

Bernstein, H.C. *et al.* (2016) Unlocking the constraints of cyanobacterial productivity: acclimations enabling ultrafast growth. *mBio* 7(4), e00949–00916.

Betterle, N. and Melis, A. (2018) Heterologous leader sequences in fusion constructs enhance expression of geranyl diphosphate synthase and yield of β-phellandrene production in cyanobacteria (*Synechocystis*). *ACS Synthetic Biology* 7, 912–921.

Betterle, N., Martinez, D.H. and Melis, A. (2020) Cyanobacterial production of biopharmaceutical and biotherapeutic proteins. *Frontiers in Plant Science* 11.

Bock, R. (2015) Engineering plastid genomes: methods, tools, and applications in basic research and biotechnology. *Annual Review of Plant Biology* 66, 211–241.

Bonente, G. *et al.* (2011) Mutagenesis and phenotypic selection as a strategy toward domestication of *Chlamydomonas reinhardtii* strains for improved performance in photobioreactors. *Photosynthesis Research* 108, 107–120.

Boynton, J.E. *et al.* (1988) Chloroplast transformation in *Chlamydomonas* with high velocity microprojectiles. *Science* 240, 1534–1538.

Brodie, J. *et al.* (2017) The algal revolution. *Trends in Plant Science* 22(8), 726–738.

Bruggeman, A.J., Kuehler, D. and Weeks, D.P. (2014) Evaluation of three herbicide resistance genes for use in genetic transformations and for potential crop protection in algae production. *Plant Biotechnology Journal* 12, 894–902.

Butler, T.O., Kapoore, R.V. and Vaidyanathan, S. (2020) *Phaeodactylum tricornutum*: a diatom cell factory. *Trends in Biotechnology* 38, 606–622.

Cerutti, H., Ma, X., Msanne, J. and Repas, T. (2011) RNA-mediated silencing in algae: biological roles and tools for analysis of gene function. *Eukaryotic Cell* 10, 1164–1172.

Chaves, J.E., Ruedaromero, P., Kirst, H. and Melis, A. (2017) Engineering isoprene synthase expression and activity in cyanobacteria. *ACS Synthetic Biology* 6, 2281–2292.

Chavez, A. et al. (2015) Highly efficient Cas9-mediated transcriptional programming. *Nature Methods* 12, 326.

Cheah, W.Y. et al. (2016) Cultivation in wastewaters for energy: a microalgae platform. *Applied Energy* 179, 609–625.

Cheah, Y.E., Albers, S.C. and Peebles, C.A.M. (2013) A novel counter-selection method for markerless genetic modification in *Synechocystis* sp. PCC 6803. *Biotechnology Progress* 29, 23–30.

Chen, B. et al. (2018a) Synthetic biology toolkits and applications in *Saccharomyces cerevisiae*. *Biotechnology Advances* 36, 1870–1881.

Chen, H., Qiu, T., Rong, J., He, C. and Wang, Q. (2015) Microalgal biofuel revisited: an informatics-based analysis of developments to date and future prospects. *Applied Energy* 155, 585–598.

Chen, W., Kallio, P.T. and Bailey, J.E. (1993) Construction and characterization of a novel cross-regulation system for regulating cloned gene expression in *Escherichia coli*. *Gene* 130, 15–22.

Chen, X., Hu, G. and Liu, L. (2018b) Hacking an algal transcription factor for lipid biosynthesis. *Trends in Plant Science* 23, 181–184.

Cui, Y., Zhao, J., Wang, Y., Qin, S. and Lu, Y. (2018) Characterization and engineering of a dual-function diacylglycerol acyltransferase in the oleaginous marine diatom *Phaeodactylum tricornutum*. *Biotechnology for Biofuels* 11, 32.

Daboussi, F. et al. (2014) Genome engineering empowers the diatom *Phaeodactylum tricornutum* for biotechnology. *Nature Communications* 5, 3831.

D'Adamo, S. et al. (2018) Engineering the unicellular alga *Phaeodactylum tricornutum* for high-value plant triterpenoid production. *Plant Biotechnology Journal* 17, 75–87.

Daniell, H., Lin, C., Yu, M. and Chang, W. (2016) Chloroplast genomes: diversity, evolution, and applications in genetic engineering. *Genome Biology* 17, 134.

Darby, A.C. and Hall, N. (2008) Fast forward genetics. *Nature Biotechnology* 26, 1248–1249.

Davies, J.P., Weeks, D.P. and Grossman, A.R. (1992) Expression of the arylsulfatase gene from the β2-tubulin promoter in *Chlamydomonas reinhardtii*. *Nucleic Acids Research* 20, 2959–2965.

De Riso, V. et al. (2009) Gene silencing in the marine diatom *Phaeodactylum tricornutum*. *Nucleic Acids Research* 37.

DeHoff, P. and Soriaga, L. (2016) *Nannochloropsis* kozak consensus sequence. *US Patent* 9, 523.

Demirer, G.S. et al. (2019) High aspect ratio nanomaterials enable delivery of functional genetic material without DNA integration in mature plants. *Nature Nanotechnology* 14, 456–464.

Demuez, M., Gonzalez-Fernandez, C. and Ballesteros, M. (2015) Algicidal microorganisms and secreted algicides: new tools to induce microalgal cell disruption. *Biotechnology Advances* 33, 1615–1625.

Dent, R. et al. (2015) Large-scale insertional mutagenesis of *Chlamydomonas* supports phylogenomic functional prediction of photosynthetic genes and analysis of classical acetate-requiring mutants. *Plant Journal* 82, 337–351.

Doron, L., Segal, N. and Shapira, M. (2016) Transgene expression in microalgae –from tools to applications. *Frontiers in Plant Science* 7, 505.

Dunahay, T.G. (1993) Transformation of *Chlamydomonas reinhardtii* with silicon carbide whiskers. *BioTechniques* 15, 452–460.

Dyo, Y.M. and Purton, S. (2018) The algal chloroplast as a synthetic biology platform for production of therapeutic proteins. *Microbiology* 164, 113–121.

Eric, P. et al. (2018) A toolkit for *Nannochloropsis oceanica* CCMP1779 enables gene stacking and genetic engineering of the eicosapentaenoic acid pathway for enhanced long-chain polyunsaturated fatty acid production. *Plant Biotechnology Journal* 16, 298–309.

Formighieri, C. and Melis, A. (2015) A phycocyanin·phellandrene synthase fusion enhances recombinant protein expression and β-phellandrene (monoterpene) hydrocarbons production in *Synechocystis* (cyanobacteria). *Metabolic Engineering* 32, 116–124.

Formighieri, C. and Melis, A. (2017) Heterologous synthesis of geranyllinalool, a diterpenol plant product, in the cyanobacterium *Synechocystis*. *Applied Microbiology and Biotechnology* 101, 2791–2800.

Freeman Rosenzweig, E.S. et al. (2017) The eukaryotic CO_2-concentrating organelle is liquid-like and exhibits dynamic reorganization. *Cell* 171, 148–162.e119.

Fujiwara, T., Ohnuma, M., Yoshida, M., Kuroiwa, T. and Hirano, T. (2013) Gene targeting in the red alga *Cyanidioschyzon merolae*: single- and multi-copy insertion using authentic and chimeric selection markers. *PLOS ONE* 8, e73608.

Gan, Q. et al. (2017) A customized contamination controlling approach for culturing oleaginous *Nannochloropsis oceanica*. *Algal Research* 27, 376–382.

Gan, Q., Jiang, J., Han, X., Wang, S. and Lu, Y. (2018) Engineering the chloroplast genome of oleaginous marine microalga *Nannochloropsis oceanica*. *Frontiers in Plant Science* 9.

Gao, F. et al. (2016) Identification of conserved and novel microRNAs in *Porphyridium purpureum* via deep sequencing and bioinformatics. *BMC Genomics* 17, 612.

Gardner, T.S., Cantor, C.R. and Collins, J.J. (2000) Construction of a genetic toggle switch in *Escherichia coli*. *Nature* 403, 339–342.

Geng, H. et al. (2015) Identification of microRNAs in the toxigenic dinoflagellate *Alexandrium catenella* by high-throughput illumina sequencing and bioinformatic analysis. *PLOS ONE* 10.

Gimpel, J.A., Henríquez, V. and Mayfield, S.P. (2015) In metabolic engineering of eukaryotic microalgae: potential and challenges come with great diversity. *Frontiers in Microbiology* 6.

Golds, T.J., Maliga, P. and Koop, H. (1993) Stable plastid transformation in PEG-treated protoplasts of *Nicotiana tabacum*. *Nature Biotechnology* 11, 95–97.

Greiner, A. et al. (2017) Targeting of photoreceptor genes in *Chlamydomonas reinhardtii* via zinc-finger nucleases and CRISPR/Cas9. *The Plant Cell* 29, 2498–2518.

Guiry, M.D. (2012) How many species of algal are there? *Journal of Phycology* 48, 1057–1063.

Gupta, S.K. and Shukla, P. (2017) Gene editing for cell engineering: trends and applications. *Critical Reviews in Biotechnology* 37, 1.

Han, X., Song, X., Li, F. and Lu, Y. (2020) Improving lipid productivity by engineering a control-knob gene in the oleaginous microalga *Nannochloropsis oceanica*. *Metabolic Engineering Communications* 11, e00142.

Hannon, G.J. (2002) RNA interference. *Nature* 418, 244–251.

Hlavova, M., Turoczy, Z. and Bisova, K. (2015) Improving microalgae for biotechnology: from genetics to synthetic biology. *Biotechnology Advances* 33, 1194–1203.

Hsu-Ching, C. and Anastasios, M. (2013) Marker-free genetic engineering of the chloroplast in the green microalga *Chlamydomonas reinhardtii*. *Plant Biotechnology Journal* 11, 818–828.

Huang, A., He, L. and Wang, G. (2011) Identification and characterization of microRNAs from *Phaeodactylum tricornutum* by high-throughput sequencing and bioinformatics analysis. *BMC Genomics* 12, 337.

Iddo, W. et al. (2018) Enhancing heterologous expression in *Chlamydomonas reinhardtii* by transcript sequence optimization. *The Plant Journal* 94, 22–31.

Ignea, C. et al. (2016) Carnosic acid biosynthesis elucidated by a synthetic biology platform. *Proceedings of the National Academy of Sciences* 113, 3681–3686.

Im, D.J. et al. (2015) Digital microfluidic approach for efficient electroporation with high productivity: transgene expression of microalgae without cell wall removal. *Analytical Chemistry* 87, 6592–6599.

Jagadevan, S. et al. (2018) Recent developments in synthetic biology and metabolic engineering in microalgae towards biofuel production. *Biotechnology for Biofuels* 11, 185.

Jeon, S. et al. (2017) Current status and perspectives of genome editing technology for microalgae. *Biotechnology for Biofuels* 10, 267.

Jeong, B., Wuscharf, D., Zhang, C. and Cerutti, H. (2002) Suppressors of transcriptional transgenic silencing in *Chlamydomonas* are sensitive to DNA-damaging agents and reactivate transposable elements. *Proceedings of the National Academy of Sciences of the United States of America* 99, 1076–1081.

Jia, B. et al. (2016) A vector for multiple gene co-expression in *Chlamydomonas reinhardtii*. *Algal Research-Biomass Biofuels and Bioproducts* 20, 53–56.

Jiang, W., Brueggeman, A.J., Horken, K.M., Plucinak, T.M. and Weeks, D.P. (2014) Successful transient expression of Cas9 and single guide RNA genes in *Chlamydomonas reinhardtii*. *Eukaryotic Cell* 13, 1465–1469.

Kao, P. and Ng, I. (2017) CRISPRi mediated phosphoenolpyruvate carboxylase regulation to enhance the production of lipid in *Chlamydomonas reinhardtii*. *Bioresource Technology* 245, 1527–1537.

Karas, B.J. et al. (2015) Designer diatom episomes delivered by bacterial conjugation. *Nature Communications* 6, 6925.

Kemmer, C. et al. (2010) Self-sufficient control of urate homeostasis in mice by a synthetic circuit. *Nature Biotechnology* 28, 355–360.

Kilian, O., Benemann, C. and Vick, B. (2011) High-efficiency homologous recombination in the oil-producing alga, *Nannochloropsis*. *Proceedings of the National Academy of Sciences of the United States of America* 108, 21265–21269.

Kim, H.S., Devarenne, T.P. and Han, A. (2018) Microfluidic systems for microalgal biotechnology: a review. *Algal Research* 30, 149–161.

Kindle, K.L. (1990) High-frequency nuclear transformation of *Chlamydomonas reinhardtii*. *Proceedings of the National Academy of Sciences of the United States of America* 87, 1228–1232.

Kirst, H., Garciacerdan, J.G., Zurbriggen, A., Ruehle, T. and Melis, A. (2012) Truncated photosystem chlorophyll antenna size in the green microalga *Chlamydomonas reinhardtii* upon deletion of the TLA3-CpSRP43 gene. *Plant Physiology* 160, 2251–2260.

Kwak, S.-Y. *et al.* (2019) Chloroplast-selective gene delivery and expression in planta using chitosan-complexed single-walled carbon nanotube carriers. *Nature Nanotechnology* 14, 447–455.

Lauersen, K.J., Kruse, O. and Mussgnug, J.H. (2015) Targeted expression of nuclear transgenes in *Chlamydomonas reinhardtii* with a versatile, modular vector toolkit. *Applied Microbiology and Biotechnology* 99, 3491–3503.

Lauersen, K.J. *et al.* (2016) Efficient phototrophic production of a high-value sesquiterpenoid from the eukaryotic microalga *Chlamydomonas reinhardtii*. *Metabolic Engineering* 38, 331–343.

Lawson, C.E. *et al.* (2019) Common principles and best practices for engineering microbiomes. *Nature Reviews Microbiology* 17, 725–741.

Leister, D. (2003) Chloroplast research in the genomic age. *Trends in Genetics* 19, 47–56.

Li, X. *et al.* (2016) An indexed, mapped mutant library enables reverse genetics studies of biological processes in *Chlamydomonas reinhardtii*. *The Plant Cell* 28, 367–387.

Li, X. *et al.* (2018) A genome-wide, mapped algal mutant library enables high-throughput genetic studies in a photosynthetic eukaryote. *Cell*, 10 April.

Li, Y., Peng, Y., Hallerman, E.M. and Wu, K. (2014) Biosafety management and commercial use of genetically modified crops in China. *Plant Cell Reports* 33, 565–573.

Lindberg, P., Park, S. and Melis, A. (2010) Engineering a platform for photosynthetic isoprene production in cyanobacteria, using *Synechocystis* as the model organism. *Metabolic Engineering* 12, 70–79.

Liu, X., Fallon, S., Sheng, J. and Curtiss, R. (2011) CO_2-limitation-inducible Green Recovery of fatty acids from cyanobacterial biomass. *Proceedings of the National Academy of Sciences* 108, 6905–6908.

Loera-Quezada, M.M. *et al.* (2016) A novel genetic engineering platform for the effective management of biological contaminants for the production of microalgae. *Plant Biotechnology Journal* 14, 2066–2076.

Lu, Y. and Xu, J. (2015) Phytohormones in microalgae: a new opportunity for microalgal biotechnology? *Trends in Plant Science* 20, 273–282.

Lu, Y. *et al.* (2014) Antagonistic roles of abscisic acid and cytokinin during response to nitrogen depletion in oleaginous microalga *Nannochloropsis oceanica* expand the evolutionary breadth of phytohormone function. *The Plant Journal* 80, 52–68.

Lu, Y. *et al.* (2020) Clade-specific sterol metabolites in dinoflagellate endosymbionts are associated with coral bleaching in response to environmental cues. *mSystems* 5(5), e00765-20.

Lu, Y. *et al.* (2021) Role of an ancient light-harvesting protein of PSI in light absorption and photoprotection. *Nature Communications* 12.

Mackinder, L.C.M. *et al.* (2017) A spatial interactome reveals the protein organization of the algal CO_2-concentrating mechanism. *Cell* 171, 133–147.e114.

Maliga, P. (2004) Plastid transformation in higher plants. *Annual Review of Plant Biology* 55, 289–313.

Matsunaga, T. and Takeyama, H. (1995) Genetic engineering in marine cyanobacteria. *Journal of Applied Phycology* 7, 77–84.

Matsunaga, T., Sakaguchi, T. and Tadakoro, F. (1991) Magnetite formation by a magnetic bacterium capable of growing aerobically. *Applied Microbiology and Biotechnology* 35, 651–655.

Mehrotra, R., Renganaath, K., Kanodia, H., Loake, G.J. and Mehrotra, S. (2017) Towards combinatorial transcriptional engineering. *Biotechnology Advances* 35, 390–405.

Melillo, J.M. *et al.* (1993) Global climate change and terrestrial net primary production. *Nature* 363, 234–240.

Melis, A. (2009) Solar energy conversion efficiencies in photosynthesis: minimizing the chlorophyll antennae to maximize efficiency. *Plant Science* 177, 272–280.

Meng, X. *et al.* (2017) Construction of a genome-wide mutant library in rice using CRISPR/Cas9. *Molecular Plant* 10, 1238–1241.

Mini, P. *et al.* (2018) Agrobacterium-mediated and electroporation-mediated transformation of *Chlamydomonas reinhardtii*: a comparative study. *BMC Biotechnology* 18, 11.

Molnar, A. *et al.* (2009) Highly specific gene silencing by artificial microRNAs in the unicellular alga *Chlamydomonas reinhardtii*. *Plant Journal* 58, 165–174.

Moog, D. *et al.* (2019) Using a marine microalga as a chassis for polyethylene terephthalate (PET) degradation. *Microbial Cell Factories* 18, 171.

Muñoz, C.F., Weusthuis, R.A., D'Adamo, S. and Wijffels, R.H. (2019) Effect of single and combined expression of lysophosphatidic acid acyltransferase, glycerol-3-phosphate acyltransferase, and diacylglycerol acyltransferase on lipid accumulation and composition in *Neochloris oleoabundans*. *Frontiers in Plant Science* 10, 1573.

Neupert, J., Karcher, D. and Bock, R. (2009) Generation of *Chlamydomonas* strains that efficiently express nuclear transgenes. *Plant Journal* 57, 1140–1150.

Ngan, C.Y. *et al.* (2015) Lineage-specific chromatin signatures reveal a regulator of lipid metabolism in microalgae. *Nature Plants* 1, 15107.

Nielsen, J.B. *et al.* (2014) Engineering synergy in biotechnology. *Nature Chemical Biology* 10, 319–322.

Nora, L.C. *et al.* (2019) Recent advances in plasmid-based tools for establishing novel microbial chassis. *Biotechnology Advances* 37, 107433.

Nymark, M., Sharma, A., Sparstad, T., Bones, A.M. and Winge, P. (2016) A CRISPR/Cas9 system adapted for gene editing in marine algae. *Scientific Reports* 6, 24951.

O'Neill, B.M. *et al.* (2011) An exogenous chloroplast genome for complex sequence manipulation in algae. *Nucleic Acids Research* 40, 2782–2792.

Parinov, S. and Sundaresan, V. (2000) Functional genomics in *Arabidopsis*: large-scale insertional mutagenesis complements the genome sequencing project. *Current Opinion in Biotechnology* 11, 157–161.

Plucinak, T.M. *et al.* (2015) Improved and versatile viral 2A platforms for dependable and inducible high-level expression of dicistronic nuclear genes in *Chlamydomonas reinhardtii*. *Plant Journal* 82, 717–729.

Poliner, E., Takeuchi, T., Du, Z.-Y., Benning, C. and Farré, E.M. (2018) Nontransgenic marker-free gene disruption by an episomal CRISPR system in the oleaginous microalga, *Nannochloropsis oceanica* CCMP1779. *ACS Synthetic Biology* 7, 962–968.

Qaim, M. (2009) The economics of genetically modified crops. *Annual Review of Resource Economics* 1, 665–694.

Qin, S., Lin, H. and Jiang, P. (2012) Advances in genetic engineering of marine algae. *Biotechnology Advances* 30, 1602–1613.

Qu, B., Eu, Y., Jeong, W. and Kim, D. (2012) Droplet electroporation in microfluidics for efficient cell transformation with or without cell wall removal. *Lab on a Chip* 12, 4483–4488.

Radakovits, R. *et al.* (2012) Draft genome sequence and genetic transformation of the oleaginous alga *Nannochloropis gaditana*. *Nature Communication* 3, 686.

Ramos-Martinez, E.M., Fimognari, L. and Sakuragi, Y. (2017) High-yield secretion of recombinant proteins from the microalga *Chlamydomonas reinhardtii*. *Plant Biotechnology Journal* 15, 1214–1224.

Rasala, B.A., Muto, M., Sullivan, J. and Mayfield, S.P. (2011) Improved heterologous protein expression in the chloroplast of *Chlamydomonas reinhardtii* through promoter and 5' untranslated region optimization. *Plant Biotechnology Journal* 9, 674–683.

Renuka, N., Guldhe, A., Prasanna, R., Singh, P. and Bux, F. (2018) Microalgae as multi-functional options in modern agriculture: current trends, prospects and challenges. *Biotechnology Advances* 36, 1255–1273.

Rosenwasser, S. *et al.* (2014) Mapping the diatom redox-sensitive proteome provides insight into response to nitrogen stress in the marine environment. *Proceedings of the National Academy of Sciences* 111, 2740–2745.

Roth, M.S. *et al.* (2017) Chromosome-level genome assembly and transcriptome of the green alga *Chromochloris zofingiensis* illuminates astaxanthin production. *Proceedings of the National Academy of Sciences* 114, E4296–E4305.

Salama, E. *et al.* (2019) Can omics approaches improve microalgal biofuels under abiotic stress? *Trends in Plant Science*.

Scaife, M.A. *et al.* (2015) Establishing *Chlamydomonas reinhardtii* as an industrial biotechnology host. *The Plant Journal* 82, 532–546.

Schroda, M., Blocker, D. and Beck, C.F. (2000) The HSP70A promoter as a tool for the improved expression of transgenes in *Chlamydomonas*. *Plant Journal* 21, 121–131.

Schulze, S. *et al.* (2017) Identification of methylated GnTI-dependent N-glycans in *Botryococcus brauni*. *New Phytologists* 215, 1361–1369.

Scotti, N., Rigano, M.M. and Cardi, T. (2012) Production of foreign proteins using plastid transformation. *Biotechnology Advances* 30, 387–397.

Scranton, M.A. *et al.* (2016) Synthetic promoters capable of driving robust nuclear gene expression in the green alga *Chlamydomonas reinhardtii*. *Algal Research* 15, 135–142.

Sengupta, A., Pakrasi, H.B. and Wangikar, P.P. (2018) Recent advances in synthetic biology of cyanobacteria. *Applied Microbiology and Biotechnology* 102, 5457–5471.

Serif, M. *et al.* (2018) One-step generation of multiple gene knock-outs in the diatom *Phaeodactylum tricornutum* by DNA-free genome editing. *Nature Communications* 9, 3924.

Shin, S. *et al.* (2016) CRISPR/Cas9-induced knockout and knock-in mutations in *Chlamydomonas reinhardtii*. *Scientific Reports* 6, 27810.

Slattery, S.S. *et al.* (2018) An expanded plasmid-based genetic toolbox enables Cas9 genome editing and stable maintenance of synthetic pathways in *Phaeodactylum tricornutum*. *ACS Synthetic Biology* 7, 328–338.

Sodeinde, O.A. and Kindle, K.L. (1993) Homologous recombination in the nuclear genome of *Chlamydomonas reinhardtii*. *Proceedings of the National Academy of Sciences of the United States of America* 90, 9199–9203.

Specht, E.A. and Mayfield, S.P. (2013) Synthetic oligonucleotide libraries reveal novel regulatory elements in *Chlamydomonas* chloroplast mRNAs. *ACS Synthetic Biology* 2, 34–46.

Tarver, J.E. *et al.* (2015) MicroRNAs and the evolution of complex multicellularity: identification of a large, diverse complement of microRNAs in the brown alga *Ectocarpus*. *Nucleic Acids Research* 43, 6384–6398.

Taylor, R.S., Tarver, J.E., Hiscock, S.J. and Donoghue, P.C. (2014) Evolutionary history of plant microRNAs. *Trends in Plant Science* 19, 175–182.

Thirietrupert, S. *et al.* (2016) Transcription factors in microalgae: genome-wide prediction and comparative analysis. *BMC Genomics* 17, 282.

Tran, M. *et al.* (2013) Production of unique immunotoxin cancer therapeutics in algal chloroplasts. *Proceedings of the National Academy of Sciences of the United States of America* 110, 14.

Trentacoste, E.M. *et al.* (2013) Metabolic engineering of lipid catabolism increases microalgal lipid accumulation without compromising growth. *Proceedings of the National Academy of Sciences of the United States of America* 110, 19748–19753.

Ungerer, J. *et al.* (2012) Sustained photosynthetic conversion of CO_2 to ethylene in recombinant cyanobacterium *Synechocystis* 6803. *Energy and Environmental Science* 5, 8998–9006.

Ungerer, J., Wendt, K.E., Hendry, J.I., Maranas, C.D. and Pakrasi, H.B. (2018) Comparative genomics reveals the molecular determinants of rapid growth of the cyanobacterium *Synechococcus elongatus* UTEX 2973. *Proceedings of the National Academy of Sciences of the United States of America* 115, E11761–e11770.

Vieler, A. *et al.* (2012) Genome, functional gene annotation, and nuclear transformation of the heterokont oleaginous alga *Nannochloropsis oceanica* CCMP1779. *PLOS Genetics* 8, e1003064.

Vila, M. *et al.* (2012) Promoter trapping in microalgae using the antibiotic paromomycin as selective agent. *Marine Drugs* 10, 2749–2765.

Wang, D. *et al.* (2014) *Nannochloropsis* genomes reveal evolution of microalgal oleaginous traits. *PLOS Genetics* 10, e1004094.

Wang, Q. *et al.* (2016) Genome editing of model oleaginous microalgae *Nannochloropsis* spp. by CRISPR/Cas9. *Plant Journal* 88, 1071–1081.

Wasmer, M. (2019) Roads forward for European GMO policy: uncertainties in wake of ECJ judgment have to be mitigated by regulatory reform. *Frontiers in Bioengineering and Biotechnology* 7, 132.

Wei, L. *et al.* (2017) RNAi-based targeted gene knockdown in the model oleaginous microalgae *Nannochloropsis oceanica*. *Plant Journal* 89, 1236–1250.

Wei, L. *et al.* (2019) Knockdown of carbonate anhydrase elevates *Nannochloropsis* productivity at high CO2 level. *Metabolic Engineering* 54, 96–108.

Werpy, T. and Petersen, G. (2004) *Top Value Added Chemicals from Biomass: Volume I – Results of Screening for Potential Candidates from Sugars and Synthesis Gas*. National Renewable Energy Lab, Golden, Colorado.

Wichmann, J., Baier, T., Wentnagel, E., Lauersen, K.J. and Kruse, O. (2018) Tailored carbon partitioning for phototrophic production of (E)-α-bisabolene from the green microalga *Chlamydomonas reinhardtii*. *Metabolic Engineering* 45, 211–222.

Wijffels, R.H., Kruse, O. and Hellingwerf, K.J. (2013) Potential of industrial biotechnology with cyanobacteria and eukaryotic microalgae. *Current Opinion in Biotechnology* 24, 405–413.

Wilson, S.A. and Roberts, S.C. (2012) Recent advances towards development and commercialization of plant cell culture processes for the synthesis of biomolecules. *Plant Biotechnology Journal* 10, 249–268.

Wu, W., Wang, P.-H., Lee, D.-J. and Chang, J.-S. (2017) Global optimization of microalgae-to-biodiesel chains with integrated cogasification combined cycle systems based on greenhouse gas emissions reductions. *Applied Energy* 197, 63–82.

Xin, Y. *et al.* (2017) Producing designer oils in industrial microalgae by rational modulation of co-evolving type-2 diacylglycerol dcyltransferases. *Molecular Plant* 10, 1523–1539.

Xue, J. *et al.* (2015) Genetic improvement of the microalga *Phaeodactylum tricornutum* for boosting neutral lipid accumulation. *Metabolic Engineering* 27, 1–9.

Yadav, V.G., De Mey, M., Giaw Lim, C., Kumaran Ajikumar, P. and Stephanopoulos, G. (2012) The future of metabolic engineering and synthetic biology: towards a systematic practice. *Metabolic Engineering* 14, 233–241.

Young, R.E.B. and Purton, S. (2016) Codon reassignment to facilitate genetic engineering and biocontainment in the chloroplast of *Chlamydomonas reinhardtii*. *Plant Biotechnology Journal* 14, 1251–1260.

Zhang, B., Shanmugaraj, B. and Daniell, H. (2017) Expression and functional evaluation of biopharmaceuticals made in plant chloroplasts. *Current Opinion in Chemical Biology* 38, 17–23.

Zhang, C. and Hu, H. (2014) High-efficiency nuclear transformation of the diatom *Phaeodactylum tricornutum* by electroporation. *Marine Genomics* 16, 63–66.

Zhang, J. *et al.* (2014) Overexpression of the soybean transcription factor GmDof4 significantly enhances the lipid content of *Chlorella ellipsoidea*. *Biotechnology for Biofuels* 7, 128–128.

Zhao, T. *et al.* (2007) A complex system of small RNAs in the unicellular green alga *Chlamydomonas reinhardtii*. *Genes & Development* 21, 1190–1203.

2 Application and Mechanism Analysis of Photosynthetic Microbial Coculture Systems for Bioproduction

Jin Wang[1,2], Xinyu Song[1,2,3]*, Lei Chen[1,2] and Weiwen Zhang[1,2,3]*

[1]Laboratory of Synthetic Microbiology, School of Chemical Engineering & Technology, Tianjin University, Tianjin, China; [2]Key Laboratory of Systems Bioengineering and Frontier Science Center of Synthetic Biology, Ministry of Education of China, Tianjin University, Tianjin, China; [3]Center for Biosafety Research and Strategy, Tianjin University, Tianjin, China

Abstract

Photosynthetic microorganisms typically live and interact with other heterotrophic microbes by establishing a stable symbiotic relationship in complex communities in nature. Natural photosynthetic microbial coculture systems have been widely used in environmental engineering, microbial cell factories and other fields due to their adaptability and stability, such as surface effluent treatment, soil remediation and biodegradation. Inspired by the commonly found symbiotic relationships between photosynthetic and heterotrophic species in nature, synthetic photosynthetic microbial coculture systems are designed and reconstructed by synthetic biology, as such systems are widely seen to hold enormous potential for foundational research as well as novel industrial, medical and environmental applications, with many proof-of-principle studies available. Although several synthetic coculture systems have been constructed for the production of value-added chemicals, such as isoprene and 3-hydroxypropionic acid, their productivities were still low for further large-scale manufacturing, partially due to the relatively poor understanding of the cell-interaction mechanism governing the coculture systems. To construct light-driven coculture systems with high efficiency and stability, omics technologies were recently employed to explore the metabolic mechanism underlying the interaction between autotrophs and heterotrophs. This chapter summarizes the recent progress in the application and mechanism analysis of photosynthetic microbial coculture systems for bioproduction.

Keywords: photosynthetic organism; coculture systems; application; mechanism; omics analysis

Abbreviations: F6P, fructose 6-phosphate; 3-HP, 3-hydroxypropionic acid; RNA, ribonucleic acid; DNA, deoxyribonucleic acid; ROS, reactive oxygen species; FDA, fluorescein diacetate; DDT, dichlorodiphenyltrichloroethane; PHB, polyhydroxybutyrate; Glk, glucokinase; FrkA, fructokinase; PHA, polyhydroxyalkanoates; NMR, nuclear magnetic resonance; GC, gas chromatography; RNA-seq, mRNA sequencing; 2-OG, 2-oxoglutarate; Rubisco, ribulose-1,5-bisphosphate carboxylase/oxygenase; NADH, nicotinamide adenine dinucleotide; PhoRB, phosphate sensor regulation; TC, two-component system; PstBS, phosphate transporters; BFR, bacterial ferritin

*Address for correspondence: wwzhang8@tju.edu.cn

1. Introduction

Symbiosis refers to close and often long-term interactions between organisms of different species, which are likely widespread and functionally relevant in diverse biological systems (Hedayat and Lapraz, 2019). Corals, together with jellyfish and sea anemones, belong to a group of animals called cnidarians that establish a symbiotic relationship with photosynthetic algae living inside their cells, which allows the building of entire reefs (Stanley and Van, 2009). Leguminous plants form a symbiotic relationship with soil bacteria rhizobia, and this process is defined as the most efficient contribution to biologically fixed nitrogen, an extremely important process for nitrogen availability in many terrestrial ecosystems (Schulte et al., 2021). Natural symbiosis systems have long been used to study the interactions between cell populations and are fundamental to cell–cell interaction studies of any kind (Goers et al., 2014). Recently, more synthetic coculture systems based on symbiosis have been designed and reconstructed by synthetic biology, as such systems are widely seen to hold enormous potential for foundational research as well as novel industrial, medical and environmental applications, with many proof-of-principle studies available (Shong et al., 2012). Compared to traditional biosynthesis that employs a clonal population of recombinant microbes, such as *Escherichia coli* (hereafter *E. coli*) or yeast, the limitations of metabolic load and the number of exogenous elements that can be cloned and optimized in a single cell can be alleviated by using coculture systems (Shong et al., 2012). Moreover, in coculture systems, different strains are divided into functional compartments, which combine the catalytic properties of different species to produce new products. Finally, relationships between cells are dynamically balanced, leading to stronger adaptability and stability to the fluctuant environment (Keller and Surette, 2006; Pai et al., 2009; Jia et al., 2016).

Photoautotrophic species, such as cyanobacteria, with the capability of producing organic matter from CO_2 using solar energy, have attracted increased attention as environmentally friendly and sustainable 'microbial cell factories' for the production of carbohydrate feedstocks to support traditional fermentation processes (Quintana et al., 2011; Hays and Ducat, 2015). For example, sucrose is an easily fermentable feedstock. Several cyanobacterial species are capable of synthesizing and secreting sucrose as an osmolyte under appropriate environmental stimuli, such as osmotic pressure (Hagemann, 2011), and this process can be sustained over a long period of time and at higher levels than that from plant feedstocks such as sugarcane and beet (Han and Watson, 1992; Niederholtmeyer et al., 2010). However, purification of sucrose from cyanobacterial cultivation supernatant is costly, and the system is easily contaminated, creating barriers to any scale-up cultivation (Chisti, 2013). In addition, any application of photosynthetic cell factories in scale-up facilities is always restricted by challenges from harsh environments, suggesting that the adaptability and compatibility of cyanobacterial cell factories should be further improved to facilitate industrial-scale biomanufacturing (Luan and Lu, 2018).

Light-driven symbiotic systems composed of photoautotrophic and heterotrophic species are widespread in nature. Lichens, for example, are a typical example of a symbiotic partnership of photoautotrophic algae and heterotrophic fungi (Rana et al., 2012). In lichens, algae synthesize organic matter using sunlight and provide organic carbon sources for fungal growth and reproduction, while fungi provide growth factors and transport inorganic nutrients for algae (Rana et al., 2012). In recent years, increasing evidence has suggested that the exchange of essential metabolites between microorganisms could be a crucial process that can significantly affect the growth, composition and structural stability of microbial communities in nature (Beliaev et al., 2014; Giri et al., 2021). For instance, in aquatic environments, the ecological interaction between photoautotrophic and heterotrophic species is based on cross-feeding and metabolite exchange (Paerl and Pinckney, 1996). To date, multiprotein complexes that cross cell membrane(s) and extracellular vesicles have been evaluated in photoautotrophs for transporting materials from the interior to the exterior of the cell (Lima et al., 2020), which facilitate the secretion of various chemicals ranging from targeted photosynthetic intermediates, such as glycolate, osmolytes and fatty acids, and extracellular polymeric substances, to the products of cell lysis, including sugars, proteins, lipids and nucleic acids (Seymour et al., 2010; Bruckner et al., 2011). These organic compounds can support the cell growth of heterotrophic species. In addition, heterotrophic species are also thought to provide essential micronutrients, such as vitamins and amino acids, which are beneficial to maintain high photosynthetic productivity (Giri et al., 2021). In addition, some positive effects on promoting cell growth were also observed, which may be attributed to the decreased oxidative stress by heterotrophs through reactive oxygen species (ROS) scavenging (Morris et al., 2008; Li et al., 2017). Inspired by the environmental adaptability and stability of symbiotic systems commonly found in nature, more efforts have been made to use symbiotic systems in the fields of environmental engineering and bioproduction (Miura et al., 1992; Kawaguchi et al., 2001; Tang et al., 2010; Markou and Georgakakis, 2011; Subashchandrabose et al., 2011; Angelis et al., 2012; Breuer et al., 2013; Fradinho et al., 2013; Liu and Vyverman, 2015; Mahdavi et al., 2015; Manjunath et al., 2016; Ryu et al., 2016; Xu et al., 2016; Ban et al., 2017; Xu et al., 2017; Li et al., 2019). Meanwhile, considering the limited flexibility and universality of natural symbiotic systems, increasing efforts have been made in recent years to design artificial routes of metabolite interchange to construct new symbiotic systems for environmental management, cell factories and other application fields (Brenner et al., 2008; Fredrickson, 2015).

Although several artificial coculture systems have been constructed for the production of value-added chemicals, such as isoprene (Liu et al., 2021) and 3-hydroxypropionic acid (Zhang et al., 2020), their productivities were still low for further large-scale manufacturing. To construct light-driven coculture systems with high efficiency and stability, it is necessary to fully understand the metabolic mechanism underlying the interaction between autotrophs and heterotrophs. Although several previous studies have shown that cyanobacterial cell growth could be improved in coculture systems (Li et al., 2017), the mechanism has yet to be determined. Moreover, while it is fully expected that the mechanism involves more than just a single gene or even a single metabolic node, to date only a few studies have utilized global-based omics techniques to explore the interaction mechanism (Nouaille et al., 2009; Bobadilla Fazzini et al., 2010; Bernstein et al., 2017). Due to the complexity of the coculture structure, the challenge of studying the interaction mechanism is also increased. Integrated omics analysis could be a good approach to obtain a 'panorama' of cells in coculture systems and reveal novel insights into the biological mechanism (Liu et al., 2021). For example, Amin et al. analyzed the signaling and interaction between diatoms and associated bacteria through integrated metabolite and transcriptomic analysis, in which tryptophan and indole-3-acetic acid were determined to be the key signaling molecules involved in the complex exchange of nutrients (Amin et al., 2015), demonstrating that the approach of integrated transcriptomics, proteomics and metabolomics should be adopted to explore microbial interactions in coculture systems.

This chapter summarizes recent progress on the application and mechanism analysis of coculture systems consisting of autotrophic and heterotrophic species.

2. Natural Photoheterotrophic Coculture Systems in Environmental Management

Natural photosynthetic microbial coculture systems have been widely used in environmental engineering, microbial cell factories and other fields due to their adaptability and stability (Table 2.1) (Miura et al., 1992; Kawaguchi et al., 2001; Tang et al., 2010; Markou and Georgakakis, 2011; Subashchandrabose et al., 2011; Angelis et al., 2012; Breuer et al., 2013; Fradinho et al., 2013; Liu and Vyverman, 2015; Mahdavi et al., 2015;

Table 2.1. Applications of photoheterotrophic coculture systems.

	Application	Co-culture system	Highlights	References
Environmental engineering	Surface sewage treatment	*Cladophora*, *Klebsormidium* and *Bacillus*, *Pseudomonas*	Efficient, clean, low-cost, and thorough treatment	Markou and Georgakakis (2011); Breuer et al. (2013); Liu and Vyverman (2015)
	Soil remediation	*Providencia* sp., *Anabaena* sp. and *Calothrix* sp.	Avoid the reduction of soil fertility; can also act on deep soil	Manjunath et al. (2016); Li et al. (2019)
	Biodegradation	*S. obliquus* GH2 and four heterotrophics	Simultaneously degrades various pollutants	Tang et al. (2010); Subashchandrabose et al. (2011); Mahdavi et al. (2015); Ryu et al. (2016)
Microbial cell factory	Lipid	Strains in activated sludge	Achieves continuous production and saves costs	Fradinho et al. (2013)
	Polysaccharide	*Spirulina platensis* (hereafter as *S. platensis*), *Chlorella vulgaris* (hereafter as *C. vulgaris*) and four fungi	For the production of new polysaccharides	Angelis et al. (2012)
	Hydrogen	*Chlamydomonas reinhardtii* (hereafter as *C. reinhardtii* MGA 161) and *Pseudomonas* sp. BS4 (hereafter as *Pseudomonas* BS4)	Increase hydrogen yield	Miura et al. (1992); Kawaguchi et al. (2001); Xu et al. (2016); Ban et al. (2017); Xu et al. (2017)

Manjunath *et al.*, 2016; Ryu *et al.*, 2016; Xu *et al.*, 2016; Ban *et al.*, 2017; Xu *et al.*, 2017; Li *et al.*, 2019), such as surface effluent treatment, soil remediation and biodegradation.

2.1 Surface effluent treatment

Large amounts of industrial production and human activities lead to excessive accumulation of nutrients in surface water, resulting in serious surface-water pollution (Dunne *et al.*, 2013). To avoid eutrophication of natural water and maintain a healthy aquatic ecosystem, the total nitrogen and phosphorus contents in water should be less than 0.3 mg/L and 0.02 mg/L, respectively (Hu *et al.*, 2010; Liu and Vyverman, 2015; Liu *et al.*, 2016; Hernández-Crespo *et al.*, 2017). In general, traditional physical and chemical sewage treatment methods have some shortcomings, such as higher energy consumption and operating cost, incomplete treatment of pollutants, and easy formation of secondary pollution. In contrast, in recent years, biological sewage treatment using the coculture system of photosynthetic microorganisms has been considered an efficient and clean remediation method (Shah, 2019).

In natural coculture systems, photosynthetic microorganisms can effectively utilize nitrogen and phosphorus in wastewater for growth and to produce oxygen necessary for the cell growth of heterotrophic bacteria; then, heterotrophic bacteria can break down organic pollutants in wastewater (Renuka et al., 2015; Liu et al., 2016). To date, many microalgae and heterotrophic bacteria have been isolated based on their ability to absorb and decompose nutrients from nature. For example, Vyverman et al. reported that *Cladophora*, *Klebsormidium* and *Anabaena* have high absorption capacities for nitrogen and phosphorus (Hernández-Crespo et al., 2017). The nitrogen uptake rates of these algae were 55.9–70.6 mg NO_3^--N/(g DW·d) under laboratory conditions (23°C). The absorption rates of phosphorus were 12.7–27.5 mg PO_4^{3-}-P/(g DW·d). In addition, these algae have good adaptability to various environmental conditions, such as temperatures ranging from 15°C to more than 35°C and pH values ranging from 3.0 to 9.0, which greatly broaden the scope of their application for effluent treatment, including purification of aquaculture wastewater, river and lake water, and urban wastewater (Markou and Georgakakis, 2011; Breuer et al., 2013; Ma et al., 2016). Loutseti et al. (2009) used a bacterial-algal coculture system to treat copper and calcium effluent, and the ion removal rates reached 80% and 100%, respectively. Subsequently, Bai and Acharya (2016) treated sulfamethoxazole and triclosan wastewater with a bacteria-algae coculture system, and the treatment efficiencies reached 32% and 74%, respectively.

2.2 Soil remediation

Both natural and human activities usually lead to soil fertility degradation and salinization; according to statistics, there is approximately 1 billion hm^2 of salinized land worldwide (Wang et al., 1993). Restoration of degraded lands affected by salinization could regain large amounts of soil resources (Li et al., 2019). Traditional soil desalinization methods mainly include leaching, removal of the surface saline-alkali layer, use of an electric field and other physical methods, which are only suitable for the upper soil and generally tend to reduce soil fertility in the process of desalination (Abou-Shady, 2016; Ullah et al., 2017). Some photosynthetic microorganisms, such as *Nostoc* and *Anabaena*, function in soil fertility recovery and desalinization, which contribute to the ability to fix nitrogen and produce extracellular matrix and extracellular polysaccharides (Li et al., 2019). During desalination, photosynthetic microorganisms usually interact with other heterotrophic microorganisms to improve soil salinity; this process cannot be accomplished by a single species. The biomass and production capacity of extracellular polysaccharide substances are indicators that identify the ability of the coculture system to remediate soil (Li et al., 2019). In a coculture system, the properties of extracellular polysaccharides produced in cyanobacteria under some stress conditions, including high salinity, high blue light and nitrogen starvation, could be affected by heterotrophic bacteria by regulating the molecular weight, branched degree, substitution base and ultramolecular structure of extracellular polysaccharides (Angelis et al., 2012; Yoshimura et al., 2012; Han et al., 2014; Tiwari et al., 2015). The features make the coculture systems suitable for different saline soil types. In addition, cyanobacteria can also interact with bacteria and fungi in the soil to improve the soil environment and soil fertility. For the first time, Prasanna et al. elucidated the positive and dynamic interaction between bacteria and cyanobacteria and its application prospects in improving crop yields such as wheat (Nain et al., 2010). The study found that wheat grown in soil with an additional cyanobacterial species and two heterotrophic bacterial species had twice as much head weight as those inoculated with full doses of phosphate and potassium. In addition, the activities of dehydrogenase, FDA hydrolase, alkaline phosphatase and microbial biomass were significantly increased in the soil inoculated with the coculture system. It was also found in field experiments that the nitrogen source cost of 40–80 kg N/hm^2 could be reduced by the application of the coculture system in soil. Nain et al. compared a common fertilizer ($N_{60}P_{60}K_{60}$) and found that the protein content of wheat grains grown in mixed soil inoculated with cyanobacteria increased by 18.6%, and the three trace elements of Fe, Mn and Cu increased by 105.3%, 36.7% and 150.0%, respectively (Manjunath et al., 2016).

2.3 Biodegradation

Although algae require several metals for normal growth and metabolism, when deprived of these metals, algae exhibit systematic, specific stress responses (Glaesener et al., 2013; Kropat et al., 2015; Malasarn et al., 2013). Metal ions can be actively ingested into cells for metabolic quenching or as a defense tool to avoid cell poisoning in algae (Zhang et al., 2013). During the growth of microalgae, some metal chelators will also be released. In addition, the pH of the living environment of algae is generally high, which will also lead to the precipitation of heavy metal ions (Muñoz and Guieysse, 2006; Subashchandrabose et al., 2013). However, in general, a single kind of photosynthetic microorganism would have a poor tolerance to metal ions and would require a long-term domestication process before application; the metal species that can be treated are also relatively simple (Shen et al., 2018). Previous studies have confirmed that many photoheterotrophic coculture systems have the ability to complex heavy metal ions and degrade organic pollutants by degrading polycyclic aromatic hydrocarbons, aliphatic chain hydrocarbons, and saturated and aromatic hydrocarbons (Boivin et al., 2007; Tang et al., 2010). Coculture systems are superior in handling a wide range of metal ions to a single species compared with a single kind of photosynthetic microorganism under the same conditions, which improves the convenience of coculture systems in practical applications (Cheloni and Slaveykova, 2018). Extracellular polymeric substances have good adsorption efficiency for heavy metals. This polymer is a viscous substance secreted by microbial cells, and its main components include proteins, polysaccharides, nucleic acids and lipids. This composition enables it to adsorb through the interaction of negatively charged coordination groups with metal ions. Therefore, the coculture system consists of algae and heterotrophic species with a remarkable ability to detoxify and assimilate metals to each other in an environment rich in metal ions.

In addition to the removal of heavy metal ions, photoheterotrophic coculture systems are expected to degrade pesticides and phenolic compounds. Subashchandrabose et al. reported that the photoheterotrophic coculture system can effectively degrade organophosphorus insecticides, such as phos, quinthiophos and methyl parathion (Subashchandrabose et al., 2011), and found that the system can degrade a variety of pesticides, including DDT, atrazine and α-endosulfan (Subashchandrabose et al., 2013). Photoheterotrophic coculture systems have superior efficiency in the degradation of phenolic compounds to single species (Muñoz and Guieysse, 2006; Subashchandrabose et al., 2013; Mahdavi et al., 2015). The removal efficiencies of organic pollutants, including black oil, acetonitrile, phenol, naphthalene, benzopyrene, dibenzofuran and azo compounds, can reach 100% by using coculture systems. However, the degradation rate of using *Scenedesmus quadricauda* alone was only 27% (Ryu et al., 2016). Tang et al. reported that multiple substances in crude oil could be simultaneously degraded when using an artificial coculture system composed of *Scenedesmus obliquus* GH2 (hereafter *S. obliquus* GH2) and four kinds of heterotrophic bacteria that can degrade different pollutants (Tang et al., 2010). Surprisingly, straight-chain alkanes were almost completely degraded using this coculture system. Ryu et al. also found that photoheterotrophic coculture systems consisting of *Thiobacillus* bacteria and *Micractinium* algae can degrade a toxic substance in wastewater called thiocyanate (Ryu et al., 2015).

3. Construction of Synthetic Photoheterotrophic Coculture Systems by Chassis Engineering

Due to hundreds of millions of years of evolution to adapt to specific living environments, natural photoheterotrophic coculture systems have some advantages in stability, but the flexibility and universality of their application are limiting. Synthetic biology aims to redesign and transform existing and natural biological systems through design and construction from scratch under the guidance of engineering ideas (Andrianantoandro et al., 2006). In recent years, the development of synthetic biology has opened up a new direction for the application of coculture systems.

3.1 Modifying the cyanobacteria chassis

In nature, there are many photosynthetic microbial coculture systems formed by long-term evolution, but they lack flexibility and universality in the application process. Sucrose, as a common carbon source, has been proved to be the most representative carbohydrate metabolite of photosynthetic microorganisms, especially cyanobacteria. Most freshwater cyanobacteria strains synthesize and accumulate sucrose as a compatible solute to resist salt stress and maintain the osmotic pressure balance between the inside and outside of the cell, thus maintaining the homeostasis of the internal environment of the cell (Hagemann, 2011). For natural cyanobacteria cells, when the concentration of sucrose accumulated inside the cell is sufficient to resist osmotic pressure, the synthesis and degradation of sucrose is in a balanced state, and this natural regulation mode of sucrose metabolism fundamentally limits the increase in sucrose production of cyanobacteria. To overcome this limitation, Ducat et al. introduced the *cscB* gene from *E. coli* W (an *Escherichia coli* that naturally utilizes sucrose) into *Synechococcus elongatus* PCC 7942 (hereafter *S. elongatus* PCC 7942), first achieving extracellular sucrose secretion in cyanobacteria. This method effectively broke through the bottleneck of the sucrose synthesis ability of cyanobacteria and laid a foundation for the subsequent artificial construction of photosynthetic microorganism coculture systems (Ducat et al., 2012). To further improve sucrose production, the sucrose biosynthesis pathway was genetically engineered by generating Δ*invA* and Δ*glgC* strains through the replacement of endogenous genes by homologous recombination (Fig. 2.1) and constructed strains

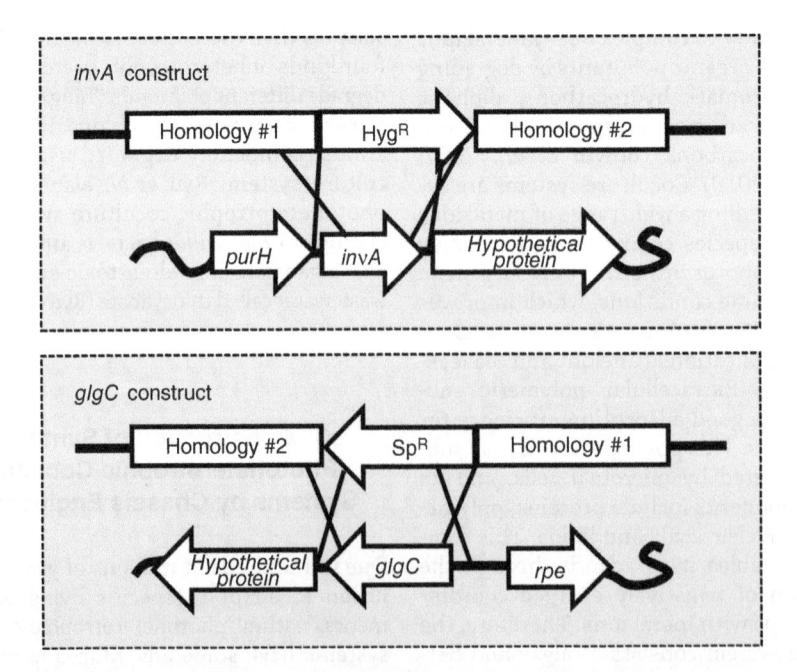

Fig. 2.1. Schematic of constructs designed for site-directed elimination of target genes (*invA* and *glgC*). Resistance cassettes were flanked by DNA homologous to sequences of neighboring target genes to allow for the selection of strains that had lost target genes following homologous recombination. Any additional/fewer base pairs recombined relative to the start or stop codons of the target gene are indicated above the construct depiction.

overexpressing *glgP* and *galU*, biasing glucose-1-phosphate flux towards sucrose and away from storage as glycogen to minimize competitive glucose or sucrose consumption reactions, which resulted in a sucrose yield up to 36.1 mg/(L·h) (Ducat *et al.*, 2012). This yield is sufficient to support the growth of certain heterotrophic bacteria. This method effectively broke through the bottleneck of the sucrose synthesis ability of cyanobacteria and laid a foundation for the subsequent generation of artificial photoheterotrophic coculture systems.

After that, Ducat's team continued trying to build an artificial coculture system consisting of heterotrophic species, including *Halomonas boliviensis* (hereafter as *H. boliviensis*) (Weiss *et al.*, 2017), *E. coli* and *Bacillus subtilis*, and the sucrose-producing *S. elongatus* PCC 7942 (Hays *et al.*, 2017). They found that coculture systems consisting of *E. coli* or yeast with cyanobacteria were able to remain relatively stable from weeks to months, and in the face of environmental variation (mainly the rhythmic change of light), the system showed very strong robustness. Ducat *et al.* further optimized the artificial coculture system using an immobilization strategy (Weiss *et al.*, 2017). The immobilized treatment has been proved by experiments to have little effect on the survival and sucrose production of cyanobacteria cells and can effectively limit cell growth so that the carbon source consumed by growth can be more effectively transferred to sucrose production. By immobilizing sucrose-producing *S. elongatus* PCC 7942 with sodium alginate, the sugar-production efficiency of *S. elongatus* PCC 7942 was increased approximately threefold. Saccharogenic *S. elongatus* PCC 7942, which had been immobilized, was more efficient than the non-immobilized cells. The coculture system with *H. boliviensis*, a natural PHB producer, can last for up to five months. The *H. boliviensis* biomass remained stable, and PHB production increased throughout the incubation process. Similarly, the system also showed strong robustness and resistance to environmental fluctuations and foreign microbial contamination, without additional antibiotics and other strategies to maintain system stability. In addition to *H. boliviensis*, Ducat *et al.* showed that cocultured strains can be replaced with *E. coli* for PHB production, which demonstrates the flexibility and universality of the synthetic bacterial coculture system. They finally realized the production of PHB, amylase and other high-value-added substances, which also proves the value of synthetic biology technologies and strategies in the application of artificial photoheterotrophic coculture systems.

3.2 Engineering a heterotrophy chassis

Using cyanobacteria to synthesize organic carbon sources to supply the growth of heterotrophic bacteria is the mainstream direction of constructing artificial photoheterotrophic coculture systems. Ducat *et al.* (2012) inoculated *S. cerevisiae* into *S. elongatus* PCC 7942, which can produce a sucrose system directly, and found that yeast cells could survive well in this system, which preliminarily confirmed the feasibility of the artificial construction of a photosynthetic microbial coculture system. Based on this, Hays *et al.* continued to explore the coculture of three classic engineered bacteria (*E. coli*, *Bacillus subtilis* and yeast) with *S. elongatus* PCC 7942, which can produce sucrose (Hays *et al.*, 2017). In previous studies, the heterotrophic bacteria used in the artificial coculture system should have the ability to utilize sucrose; however, many heterotrophic bacteria cannot naturally use sucrose as a carbon source. Therefore, it is necessary to modify heterotrophic microorganisms for sucrose utilization.

Sucrose breakdown is catalyzed by an invertase (Inv) (Kirsch *et al.*, 2018), resulting in the production of glucose and fructose, which are likely phosphorylated to glucose-6-P by glucokinase (Glk) and fructose-6-P by fructokinase (FrkA) and cycled back into glycolysis (Mills *et al.*, 2020). Löwe *et al.* introduced the *cscA-cscB* system of *E. coli* W from naturally available sucrose into *Pseudomonas putida* (hereafter *P. putida*), in which *cscA* is a sucrose hydrolase that can hydrolyse sucrose into glucose and fructose (Löwe *et al.*, 2017).

Transporter encoded by *cscB* can promote the extracellular to intracellular transfer of sucrose, so *P. putida* has the ability to use sucrose and thus construct a light-activated hybrid system for PHA synthesis (Sahin-Tóth et al., 1995). To ensure that *E. coli* BL21 utilizes sucrose as the sole carbon source, Zhang et al. cloned and expressed the essential genes for sucrose metabolism, namely, *cscB* (*ECW_m2594*), *cscK* (*ECW_m2595*) and *cscA* (*ECW_m2596*), into *E. coli* BL21 to generate an engineered strain, *E. coli* cscN. In addition, to synthesize 3-HP, the malonyl-CoA reductase-coding gene *mcr* (*Caur_2614*) was introduced into *E. coli* cscN, resulting in the engineered strain *E. coli* ABKm (Zhang et al., 2020) (Fig. 2.2).

4. Application of Synthetic Photoheterotrophic Coculture Systems for the Production of Value-added Chemicals

4.1 Lipid production

Polyhydroxyalkanoate (PHA) is the most abundant neutral lipid and is stored as a carbon and energy compound in bacterial cells (Steinbüchel, 1991). Microbial PHA have been developed as biodegradable plastics for the past many years (Hao and Zhu, 2005). Although the technologies for the production of PHA in coculture systems are mature, most coculture systems consist of heterotrophic species (Reis et al., 2003; Salehizadeh and Loosdrecht, 2004; Albuquerque et al., 2010; Moita et al., 2014). In these cases, electron acceptors (such as oxygen) must be provided to the system, and aeration is the most commonly used. Aeration is generally considered to be the most energy-consuming demand in mixed-culture bioreactors, such as bioreactors used in activated sludge wastewater treatment plants (Rosso et al., 2008). The need for aeration substantially increases the operational costs of the MMC process for PHA production. The first coculture system for PHA production consisting of bacteria and algae was published by Fradinho et al. (2013). In this system, PHA production mainly consists of two stages, 'PHA-enriched synthetic strains' and 'PHA batch synthesis' (Fradinho et al., 2013). The first phase uses the 'Feast and Famine' regime. The accumulation of PHA requires acetic acid as the substrate, so the heterotrophic species in activated sludge first accumulate PHA at the stage when acetic acid is sufficient. Once the substrate is exhausted, the heterotrophic species can use the oxygen produced by algae to consume the PHA synthesized to maintain survival. Through the repeated cycle of this stage, the PHA-producing heterotrophic species can survive in the presence of algae, which results in the enrichment of PHA-producing strains. In the second stage, PHA production is achieved by continuously supplementing acetic acid so that the heterotrophic species can continuously produce without consuming PHA. In the study, the

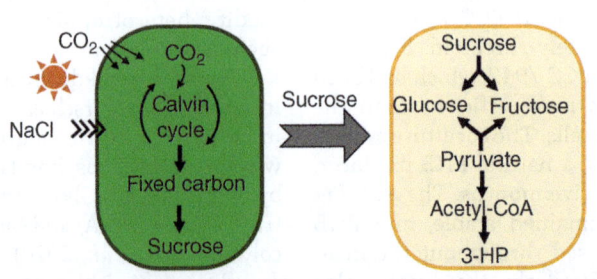

Synechococcus elongatus UTEX 2973   *Escherichia coli* BL21

Fig. 2.2. Schematic diagram of the artificial consortium system. The engineered *S. elongatus* UTEX 2973 secreted sucrose under osmotic stress to support *E. coli* growth and the synthesis of 3-HP under photoautotrophic growth conditions.

maximum content of PHA in the bacteria was as high as 20% (Fradinho et al., 2013). After screening, the coculture system of bacteria and algae had a higher PHA yield than the pure heterotrophic system, and the mutual relationship between bacteria and algae made the system stable over a long period of time to achieve the continuous production of PHA.

4.2 Polysaccharide production

Microbial extracellular polysaccharides are polymers that can serve as renewable sources for hydrocolloids used in food, pharmaceutical and other industrial applications (Zhang et al., 2010a). For example, β-D-glucan has immunomodulatory and tumor-suppressive functions, bacterial cellulose can be used in the manufacture of sound card film, and hyaluronic acid can be used in cosmetics (Sutherland, 1998). According to statistics, approximately 76 species of microorganisms belonging to 49 genera can produce extracellular polysaccharides (Ding et al., 2014). Microbial polysaccharides include intracellular, cell wall and extracellular polysaccharides. Extracellular polysaccharides are those produced by microorganisms in large quantities and are easily separated from bacteria. They can also be industrialized through deep fermentation. Generally, microbial polysaccharides are produced by fermentation with starch hydrolysis as a carbon source or can be directly produced by using soluble starch through microbial enzymes. To date, microbial extracellular polysaccharides have been produced in large quantities, mainly including xanthan gum, gellan gum, sclerotinia glucan, peduncle polysaccharide, and thermosetting polysaccharide (Wei, 2002).

Photoheterotrophic coculture systems are also suitable for the production of extracellular polysaccharides. Coculture systems are composed of at least two microorganisms; the interaction between these microorganisms may lead to stress conditions, and the response of microorganisms to stress conditions may generate a large amount of EPS, such as growing under these adverse conditions as a metabolic strategy (Angelis et al., 2012). Angelis et al. reported a coculture system consisting of microalgae, cyanobacteria and macrofungi for the production of extracellular polysaccharides (Angelis et al., 2012). In this study, two photosynthetic species, S. platensis and C. vulgaris, and five fungi, Ganoderma, Lentinus, Trametes, Pleurotus and Agaricus, which were screened based on biomass and extracellular polysaccharide productivity in a single culture system, were constructed as a new coculture system. By using ^{13}C NMR and GC spectroscopy, the authors characterized and compared the water-soluble extracellular polysaccharides in single and coculture systems. The results showed that a variety of extracellular polysaccharides were produced in the coculture system and were quite different from those produced by pure culture, and many new polysaccharides were identified from the molecular structure. On this basis, to achieve the maximum production of polysaccharides, Angelis et al. further optimized these coculture conditions, such as temperature, stirring speed, inoculation rates of fungi and microalgae, initial pH, light intensity and glucose concentration, and ultimately increased the polysaccharide and biomass yields by 33% and 61% compared with a single culture of cyanobacteria, respectively (Angelis et al., 2012).

4.3 Biological hydrogen production

Today, the global energy demand mainly depends on fossil fuels (accounting for approximately 80% of the current world energy demand). This status will eventually lead to the foreseeable depletion of limited fossil-energy resources. Carbon dioxide and other pollutants produced by these fossil fuels are causing global climate change. To compensate for the depletion of fossil fuels and their adverse impact on the environment, it is first proposed to use hydrogen as a future energy carrier. Hydrogen has various other uses (Veziroğlu, 1995; Ramachandran and Menon, 1998). For example, hydrogen can be a reactant in hydrogenation processes or an O_2 scavenger; it can also be a fuel in rocket engines and a coolant in electrical generators to take advantage of its unique physical properties (Das and Veziroğlu, 2001). The biological hydrogen production process can be divided into various categories.

The first is the biological photolysis of water by algae and cyanobacteria. This method uses the same process as photosynthesis of plants and algae but is suitable for producing hydrogen instead of carbon-containing biomass. Photosynthesis involves the absorption of light by two different photosynthetic systems operating in series, with two photons used to remove each electron from water and for CO_2 reduction or H_2 formation. In green plants, only CO_2 reduction occurs because no hydrogenase is present (Ramachandran and Menon, 1998). Microalgae produce hydrogenase, whether eukaryotes (such as green algae) or prokaryotes (such as cyanobacteria), and can produce hydrogen under specific conditions (Benemann, 1997). Hydrogen photo-evolution by unicellular algae was first demonstrated by Gaffron and Rubin (1942). Hydrogen production by microalgae has been reviewed by several researchers (Healey, 1970; Benemann et al., 1980; Kumazawa and Mitsui, 1981).

The second is photolysis of organic matter by photosynthetic bacteria. Currently, the most promising biological hydrogen production microbial system in the literature is phototrophic bacteria (Kim, 1981; Vincenzini et al., 1982; Miyake and Kawamura, 1987; Fascetti and Todini, 1995; Fascetti et al., 1998). These bacteria have a high theoretical conversion rate, the ability to use broad-spectrum light, and the ability to consume organic substrates derived from waste.

The third is hydrogen production by the fermentation of organic compounds. There have been many reports on the application of coculture bacterial systems for biological hydrogen production. For example, Miura et al. cocultivated C. reinhardtii MGA 161 with the marine photosynthetic bacterial strain W-1S for hydrogen production (Miura et al., 1992). In this study, C. reinhardtii MGA 161 was used as an electron donor for hydrogen photolysis, and the hydrogen yield in the optimized system reached 0.083 μmol/h. Kawaguchi et al. metabolized algal starch into lactate as an electron donor by coculturing Rhodobium marinum A-501 with Lactobacillus amylovorus and algal biomass and further improved it to make it more conducive to hydrogen production by adding algal extracts to enhance the nitrogenase activity of this strain (Kawaguchi et al., 2001). In this coculture system, the hydrogen yield increased 1.5 times, and the highest yield reached 100.75 μmol/h, which was 1.4 times higher than that under single-culture conditions. In recent years, Xu et al. cocultivated Chlamydomonas rheiniscens with Bradyrhizobium japonicum or Azotobacter chroococcum to produce hydrogen and obtained hydrogen yields of 0.42 μmol/h and 0.36 μmol/h, respectively (Xu et al., 2016, 2017). Luo et al. found high hydrogen accumulation in contaminated C. reinhardtii bacterial fluid and confirmed that Pseudomonas BS4 could promote hydrogen production (Ban et al., 2017). The hydrogen production of this optimized system containing Pseudomonas BS4 was 811 μmol/h. This kind of heterotrophic bacteria can not only assist Chlamydomonas reischlamydomonas in hydrogen production but also promote the hydrogen production of photosynthetic microorganisms such as Chlorella and Scenedesmus.

5. Mechanism Analysis of Photoheterotrophic Coculture Systems by Omics Analysis

5.1 Tools

In recent decades, systems biology methods, including metagenomics, transcriptomics, proteomics and metabolomics, have been applied to analyze the interaction mechanisms in coculture systems (Faust and Raes, 2012; Ponomarova and Patil, 2015; Khan et al., 2018; Ravikrishnan et al., 2019). Moreover, the multiomics integrated bioinformatics method provides a large amount of data and information for an in-depth understanding of the interactions between the members of the coculture system. With the rapid development of omics technology, the interaction mechanisms in natural and artificial coculture systems have become increasingly clear. This section examines the four main technologies used for mechanism analysis of coculture systems.

5.1.1 Metagenomics

Metagenomics enables the sequencing of total microbial DNA sampled from environmental niches, which transforms the traditional technologies for natural microbial communities and provides a rich set of biological insights. Metagenomics is considered a useful technique to study the functions and interaction mechanisms of complex ecosystems, such as the human gut microbiota and marine microbes (DeLong et al., 2006; Gill et al., 2006). In addition, as a cell culture-independent method, metagenomics provides a new method to characterize uncultured microorganisms at higher resolution from their natural habitats and enables researchers to understand the potential functions of these microorganisms in communities (Mick and Sorek, 2014). Therefore, metagenomics can be used to analyze natural and artificial coculture systems consisting of various microbial species.

5.1.2 Transcriptomics

Transcriptomics is a discipline that studies gene transcription and transcriptional regulation profiles in cells; the study of gene expression at the RNA level is a key element in functional genomics (Bhadauria et al., 2007). Microarray and high-throughput mRNA sequencing (RNA-seq) are commonly used transcriptomic techniques (Heller, 2002). Recent advances in next-generation sequencing technologies such as RNA-seq have helped decipher the functional complexity of the entire transcriptomes of organisms (Chandhini and Rejish Kumar, 2019). Moreover, RNA-seq has distinct advantages over other technologies in mapping and quantifying transcriptomes (Wang et al., 2009; Xi et al., 2014).

5.1.3 Proteomics

The term 'proteome' was first used to describe the collection of proteins encoded by the genome (Wilkins et al., 1996). Proteins are the molecular products of genes and are vital to an organism because they are the essential building blocks for switching on and regulating metabolic pathways to function properly. Genomics provides a blueprint of possible gene products for proteomic research priorities (Tyers and Mann, 2003). Proteomics refers to the study of proteins on a large scale, especially protein expression, structure and function, and this increasingly sophisticated technology aims to describe and characterize all expressed proteins in biological systems (Zhang et al., 2010b). For example, using proteomic analysis, Christie-Oleza et al. found that it is not the concentration of nutrients that maintains stable interactions and dynamic systems but the cycling of nutrients that underlies the maintenance of the long-term stable coculture of *Synechococcus* sp. WH7803 (hereafter *Synechococcus* WH7803) with heterotrophic microorganisms (Christie-Oleza et al., 2017).

5.1.4 Metabolomics

Metabolomics is a method for investigating the overall metabolite profile in a system under a given set of conditions. Metabolites are the result of the interaction between the genome and its environment in a certain system, not only the final product of gene expression but also the important components in regulating the entire system. (Rochfort, 2005). Metabolic crosstalk plays a vital role in determining microbes in the interaction mechanism within the biological community (Straight and Kolter, 2009). Microbes communicate with each other by exchanging various metabolites, including primary metabolites, such as organic acids and amino acids, and secondary metabolites, such as autoinducers, in quorum sensing. Therefore, investigation of these specific metabolites is a focus of research on microbial interaction mechanisms (Yang et al., 2011).

5.1.5 Others

Metagenomics, transcriptomics, proteomics and metabolomics are important methods to study the interaction mechanisms in coculture systems. All methods can analyze the mechanisms of coculture systems at various levels; however, due to the incompleteness and complementarity of these different approaches, multiomics analysis can achieve a panoramic view of cells in mixed bacterial systems and reveal new insights into biological mechanisms (Yu *et al.* n.d.). To further improve the stability and efficiency in artificial coculture systems, it is necessary to investigate complicated metabolism networks, including carbohydrate, lipid, nucleic acid, amino acid, and vitamin and cofactor metabolism networks (Chunmei, 2015).

5.2 Deciphering metabolic interactions between phototrophic and heterotrophic cells

Omics technologies are valuable tools for the mechanism analysis of the interaction relationships in synthetic coculture systems. Omics findings can provide an important theoretical basis for the redesign and industrial application of synthetic coculture systems. Below, we summarize the progress from different aspects of cell metabolism.

5.2.1 Energy metabolism

Energy metabolism generally includes oxidative phosphorylation, photosynthesis, photosynthesis-antenna protein, nitrogen metabolism, carbon fixation, methane metabolism and sulfur metabolism (Wang *et al.*, 2020), which have been proven to play important roles in the interactions among different species in coculture systems. Carbon and nitrogen are the most abundant nutrient elements in organisms. Therefore, the metabolic balance of carbon and nitrogen is the basis of various lifestyles to adapt to fluctuating conditions and needs to be strictly controlled through various mechanisms. In autotrophic microorganisms such as cyanobacteria, the control of carbon digestion and metabolism is particularly important because they have limited choices for carbon source and nitrogen assimilation. Cyanobacteria are capable of producing oxygen for photosynthesis and use different forms of nitrogen sources, such as nitrate, ammonium and/or N_2, for diazotrophic strains. These nitrogen sources are converted to ammonium, which is then incorporated into the carbon skeleton 2-oxoglutarate (2-OG) to synthesize various biomolecules (Flores and Herrero, 2005). Thus, the accumulation of 2-OG *in vivo* constitutes a nitrogen starvation signal and causes a series of cellular reactions, including the formation of nitrogen-fixing heterovesicles in some filamentous cyanobacteria strains (Laurent *et al.*, 2005). Beliaev *et al.* analyzed the transcriptional response of *Synechococcus* PCC 7002 (hereafter *Synechococcus* 7002) in a coculture system consisting of *Synechococcus* 7002 and the marine facultative aerobe *Shewanella putrefaciens* W3-18-1 through transcriptomics and found that there were 473 transcripts showing ≥ twofold changes in *Synechococcus* 7002 (Beliaev *et al.*, 2014). Among them, transcripts related to energy metabolism accounted for 8%. For example, transcripts related to nitrogen assimilation, aminoamide synthase (*glnA*) and ferredoxin-nitrite reductase (*nirA*) were upregulated. The assimilation reduction of nitrate nitrogen to ammonium nitrogen is a key step in the biosphere nitrogen cycle, and nitrite reductase reduces nitrite to ammonium in a six-electron reaction (Pierre *et al.*, 2009). In addition, increased mRNA abundance levels of genes encoding carbon fixation-related functions (i.e. RuBisCo, carboxylsome components, bicarbonate transporter, NADH-quinone dehydrogenase) were also found in this study, which might be caused by the limitation of inorganic carbon in the coculture system. Using transcriptomic analysis, Bernstein *et al.* found that genes associated with photosystem II (*psbV*, *psbX* and *psbV2*; *tll1285*, *tsr2013* and *tll1284*) and carboxylsomes (*ccmK1* and *ccmL*; *tll0946* and *tll0945*) were significantly increased in *Thermosynechococcus elongatus* BP-1, which was cocultured with *Thermus ruber* strain A (Bernstein *et al.*, 2017). All these findings enhanced the understanding of energy metabolism in coculture systems at the transcriptional level.

5.2.2 Signal transduction

Signal transduction is an important mechanism for continuously responding to changes in the external environment and responding to external stimuli by regulating the expression of related genes in bacteria (Wang Xueying, 2021). A two-component system comprising sensor kinases and response regulators is a ubiquitous signal transduction pathway and a ubiquitous signaling mechanism in bacteria (West and Stock, 2001; Palenik et al., 2003; Vidal et al., 2009). Compared with pure culture, in coculture systems, in addition to dealing with the external environment, the strains in the coculture systems also need to resolve the relationship between the constantly changing members in the system. Because of the existence of multiple strains, the established interaction relationship between members might be changed with the proportion of members in the process of growth and reproduction, which highlights the role of signal transduction in coculture systems (Wang et al., 2020). The upregulation of genes or proteins related to signal transduction in cocultured strains reflects the important role of the signal transduction system in coculture systems (Palenik et al., 2003; Thevenard et al., 2011; Zheng et al., 2020) Zheng et al. investigated the interaction between marine *Synechococcus* sp. YX04-3 (hereafter *Synechococcus* YX04-3) and its cocultured heterotrophic bacteria through metagenomics and proteomics (Zheng et al., 2020). The omics data demonstrated that three sensor histidine kinases and six response regulators were significantly changed in the *Synechococcus* YX04-3 proteome, two of which were predicted to be involved in the phosphate sensor regulation (PhoRB) system. Significant changes were also detected in eight sensor histidine kinases and 11 response regulators in the *Synechococcus* YX04-3 genome. In addition, the abundance of sensors is generally lower than that of response regulators, which may represent an efficient regulatory strategy, and there appears to be an economy of regulation in which certain sensors can transmit signals to multiple response regulators (Palenik et al., 2003, 2006). In addition, using transcriptomic analysis, Thevenard et al. found that all response regulator genes of *Streptococcus thermophilus* LMD-9 were expressed in milk, and in the presence of *Lactobacillus delbrueckii* subsp. *Bulgaricus*, *S. thermophilus* LMD-9 significantly increased the expression levels of four response regulators (Thevenard et al., 2011). This finding suggested that the presence of *L. bulgaricus* alters the regulatory system of *S. thermophilus*, providing evidence for the contribution of each two-component system to the growth of *S. thermophilus* LMD-9 in milk.

5.2.3 Membrane transport

Transport systems enable bacteria to accumulate needed nutrients, extrude unwanted byproducts and modify their cytoplasmic contents of protons and salts to maintain a composition conducive to growth and development (Padan, 2009). A stable synthetic coculture system usually requires the membrane transport system to exchange and circulate nutrients to maintain the mutualistic relationship between cocultured species (Wang et al., 2015). Compared with pure culture conditions, the upregulation of genes or proteins related to membrane transport indicated that the uptake of various nutrients in cocultured member species was improved, which facilitated the growth of strains and metabolic interactions between member bacteria (Aharonovich and Sher, 2016; Christie-Oleza et al., 2017). Christie-Oleza et al. designed long-term coculture experiments using *Synechococcus* WH7803 and heterotrophic microorganisms under eutrophic and natural oligotrophic seawater conditions (Christie-Oleza et al., 2017). Through proteomic analysis, it was found that *Synechococcus* has a small number of differentially expressed proteins when cocultured with *Roseobacter*, and most of the differentially expressed proteins have unknown functions, which is consistent with another short-term photoautotrophic organism (Aharonovich and Sher, 2016). At the functional level, ~61% of the proteome in *Synechococcus* WH7803 was allocated to the membrane transport pathway when cocultured with heterotrophic bacteria, which was mainly due to increases in the abundances of two periplasmic binding proteins targeting amino acids and iron. The proteomics analysis by Zheng et al. also provided a reference for the analysis of membrane transport pathways (Zheng et al., 2020). In this study, the proteins related to nitrate/nitrite transporters, phosphate transporters (PstBS), urea transport (UrtABCDE) and iron-binding ABC transporters showed significantly different expression in *Synechococcus* YX04-3, which suggests that the uptake of nutrients such as nitrogen and phosphorus was enhanced under coculture conditions.

5.2.4 Stress responses

ROS can cause damage to cells and lead to oxidative stress by reacting with many biomolecules, including nucleic acids, proteins and lipids (Gallego et al., 2012; Fischer et al., 2013). The production of ROS is an inevitable result of aerobic metabolism; thus, the oxidative stress response is one of the main components of the interaction mechanism in coculture systems (Halliwell, 2006). It is generally believed that the better robustness and stress resistance of coculture systems is due to the existence of heterotrophic species helping to remove the ROS-generated compared with pure cultured strains. Previous studies pointed out the possibility that heterotrophic bacteria can reduce oxidative stress in the medium to improve the cell growth of cyanobacteria in a coculture system, eventually enhancing the stability of the coculture system (Morris et al., 2008; Li et al., 2017). Ren et al. selected ascorbate as an antioxidant to quench intracellular ROS in the microalga *Schizochytrium* sp., which has great significance in improving the production of DHA (Ren et al., 2017). Liu et al. investigated the effect of cyanobacteria on *Escherichia coli* under coculture conditions using an isoprene-producing coculture system through a combination of transcriptomics, proteomics and metabolomics and found that the response of *E. coli* BL21(DE3) to *S. elongatus* PCC 7942 was triggered by the oxidative stress generated by photosynthesis under coculture conditions (Liu et al., 2021). In this process, to stop the Fenton reaction, *E. coli* first upregulated bacterial ferritin (BFR) to reduce the Fe^{2+} concentration, while thiocysteine was converted to cystine to reduce H_2O_2. Second, YtfE was upregulated to repair damaged Fe-S protein. Third, the phage shock proteins PspB, PspD and PspE were upregulated in response to oxidative stress. Consistent with the above-mentioned stress-response mechanisms, Christie-Oleza et al. found that the proteins related to oxidative stress were decreased when *Synechococcus* WH7803 was cocultured with heterotrophic bacteria (Christie-Oleza et al., 2017). These studies indicated that the heterotrophic bacteria present in coculture systems provide cyanobacteria with protection against oxidative stress. More recently, Ma et al. explored how the relieved oxidative stress affected cyanobacterial cell growth in a coculture system by supplementing with additional ascorbic acid (Ma et al., 2022). They found that the cell growth of cyanobacteria was clearly improved with an additional 1 mM ascorbic acid under axenic culture; however, its growth was still slower than that in the coculture system. This result suggests that the improved cell growth of cyanobacteria may be caused by multiple factors, including reduced oxidative stress. Therefore, they further explored the cellular responses of cyanobacteria in the system through quantitative transcriptomics, proteomics and metabolomics, and the integrated omics results demonstrated that in the coculture system, the relieved oxidative stress and increased CO_2 availability improved the cell growth of cyanobacteria.

6. Conclusions and Prospective

Symbiosis, especially photoheterotrophic systems, is widespread and functionally relevant in diverse biological systems. After hundreds of millions of years of evolution, each natural symbiotic system possesses its own unique advantages and characteristics owing to its unique living conditions. Although some natural bacterial symbiotic systems have been used in effluent treatment and biosynthesis through simple modification, the scope of application is limited. In recent years, inspired by the interaction of different species in natural symbiotic systems, synthetic coculture systems consisting of photosynthetic and heterotrophic species have been constructed by synthetic biology. For example, many studies have demonstrated the feasibility of synthetic photoheterotrophic coculture systems to achieve the one-step conversion of carbon feedstocks, such as sucrose, to value-added chemicals, such as 3-HP and isoprene. However, to date, the yield of chemicals produced in synthetic coculture systems is still low, severely limiting their use for large-scale production. To further improve the stability and efficiency of synthetic coculture systems, omics technology was employed to analyze the interaction relationships. Multiple omics analyses provide strong evidence that mutual interactions can be achieved through cross-feeding and competition between phototrophs and prokaryotic heterotrophs, which is indispensable for developing more intelligent artificial consortia.

Acknowledgement

This chapter was supported by grants from the National Key Research and Development Program of China (Grant nos 2019YFA0904600, 2018YFA0903000, 2018YFA0903600, 2020YFA0906800) and the National Natural Science Foundation of China (Grant nos 31901016, 31770035, 31972931, 21621004).

References

Abou-Shady, A. (2016) Reclaiming salt-affected soils using electro-remediation technology: PCPSS evaluation. *Electrochimica Acta* 190, 511–520.

Aharonovich, D. and Sher, D. (2016) Transcriptional response of *Prochlorococcus* to co-culture with a marine *Alteromonas*: differences between strains and the involvement of putative infochemicals. *Isme Journal* 10, 2892–2906.

Albuquerque, M., Torres, C. and Reis, M. (2010) Polyhydroxyalkanoate (PHA) production by a mixed microbial culture using sugar molasses: effect of the influent substrate concentration on culture selection. *Water Research* 44, 3419–3433.

Amin, S., Hmelo, L., Van Tol, H., Durham, B., Carlson, L. et al. (2015) Interaction and signalling between a cosmopolitan phytoplankton and associated bacteria. *Nature* 522, 98–101.

Andrianantoandro, E., Basu, S., Karig, D.K. and Weiss, R. (2006) Synthetic biology: new engineering rules for an emerging discipline. *Molecular Systems Biology* 2, 2006.0028.

Angelis, S., Novak, A.C., Sydney, E.B., Soccol, V.T., Carvalho, J.C. et al. (2012) Co-culture of microalgae, cyanobacteria, and macromycetes for exopolysaccharides production: process preliminary optimization and partial characterization. *Applied Biochemistry and Biotechnology* 167, 1092–1106.

Bai, X. and Acharya, K. (2016) Removal of trimethoprim, sulfamethoxazole, and triclosan by the green alga *Nannochloris* sp. *Journal of Hazardous Materials* 315, 70–75.

Ban, S., Lin, W., Wu, F. and Luo, J. (2017) Algal-bacterial cooperation improves algal photolysis-mediated hydrogen production. *Bioresource Technology* 251, 350.

Beliaev, A.S., Romine, M.F., Serres, M., Bernstein, H.C., Linggi, B.E. et al. (2014) Inference of interactions in cyanobacterial-heterotrophic co-cultures via transcriptome sequencing. *Isme Journal* 8, 2243–2255.

Benemann, J.R. (1997) Feasibility analysis of photobiological hydrogen production. *International Journal of Hydrogen Energy* 22, 979–987.

Benemann, J.R., Miyamoto, K. and Hallenbeck, P.C. (1980) Bioengineering aspects of biophotolysis. *Enzyme & Microbial Technology* 2, 103–111.

Bernstein, H.C., McClure, R.S., Thiel, V., Sadler, N.C., Kim, Y.M. et al. (2017) Indirect interspecies regulation: transcriptional and physiological responses of a cyanobacterium to heterotrophic partnership. *mSystems* 2, e00181–00216.

Bhadauria, V., Popescu, L., Zhao, W.S. and Peng, Y.L. (2007) Fungal transcriptomics. *Microbiological Research* 162, 285–298.

Bobadilla Fazzini, R.A., Preto, M.J., Quintas, A.C., Bielecka, A. and Dos Santos, V.A. (2010) Consortia modulation of the stress response: proteomic analysis of single strain versus mixed culture. *Environmental Microbiology* 12, 2436–2449.

Boivin, M., Greve, G.D., Garcia-Meza, J.V., Massieux, B., Sprenger, W. et al. (2007) Algal-bacterial interactions in metal contaminated floodplain sediments. *Environmental Pollution* 145, 884–894.

Brenner, K., You, L. and Arnold, F.H. (2008) Engineering microbial consortia: a new frontier in synthetic biology. *Trends in Biotechnology* 26, 483–489.

Breuer, G., Lamers, P.P., Martens, D.E., Draaisma, R.B. and Wijffels, R.H. (2013) Effect of light intensity, pH, and temperature on triacylglycerol (TAG) accumulation induced by nitrogen starvation in *Scenedesmus obliquus*. *Bioresource Technology* 143, 1–9.

Bruckner, C.G., Rehm, C., Grossart, H.P. and Kroth, P.G. (2011) Growth and release of extracellular organic compounds by benthic diatoms depend on interactions with bacteria. *Environmental Microbiology* 13, 1052–1063.

Chandhini, S. and Rejish Kumar, V.J. (2019) Transcriptomics in aquaculture: current status and applications. *Reviews in Aquaculture* 11, 1379–1397.

Cheloni, G. and Slaveykova, V.I. (2018) Combined effects of trace metals and light on photosynthetic microorganisms in aquatic environment. *Environments* 5, 81.

Chisti, Y. (2013) Constraints to commercialization of algal fuels. *Journal of Biotechnology* 167, 201–214.

Christie-Oleza, J.A., Sousoni, D., Lloyd, M., Armengaud, J. and Scanlan, D.J. (2017) Nutrient recycling facilitates long-term stability of marine microbial phototroph–heterotroph interactions. *Nature Microbiology* 2, 17100.

Chunmei, X.D.W.L.D. (2015) Research progress of microbial co-culture technology. *Journal of Microbiology* 55, 1089–1096.

Das, D. and Veziroǧlu, T.N. (2001) Hydrogen production by biological processes: a survey of literature. *International Journal of Hydrogen Energy* 26(1), 13–28.

DeLong, E.F., Prestonc, M., Mincer, T., Rich, V., Hallam, S.J. et al. (2006) Community genomics among stratified microbial assemblages in the ocean's interior. *Science* 311, 496–503.

Ding, M.Z., Zou, Y., Song, H. and Yuan, Y.J. (2014) Metabolomic analysis of cooperative adaptation between co-cultured *Bacillus cereus* and *Ketogulonicigenium vulgare*. *PLOS ONE* 9, e94889.

Ducat, D.C., Avelar-Rivas, J.A., Way, J.C. and Silver, P.A. (2012) Rerouting carbon flux to enhance photosynthetic productivity. *Applied and Environmental Microbiology* 78, 2660–2668.

Dunne, E.J., Coveney, M.F. and Marzolf, E.R. (2013) Nitrogen dynamics of a large-scale constructed wetland used to remove excess nitrogen from eutrophic lake water. *Ecological Engineering* 61, 224–234.

Fascetti, E. and Todini, O. (1995) *Rhodobacter sphaeroides* RV cultivation and hydrogen production in a one- and two-stage chemostat. *Applied Microbiology and Biotechnology* 44, 300–305.

Fascetti, E., D'Addario, E., Todini, O. and Robertiello, A. (1998) Photosynthetic hydrogen evolution with volatile organic acids derived from the fermentation of source selected municipal solid wastes. *International Journal of Hydrogen Energy* 23, 753–760.

Faust, K. and Raes, J. (2012) Microbial interactions: from networks to models. *Nature Reviews Microbiology* 10, 538–550.

Fischer, B.B., Hideg, É. and Krieger-Liszkay, A. (2013) Production, detection, and signaling of singlet oxygen in photosynthetic organisms. *Antioxidants & Redox Signaling* 18, 2145–2162.

Flores, E. and Herrero, A. (2005) Nitrogen assimilation and nitrogen control in cyanobacteria: figure 1. *Biochemical Society Transactions* 33, 164–167.

Fradinho, J.C., Domingos, J., Carvalho, G., Oehmen, A. and Reis, M. (2013) Polyhydroxyalkanoates production by a mixed photosynthetic consortium of bacteria and algae. *Bioresource Technology* 132, 146–153.

Fredrickson, J.K. (2015) Ecological communities by design. *Science* 348, 1425–1427.

Gaffron, H. and Rubin, J. (1942) Fermentative and photochemical production of hydrogen in algae. *The Journal of General Physiology* 26, 219–240.

Gallego, S.M., Pena, L.B., Barcia, R.A., Azpilicueta, C.E. and Benavides, M.P. (2012) Unravelling cadmium toxicity and tolerance in plants: insight into regulatory mechanisms. *Environmental & Experimental Botany* 83, 33–46.

Gill, S.R., Pop, M., Deboy, R.T., Eckburg, P.B., Turnbaugh, P.J. et al. (2006) Metagenomic analysis of the human distal gut microbiome. *Science* 312, 1355–1359.

Giri, S., Ona, L., Waschina, S., Shitut, S., Yousif, G. et al. (2021) Metabolic dissimilarity determines the establishment of cross-feeding interactions in bacteria. *Current Biology* 31, 5547–5557.

Glaesener, A.G., Merchant, S.S. and Blaby-Haas, C.E. (2013) Iron economy in *Chlamydomonas reinhardtii*. *Frontiers in Plant Science* 4, 337.

Goers, L., Freemont, P. and Polizzi, K.M. (2014) Co-culture systems and technologies: taking synthetic biology to the next level. *Journal of the Royal Society Interface* 11, 1058–1069.

Hagemann, M. (2011) Molecular biology of cyanobacterial salt acclimation. *FEMS Microbiology Reviews* 35, 87–123.

Halliwell, B. (2006) Reactive species and antioxidants: redox biology is a fundamental theme of aerobic life. *Plant Physiology* 141, 312–322.

Han, P.P., Ying, S., Jia, S.R., Cheng, Z. and Tan, Z.L. (2014) Effects of light wavelengths on extracellular and capsular polysaccharide production by *Nostoc flagelliforme*. *Carbohydrate Polymers* 105, 145–151.

Han, Y. and Watson, M. (1992) Production of microbial levan from sucrose, sugarcane juice and beet molasses. *Journal of Industrial Microbiology* 9, 257–260.

Hao, X.D. and Zhu, J.Y. (2005) Production of biodegradable plastics PHA by using mixed cultures and activated sludge. *Ecological Environment* 2005(6), 967–971.

Hays, S.G. and Ducat, D.C. (2015) Engineering cyanobacteria as photosynthetic feedstock factories. *Photosynthesis Research* 123, 285–295.

Hays, S.G., Yan, L., Silver, P.A. and Ducat, D.C. (2017) Synthetic photosynthetic consortia define interactions leading to robustness and photoproduction. *Journal of Biological Engineering* 11, 4.

Healey, F.P. (1970) Hydrogen evolution by several algae. *Planta* 91, 220–226.

Hedayat, K.M. and Lapraz, J.-C. (2019) Symbiosis. In: Kamyar, M. and Hedayat, J.-C.L. (eds) *The Theory of Endobiogeny*. Academic Press, pp. 63–75.

Heller, M.J. (2002) DNA microarray technology: devices, systems, and applications. *Annual Review of Biomedical Engineering* 4, 129.

Hernández-Crespo, C., Gargallo, S., Benedito-Durá, V., Nácher-Rodríguez, B., Rodrigo-Alacreu, M.A. et al. (2017) Performance of surface and subsurface flow constructed wetlands treating eutrophic waters. *Science of the Total Environment* 595, 584.

Hu, G.J., Zhou, M., Hou, H.B., Zhu, X. and Zhang, W.H. (2010) An ecological floating-bed made from dredged lake sludge for purification of eutrophic water. *Ecological Engineering* 36, 1448–1458.

Jia, X., Liu, C., Song, H., Ding, M., Du, J. et al. (2016) Design, analysis and application of synthetic microbial consortia. *Synthetic & Systems Biotechnology* 1, 109–117.

Kawaguchi, H., Hashimoto, K., Hirata, K. and Miyamoto, K. (2001) H_2 production from algal biomass by a mixed culture of *Rhodobium marinum* A-501 and *Lactobacillus amylovorus*. *Journal of Bioscience & Bioengineering* 91, 277–282.

Keller, L. and Surette, M.G. (2006) Communication in bacteria: an ecological and evolutionary perspective. *Nature Reviews Microbiology* 4, 249–258.

Khan, N., Maezato, Y., McClure, R.S., Brislawn, C.J., Mobberley, J.M. et al. (2018) Phenotypic responses to interspecies competition and commensalism in a naturally-derived microbial co-culture. *Scientific Reports* 8, 297.

Kim, J.S. (1981) Production of molecular hydrogen by *Rhodopseudomonas* sp. *Journal of Fermentation Technology* 59, 185–190.

Kirsch, F., Luo, Q., Lu, X. and Hagemann, M. (2018) Inactivation of invertase enhances sucrose production in the cyanobacterium *Synechocystis* sp. PCC 6803. *Microbiology* 164, 1220–1228.

Kropat, J., Gallaher, S.D., Urzica, E.I., Nakamoto, S.S. and Merchant, S.S. (2015) Copper economy in *Chlamydomonas*: prioritized allocation and reallocation of copper to respiration vs. photosynthesis. *Proceedings of the National Academy of Sciences of the United States of America* 112, 2644–2651.

Kumazawa, S. and Mitsui, A. (1981) Characterization and optimization of hydrogen photoproduction by a saltwater blue-green alga, *Oscillatoria* sp. Miami BG7. I. Enhancement through limiting the supply of nitrogen nutrients. *International Journal of Hydrogen Energy* 6, 339–348.

Laurent, S., Chen, H., Bédu, S., Ziarelli, F., Peng, L. et al. (2005) Nonmetabolizable analogue of 2-oxoglutarate elicits heterocyst differentiation under repressive conditions in *Anabaena* sp. PCC 7120. *Proceedings of the National Academy of Sciences of the United States of America* 102, 9907–9912.

Li, H., Zhao, Q. and Huang, H. (2019) Current states and challenges of salt-affected soil remediation by cyanobacteria. *Science of the Total Environment* 669, 258–272.

Li, T., Li, C.T., Butler, K., Hays, S.G., Guarnieri, M.T. et al. (2017) Mimicking lichens: incorporation of yeast strains together with sucrose-secreting cyanobacteria improves survival, growth, ROS removal, and lipid production in a stable mutualistic co-culture production platform. *Biotechnology Biofuels* 10, 55.

Lima, S., Matinha-Cardoso, J., Tamagnini, P. and Oliveira, P. (2020) Extracellular vesicles: an overlooked secretion system in cyanobacteria. *Life (Basel)* 10, 129.

Liu, H., Cao, Y., Guo, J., Xu, X., Long, Q. et al. (2021) Study on the isoprene-producing co-culture system of *Synechococcus elongates-Escherichia coli* through omics analysis. *Microbial Cell Factories* 20, 1–18.

Liu, J. and Vyverman, W. (2015) Differences in nutrient uptake capacity of the benthic filamentous algae *Cladophora* sp., *Klebsormidium* sp. and *Pseudanabaena* sp. under varying N/P conditions. *Bioresource Technology* 179, 234–242.

Liu, J., Danneels, B., Vanormelingen, P. and Vyverman, W. (2016) Nutrient removal from horticultural wastewater by benthic filamentous algae *Klebsormidium* sp., *Stigeoclonium* spp. and their communities: from laboratory flask to outdoor algal turf scrubber (ATS). *Water Research* 92, 61–68.

Loutseti, S., Danielidis, D.B., Economou-Amilli, A., Katsaros, C., Santas, R. *et al.* (2009) The application of a micro-algal/bacterial biofilter for the detoxification of copper and cadmium metal wastes. *Bioresource Technology* 99, 2099–2105.

Löwe, H., Hobmeier, K., Moos, M., Kremling, A. and Pflüger-Grau, K. (2017) Photoautotrophic production of polyhydroxyalkanoates in a synthetic mixed culture of *Synechococcus elongatus cscB* and *Pseudomonas putida cscAB*. *Biotechnology for Biofuels* 10, 190.

Luan, G. and Lu, X. (2018) Tailoring cyanobacterial cell factory for improved industrial properties. *Biotechnology Advances* 36, 430–442.

Ma, B.Y., Zhang, G.M., Wang, H.Y., Xu, H.Z., Peng, M. *et al.* (2016) Research progress on application of photosynthetic microbial mixed culture. *China Biotechnol* 36, 113–122.

Ma, J., Guo, T., Ren, M., Chen, L., Song, X. *et al.* (2022) Cross-feeding between cyanobacterium *Synechococcus* and *Escherichia coli* in an artificial autotrophic–heterotrophic coculture system revealed by integrated omics analysis. *Biotechnology for Biofuels and Bioproducts* 15, 69.

Mahdavi, H., Prasad, V., Liu, Y. and Ulrich, A.C. (2015) In situ biodegradation of naphthenic acids in oil sands tailings pond water using indigenous algae–bacteria consortium. *Bioresource Technology* 187, 97–105.

Malasarn, D., Kropat, J., Hsieh, S.I., Finazzi, G., Casero, D. *et al.* (2013) Zinc deficiency impacts CO_2 assimilation and disrupts copper homeostasis in *Chlamydomonas reinhardtii*. *Journal of Biological Chemistry* 288, 10672–10683.

Manjunath, M., Kanchan, A., Ranjan, K., Venkatachalam, S., Prasanna, R. *et al.* (2016) Beneficial cyanobacteria and eubacteria synergistically enhance bioavailability of soil nutrients and yield of okra. *Heliyon* 2, e00066.

Markou, G. and Georgakakis, D. (2011) Cultivation of filamentous cyanobacteria (blue-green algae) in agro-industrial wastes and wastewaters: a review. *Applied Energy* 88, 3389–3401.

Mick, E. and Sorek, R. (2014) High-resolution metagenomics. *Nature Biotechnology* 32, 750–751.

Mills, L.A., Mccormick, A.J. and Lea-Smith, D.J. (2020) Current knowledge and recent advances in understanding metabolism of the model cyanobacterium *Synechocystis* sp. PCC 6803. *Bioscience Reports* 40(4), BSR20193325.

Miura, Y., Saitoh, C., Matsuoka, S. and Miyamoto, K. (1992) Stably sustained hydrogen production with high molar yield through a combination of a marine green alga and a photosynthetic bacterium. *Journal of the Agricultural Chemical Society of Japan* 56, 751–754.

Miyake, J. and Kawamura, S. (1987) Efficiency of light energy conversion to hydrogen by the photosynthetic bacterium *Rhodobacter sphaeroides*. *International Journal of Hydrogen Energy* 12, 147–149.

Moita, R., Freches, A. and Lemos, P.C. (2014) Crude glycerol as feedstock for polyhydroxyalkanoates production by mixed microbial cultures. *Water Research* 58, 9–20.

Morris, J.J., Kirkegaard, R., Szul, M.J., Johnson, Z.I. and Zinser, E.R. (2008) Facilitation of robust growth of *Prochlorococcus* colonies and dilute liquid cultures by 'helper' heterotrophic bacteria. *Applied and Environmental Microbiology* 74, 4530–4534.

Muñoz, R. and Guieysse, B. (2006) Algal-bacterial processes for the treatment of hazardous contaminants: a review. *Water Research* 40, 2799–2815.

Nain, L., Rana, A., Joshi, M., Jadhav, S.D., Kumar, D. *et al.* (2010) Evaluation of synergistic effects of bacterial and cyanobacterial strains as biofertilizers for wheat. *Plant & Soil* 331, 217–230.

Niederholtmeyer, H., Wolfstadter, B.T., Savage, D.F., Silver, P.A. and Way, J.C. (2010) Engineering cyanobacteria to synthesize and export hydrophilic products. *Applied and Environmental Microbiology* 76, 3462–3466.

Nouaille, S., Even, S., Charlier, C., Le Loir, Y., Cocaign-Bousquet, M. *et al.* (2009) Transcriptomic response of *Lactococcus lactis* in mixed culture with *Staphylococcus aureus*. *Applied and Environmental Microbiology* 75, 4473–4482.

Padan, E. (2009) Bacterial membrane transport: organization of membrane activities. In: *Encyclopedia of Life Sciences*. Wiley.

Paerl, H.W. and Pinckney, J. (1996) A mini-review of microbial consortia: their roles in aquatic production and biogeochemical cycling. *Microbial Ecology* 31, 225–247.

Pai, A., Tanouchi, Y., Collins, C.H. and You, L. (2009) Engineering multicellular systems by cell–cell communication. *Current Opinion in Biotechnology* 20, 461–470.

Palenik, B., Brahamsha, B., Larimer, F.W., Land, M., Hauser, L. et al. (2003) The genome of a motile marine *Synechococcus*. *Nature* 424, 1037–1042.

Palenik, B., Ren, Q., Dupont, C.L., Myers, G.S., Heidelberg, J.F. et al. (2006) Genome sequence of *Synechococcus* CC9311: insights into adaptation to a coastal environment. *Proceedings of the National Academy of Sciences of the United States of America* 103, 13555–13559.

Pierre, S., Hirasawa, M., Cassan, N., Lagoutte, B., Tripathy, J.N. et al. (2009) New insights into the catalytic cycle of plant nitrite reductase: electron transfer kinetics and charge storage. *Biochemistry* 48, 2828–2838.

Ponomarova, O. and Patil, K.R. (2015) Metabolic interactions in microbial communities: untangling the gordian knot. *Current Opinion in Microbiology* 27, 37–44.

Quintana, N., van der Kooy, F., van de Rhee, M.D., Voshol, G.P. and Verpoorte, R. (2011) Renewable energy from cyanobacteria: energy production optimization by metabolic pathway engineering. *Applied and Environmental Microbiology* 91, 471–490.

Ramachandran, R. and Menon, R.K. (1998) An overview of industrial uses of hydrogen. *International Journal of Hydrogen Energy* 23, 593–598.

Rana, A., Joshi, M., Prasanna, R., Shivay, Y.S. and Nain, L. (2012) Biofortification of wheat through inoculation of plant growth promoting rhizobacteria and cyanobacteria. *European Journal of Soil Biology* 50, 118–126.

Ravikrishnan, A., Blank, L.M., Srivastava, S. and Raman, K. (2019) Investigating metabolic interactions in a microbial co-culture through integrated modelling and experiments. *Cold Spring Harbor Laboratory* 18, 1249–1258.

Reis, M.A.M., Serafim, L.S., Lemos, P.C., Ramos, A.M., Aguiar, F.R. et al. (2003) Production of polyhydroxyalkanoates by mixed microbial cultures. *Bioprocess and Biosystems Engineering* 25, 377–385.

Ren, L.J., Sun, X.M., Ji, X.J., Chen, S.L., Guo, D.S. et al. (2017) Enhancement of docosahexaenoic acid synthesis by manipulation of antioxidant capacity and prevention of oxidative damage in *Schizochytrium* sp. *Bioresource Technology* 223, 141–148.

Renuka, N., Sood, A., Prasanna, R. and Ahluwalia, A.S. (2015) Phycoremediation of wastewaters: a synergistic approach using microalgae for bioremediation and biomass generation. *International Journal of Environmental Science and Technology* 12, 1443–1460.

Rochfort, S. (2005) Metabolomics reviewed: a new 'omics' platform technology for systems biology and implications for natural products research. *Journal of Natural Products* 68, 1813–1820.

Rosso, D., Stenstrom, M.K. and Larson, L.E. (2008) Aeration of large-scale municipal wastewater treatment plants: state of the art. *Water Science and Technology* 57, 973–978.

Ryu, B.G., Kim, W., Nam, K., Kim, S., Lee, B. et al. (2015) A comprehensive study on algal–bacterial communities shift during thiocyanate degradation in a microalga-mediated process. *Bioresource Technology* 191, 496–504.

Ryu, B.G., Kim, J., Han, J.I. and Yang, J.W. (2016) Feasibility of using a microalgal-bacterial consortium for treatment of toxic coke wastewater with concomitant production of microbial lipids. *Bioresource Technology* 225, 58–66.

Sahin-Tóth, M., Frillingos, S., Lengeler, J.W. and Kaback, H.R. (1995) Active transport by the *cscB* permease in *Escherichia coli* K-12. *Biochemical and Biophysical Research Communications* 208, 1116–1123.

Salehizadeh, H. and Loosdrecht, M. (2004) Production of polyhydroxyalkanoates by mixed culture: recent trends and biotechnological importance. *Biotechnology Advances* 22, 261–279.

Schulte, C., Borah, K., Wheatley, R.M., Terpolilli, J.J., Saalbach, G. et al. (2021) Metabolic control of nitrogen fixation in rhizobium-legume symbioses. *Science Advances* 7, eabh2433.

Seymour, J.R., Ahmed, T., Durham, W.M. and Stocker, R. (2010) Chemotactic response of marine bacteria to the extracellular products of *Synechococcus* and *Prochlorococcus*. *Aquatic Microbial Ecology* 59, 161–168.

Shah, M.P. (2019) Bioremediation of azo dye. In: Shah, M.P. and Rodriguez-Couto, S. (eds) *Microbial Wastewater Treatment*. Elsevier, pp. 103–126.

Shen, L., Li, Z., Wang, J., Liu, A., Li, Z. et al. (2018) Characterization of extracellular polysaccharide/protein contents during the adsorption of Cd(II) by *Synechocystis* sp. PCC 6803. *Environmental Science and Pollution Research* 25, 20713–20722.

Shong, J., Diaz, M.R.J. and Collins, C.H. (2012) Towards synthetic microbial consortia for bioprocessing. *Current Opinion in Biotechnology* 23, 798–802.

Stanley, G.D. and Van, D. (2009) *The Evolution of the Coral–algal Symbiosis*. Springer, Berlin Heidelberg, 7–19.

Steinbüchel, A. (1991) *Polyhydroxyalkanoic Acids*. Palgrave Macmillan, pp. 123–213

Straight, P.D. and Kolter, R. (2009) Interspecies chemical communication in bacterial development. *Annual Review of Microbiology* 2009, 99–118.

Subashchandrabose, S.R., Ramakrishnan, B., Megharaj, M., Venkateswarlu, K. and Naidu, R. (2011) Consortia of cyanobacteria/microalgae and bacteria: biotechnological potential. *Biotechnology Advances* 29, 896–907.

Subashchandrabose, S.R., Ramakrishnan, B., Megharaj, M., Venkateswarlu, K. and Naidu, R. (2013) Mixotrophic cyanobacteria and microalgae as distinctive biological agents for organic pollutant degradation. *Environment International* 51, 59–72.

Sutherland, I.W. (1998) Novel and established applications of microbial polysaccharide. *Trends in Biotechnology* 16, 41–46.

Tang, X., He, R.Y., Tao, R.Q., Dang, R., Guo, R.L. et al. (2010) Construction of an artificial microalgalbacterial consortium that efficiently degrades crude oil. *Journal of Hazardous Materials* 181, 1158–1162.

Thevenard, B., Rasoava, N., Fourcassié, P., Monnet, V., Boyaval, P. et al. (2011) Characterization of *Streptococcus thermophilus* two-component systems: in silico analysis, functional analysis and expression of response regulator genes in pure or mixed culture with its yogurt partner, *Lactobacillus delbrueckii* subsp. *bulgaricus*. *International Journal of Food Microbiology* 151, 171–181.

Tiwari, O.N., Khangembam, R., Shamjetshabam, M., Sharma, A.S., Oinam, G. et al. (2015) Characterization and optimization of bioflocculant exopolysaccharide production by cyanobacteria *Nostoc* sp. BTA97 and *Anabaena* sp. BTA990 in culture conditions. *Applied Biochemistry and Biotechnology, Part A. Enzyme Engineering and Biotechnology* 176, 1950–1963.

Tyers, M. and Mann, M. (2003) From genomics to proteomics. *Nature* 422, 193–197.

Ullah, S., Dahlawi, S., Naeem, A., Rangel, Z. and Naidu, R. (2017) Biochar application for the remediation of salt-affected soils: challenges and opportunities. *Science of the Total Environment* 625, 320.

Veziroğlu, T.N. (1995) Twenty years of the hydrogen movement 1974–1994. *International Journal of Hydrogen Energy* 20, 1–7.

Vidal, R., López-Maury, L., Guerrero, M.G. and Florencio, F.J. (2009) Characterization of an alcohol dehydrogenase from the cyanobacterium *Synechocystis* sp. strain PCC 6803 that responds to environmental stress conditions via the Hik34-Rre1 two-component system. *Journal of Bacteriology* 191, 4383–4391.

Vincenzini, M., Materassi, R., Tredici, M.R. and Florenzano, G. (1982) Hydrogen production by immobilized cells. I. Light dependent dissimilation of organic substances by *Rhodopseudomonas palustris*. *International Journal of Hydrogen Energy* 7, 231–236.

Wang, V.B., Sivakumar, K., Yang, L., Zhang, Q. and Cao, B. (2015) Metabolite-enabled mutualistic interaction between *Shewanella oneidensis* and *Escherichia coli* in a co-culture using an electrode as electron acceptor. *Scientific Reports* 5, 11222.

Wang, X. (2021) Bacterial transcription factors and gene expression regulation. *Chinese Journal of Biochemistry and Molecular Biology* 37, 697–703.

Wang, X., Li, Z., Policarpio, L., Koffas, M.A.G. and Zhang, H. (2020) De novo biosynthesis of complex natural product sakuranetin using modular co-culture engineering. *Applied Microbiology and Biotechnology* 104, 4849–4861.

Wang, Z., Z.S.Q. and Yu, R.P. (1993) *Saline Soil of China*. Science Press, Beijing.

Wang, Z., Gerstein, M. and Snyder, M. (2009) RNA-Seq: a revolutionary tool for transcriptomics. *Nature Reviews Genetics* 10, 57–63.

Wei, P. (2002) A summary of microbial extra polysaccharides. *Journal of Zhejiang University of Science and Technology* 14, 8–14.

Weiss, T.L., Young, E.J. and Ducat, D.C. (2017) A synthetic, light-driven consortium of cyanobacteria and heterotrophic bacteria enables stable polyhydroxybutyrate production. *Metabolic Engineering* 44, 236–245.

West, A.H. and Stock, A.M. (2001) Histidine kinases and response regulator proteins in two-component signaling systems. *Trends in Biochemical Sciences* 26, 369–376.

Wilkins, M.R., Pasquali, C., Appel, R.D., Ou, K., Golaz, O. et al. (1996) From proteins to proteomes: large scale protein identification by two-dimensional electrophoresis and amino acid analysis. *Nature Biotechnology* 14, 61–65.

Xi, Q., Yi, B., Zhuang, Q. and Zhong, G. (2014) RNA-Seq technology and its application in fish transcriptomics. *Omics A Journal of Integrative Biology* 18, 98–110.

Xu, L., Li, D., Wang, Q. and Wu, S. (2016) Improved hydrogen production and biomass through the co-cultivation of *Chlamydomonas reinhardtii* and *Bradyrhizobium japonicum*. *International Journal of Hydrogen Energy* 41, 9276–9283.

Xu, L., Cheng, X., Wu, S. and Wang, Q. (2017) Co-cultivation of *Chlamydomonas reinhardtii* with *Azotobacter chroococcum* improved H$_2$ production. *Biotechnology Letters* 39, 731–738.

Yang, J.Y., Karr, J.R., Watrous, J.D. and Dorrestein, P.C. (2011) Integrating '-omics' and natural product discovery platforms to investigate metabolic exchange in microbiomes. *Current Opinion in Chemical Biology* 15, 79–87.

Yoshimura, H., Kotake, T., Aohara, T., Tsumuraya, Y., Ikeuchi, M. *et al.* (2012) The role of extracellular polysaccharides produced by the terrestrial cyanobacterium *Nostoc* sp. strain HK-01 in NaCl tolerance. *Journal of Applied Phycology* 24, 237–243.

Yu, K., Yi, S., Li, B., Guo, F., Peng, X. *et al.* An integrated meta-omics approach reveals substrates involved in synergistic interactions in a bisphenol A (BPA)-degrading microbial community. *Microbiome* 7, 1–13.

Zhang, H.O., Zhou, W.Z., Ma, Y.H., Zhao, H.X. and Zhang, Y.Z. (2013) FTIR spectrum and detoxication of extracellular polymeric substances secreted by microorganism. *Spectroscopy and Spectral Analysis* 33, 3041–3043.

Zhang, L., Chen, L., Diao, J., Song, X., Shi, M. and Zhang, W. (2020) Construction and analysis of an artificial consortium based on the fast-growing cyanobacterium *Synechococcus elongatus* UTEX 2973 to produce the platform chemical 3-hydroxypropionic acid from CO$_2$. *Biotechnology for Biofuels* 13, 1–14.

Zhang, W., Shen, N.H. and Chen, X.F. (2010a) Properties of several microbial polysaccharides and their applications in food industry. *Science and Technology of Food Industry*.

Zhang, X., Fang, A., Riley, C.P., Wang, M., Regnier, F.E. *et al.* (2010b) Multi-dimensional liquid chromatography in proteomics: a review. *Analytica Chimica Acta* 664, 101–113.

Zheng, Q., Wang, Y., Lu, J., Lin, W., Chen, F. *et al.* (2020) Metagenomic and metaproteomic insights into photoautotrophic and heterotrophic interactions in a *Synechococcus* culture. *mBio* 11(1).

3 Genome Editing in Diatoms: Current Progress and Challenges

Xiahui Hao[1], Fan Hu[2], Yufang Pan[1], Wenxiu Yin[1] and Hanhua Hu[1*]

[1]Institute of Hydrobiology, Chinese Academy of Sciences, Wuhan, China; [2]School of Foreign Languages, China University of Geosciences, Wuhan, China

Abstract

Genome editing technology, as one of the most important biotechnologies in the last decade, has greatly promoted the development of reverse genetics and synthetic biology. Rapid development of genome editing technology enables different kinds of mutation at targeted sequences like deletion, insertion, epigenetic modification and even alteration of chromosomes. TALEN and CRISPR/Cas9, two genome editing technologies have been successfully applied to model diatoms; however, the corresponding researches are still very limited compared with other model organisms at present. This review summarizes the current published researches on diatom genome editing technology, and introduces the essential factors in the practical application of genome editing technology, including the delivery of genome editing elements, screening of genome edited strains, and bioinformatics tools. Through the published literature, the potential of genome editing technology in diatom genetic research and diatom synthetic biology is revealed.

1. Genome Editing Technology

Genome editing technology, the 'superstar' of bioengineering researches in recent decades, generally refers to a series of approaches that target specific genome nucleic acid sequences through active complexes in order to modify the target sequences as desired. In a few years, multiple technologies emerged one after another, such as zinc-finger nucleases, meganucleases, transcription activator-like effector nucleases (TALEN) and clustered regularly interspaced short palindromic repeats (CRISPR) system. TALEN and CRISPR systems, in particular, are the most widely used genome editing technologies at present.

TALEN is a transcription activator-like effector (TALE) from *Xanthomonas* plant pathogens with a *Fok*I nuclease domain (Sanjana *et al.*, 2012). TALE is composed of tandemly repeated 34-amino acid sequence with two polymorphic amino acids called repeat variable di-residue (RVD) that recognizes specific DNA sequences (Kay *et al.*, 2007), so the *Fok*I nuclease domain can be aligned to the target locus. As *Fok*I functions only as a dimer, two independent TALENs are required to generate DNA double-strand breaks (DSBs) (Li *et al.*, 2012).

CRISPR system was first identified as 'immune systems' in different Eubacteria and Archaea genomes (Horvath and Barrangou, 2010). There are two major classes and six types in CRISPR-Cas systems, and Class 2 type II Cas system only utilizes one large single CRISPR-associated protein 9 (Cas9) to achieve gRNA binding, locus recognition and DNA

*Address for correspondence: hanhuahu@ihb.ac.cn

cleavage (Jiang and Doudna, 2017). CRISPR RNA (crRNA) and trans-activating crRNA (tracrRNA), the two RNA elements that form a crRNA:tracrRNA duplex, can assist Cas9 in targeting and cleaving dangerous exogenous nucleic acids, which inspires researchers to use a single guide RNA (sgRNA) to direct Cas protein to create DSBs in target loci. Therefore, Cas9 was chosen to be the first attempt in CRISPR genome editing technology, but with the limitation of protospacer adjacent motif (PAM) reading NGG at the end of sequence (Sander and Joung, 2014). Well known for its simple structure and easy accessibility, Cas9 and other CRISPR/Cas systems have become powerful tools with countless variation developed by researchers around the world, and the restrictions of PAM are no longer a crushing issue (Sukegawa et al., 2021).

TALEN and CRISPR systems both act as molecular scissors; however, in order to delete, insert or modify sequences, they also require genome repair mechanism, namely non-homologous end joining (NHEJ) and homology-directed recombination (HDR). NHEJ mechanism is a universal repair mechanism that creates random indels in target loci (Lieber, 2010). HDR repair mechanism can create precise genome editing when a donor DNA is provided, but it is often confined to the S and G2 phases of the cell cycle (Cuozzo et al., 2007).

Even though the procedure of genome editing is rather easy to understand, to achieve different goals of genome editing demands optimized methods of these technologies. The most simple and common use of genome editing technology is gene knockout, which can be accomplished by indels of NHEJ or HDR. Precise insertion of genes or any DNA sequences, however, relies on HDR mechanism, therefore mutation repair templates are indispensable when genome editing elements are introduced into cells. It should be emphasized that the key of genome editing is the ability to target any desired sequences, and the subsequent modification is varied. For instance, base editing technology is also achieved by using nuclease deactivated Cas9, like dCas9 or nCas9, with cytidine deaminase or adenosine deaminase to catalyze the conversion of DNA base (Komor et al., 2016; Gaudelli et al., 2017). CRISPR activation (CRISPRa) uses the locus recognition function of nuclease deactivated Cas9 with activator domain to activate gene expression, and, contrarily, CRISPR interference (CRISPRi) uses dCas9 to target DNA to sterically block transcription initiation and elongation (Ameruoso et al., 2022). Cas9 protein can be fused with other epigenetic modifiers, such as DNA methyltransferase and demethylase, to achieve CRISPR/Cas-based epigenetic regulation, which is considered as another great application of accurate position genome editing (Gallego-Bartolomé et al., 2018; Papikian et al., 2019). Moreover, remote DSBs can create large fragment deletion or even alteration of chromosomes (Zhou et al., 2014; Beying et al., 2020).

It is an uplifting fact that the promising future of genome editing technology brings more possibilities in reverse genetics and synthetic biology, and thus it is natural to have in-depth researches developed in a wide variety of species including diatoms. Diatom researchers are always keeping pace with the trend. TALEN and CRISPR/Cas9 have been successfully applied to model diatoms with different reliable workflows, providing great examples for future diatom research. Currently, diatom genome editing studies mainly focus on gene knockout and gene function analysis, which is not comparable to numerous studies in mammals, plant and yeast. It should be noted that establishing a stable genome editing protocol for certain species takes great effort, and modifying certain species into a synthetic chassis requires fast, efficient and maneuverable genome editing protocol. Today, diatom synthetic biology is ready to cut a striking figure, and more and more in-depth research is focused on key genes of diatom metabolic pathways. The development and application of gene editing technology is no doubt one of the major focuses of diatom research in the near future.

2. Approaches to Deliver Genome Editing Elements into Diatom Cells

To introduce TALEN or CRISPR/Cas9 into diatom cells, plasmid expression systems are straightforward choices. After years of innovation of key components such as promoters and select markers, efficient plasmid systems in diatoms can be applied in genome editing technology without obstacles.

In reported diatom genome editing researches, TALEN and Cas9 proteins are usually driven by *fucoxanthin chlorophyll a/c binding protein (FCP/LHCF)* or *nitrate reductase (NR)* promoters. Series of *FCP/LHCF* promoters are often the first choices for protein expression in diatoms, as *FCP/LHCF* are constitutively and robustly expressed in diatoms. *NR* promoter, an inducible promoter, can be used to control expression of introduced genes by sensing nitrogen source in the culture, and it switches off in ammonium medium and is induced in nitrate medium (Poulsen et al., 2006; Miyagawa et al., 2009). Therefore, *NR* promoter is used to regulate expression of Cas9 to reduce the off-target effects. Unlike TALEN and Cas9 protein, sgRNA is driven by RNA polymerase III promoters, like U6 small nuclear RNA promoter or sRNAi promoter. U6 promoter is proved to be a priority choice for sgRNA expression in different species (Sander and Joung, 2014; Aksoy et al., 2022). sRNAi promoter has also been successfully applied to sgRNA expressions (Moosburner et al., 2020), and it was reported to drive a most highly expressed small, non-coding, intergenic RNA in *Phaeodactylum tricornutum* (Rogato et al., 2014). Besides RNA polymerase III promoters, RNA polymerase II promoter has also been used to drive sgRNA expression with defined length by combining a self-cleaving ribozyme and an artificial poly(A) tail, for it not only shortened the expression array of multiplex sgRNAs cassette but also released multiple sgRNA at the same time. Taparia et al. (2022) demonstrated that a tandem array of sgRNA targets, flanked by Hammer Head and Hepatitis Delta Virus ribozymes, could be expressed as a single transcript using *P. tricornutum* promoter-49202 or promoter H4-1b.

As for selection markers, there are indeed few choices for protist diatoms, and zeocin and nourseothricin are the most used antibiotics in diatom researches. Besides, genome editing is considered to be the method to create auxotroph for selection marker. For example, PtUMPS knockout *P. tricornutum* strain cannot live in media without uracil, and other endogenous counter-selectable marker genes like PtAPT (adenine) and PtPRA-PH/CH (histidine) are also applied to antibiotic-free selection (Serif et al., 2018; Slattery et al., 2018). A stable auxotroph-based selection procedure will be a great starting point for diatom synthetic biology.

Transformation is critical for plasmids to enter diatom cells and express successfully. Biolistic bombardment and electroporation, the well-established methods in diatom for years (Zaslavskaia et al., 2000; Zhang and Hu, 2014; Yin and Hu, 2021), are unquestionably suitable for delivering TALEN or CRISPR/Cas9 plasmids. Bacterial conjugation, first reported by Karas et al. (2015), is a rather new technique in diatoms and is still not as popular as biolistic bombardment and electroporation, but it is reported to have several advantages over other methods. Firstly, episomal vectors delivered by conjugation can be maintained as an episome outside the genome due to a yeast centromeric CEN6-ARSH4-HIS3 sequence, which therefore greatly reduces the probability of random integration of the transgene in the genome. Secondly, the episomal vectors can be cured upon the removal of selection pressure, allowing the exclusion of the genome editing elements from the cells once the desirable modification has been obtained. Lastly, the colonial period is short. Colonies of *P. tricornutum* and *Thalassiosira pseudonana* appeared 10–14 and 7–14 days after transformation by bacterial conjugation respectively (Karas et al., 2015; Moosburner et al., 2020). Overall, no matter what methods are used, stable and effective protocols are fundamental to genome editing experiment practices.

DNA-free genome editing has drawn much attention because of the feature of minimizing the unwanted effects and random integrations caused by the introduction of vector DNA. The key of DNA-free genome editing is the direct introduction of Cas9/sgRNA ribonucleoprotein (RNP) complexes. Serif et al. (2018) successfully demonstrated the procedure of biolistic bombardment of Cas9 RNP complexes into the diatom. According to their description, different crRNAs and tracrRNA were purchased from IDT and assembled into Cas9-gRNA RNP complex using HiFi Cas9 (IDT #1074181). Each RNP complex targeting different loci of corresponding genes was coated onto gold particles and bombarded into *P. tricornutum* cells. Without additional selection marker, gene ptUMPS or ptAPT was chosen to be knocked out along with other target genes; therefore clones can be obtained by using a selective medium containing 5-FOA and uracil or a medium containing 2-FA and adenine. The study represents a true DNA-free method to knock out three genes simultaneously using Cas9-gRNA RNP complex, thus extending a new strategy for diatom bioengineering. But the usage of endogenous counter-selectable marker genes is not sustainable for multiple rounds of transformation, and additional gene knockout may interfere with the analysis of target genes.

3. Screening of Genome Edited Strains

Screening of diatom cells is far more difficult compared to other fast-growing and monoploid microorganisms. *P. tricornutum* and *T. pseudonana*, two species widely used in diatom genome editing research, are both diploids, whereas heterozygous mutations (only one allele was mutated) or biallelic mutations (both alleles were mutated but the sequence of each allele was distinct) are likely to be created when different editing events happen in two homologous chromosomes. What is more, several different mutated alleles were usually found within a colony, as a few cell divisions have occurred before the first mutagenic event. It is understandable that homozygous mutates are difficult to obtain. Several experiments are designed to detect mutations of transformations. Fragment size difference of PCR products with specific primers is simple and reliable to detect large deletions or insertions in mutations. For example, gene knock-in and knockout will lead to a large and small PCR fragment, respectively. However, small indels cannot be distinguished by the difference of PCR product size directly. Mismatch detection assays are also simple, rapid and cost-effective methods based on PCR to detect targeted mutations. These methods include the mismatch cleavage assay, high-resolution melting analysis (HRMA) and heteroduplex mobility assay (Daboussi et al., 2014; Nymark et al., 2016; Zischewski et al., 2017; Slattery et al., 2018). Southern blot is another useful method to reveal the alteration of DNA sequences compared to the wild-type DNA (Weyman et al., 2015). Unless the mutation is homozygous, multiple traces with overlapping peaks starting at the mutation site will be generated by direct Sanger sequencing (Stukenberg et al., 2018). Fortunately, there are bioinformatic tools like DSDecodeM (http://skl.scau.edu.cn/dsdecode/ [accessed 12/6/23]; Liu et al., 2015), ICE (Inference of CRISPR Edits) Analysis Tool (http://ice.synthego.com [accessed 12/6/23]; Conant et al., 2022) and TIDE (Tracking of Indels by Decomposition: http://tide.nki.nl [accessed 12/6/23]; Brinkman et al., 2014) to assist in screening clones with higher mutation frequency, which is conducive to obtaining homozygotes efficiently in subsequent subcloning of this mutant.

4. Tools for Diatom Genome Editing

Genome editing technology is advancing alongside bioinformatics. Popularization of genome editing technology is unrealizable without the peerless bioinformatic tools developed by scientists around the world, and a lot of tools are available for diatom genome editing. TALE-NT 2.0 (TAL Effector Nucleotide Targeter 2.0, https://tale-nt.cac.cornell.edu/ [accessed 12/6/23]) is a handy tool for designing pairs of TALEs for TALENs to target a specific gene sequence and also for designing custom TALEs for single sites (Cermak et al., 2011; Doyle et al., 2012). SAPTA (Scoring Algorithm for Predicting TALE(N) Activity, http://baolab.bme.gatech.edu/Research/BioinformaticTools [accessed 12/6/23]) is also used for the selection of optimal TALEN binding sites and estimation of cutting efficiency of TALENs (Lin et al., 2014). Selection of sgRNA is vital for increasing editing frequency, and many bioinformatic programs are specialized to predict a better sgRNA sequence pattern. There are plenty of user-friendly tools for the optimization of sgRNA and off-target prediction: CRISPR RGEN Tools (Bae *et al.*, 2014; Park *et al.*, 2015, http://www.rgenome.net/ [accessed 12/6/23]), CRISPR-GE (Xie *et al.*, 2017, http://skl.scau.edu.cn/), CRISPick (Sanson *et al.*, 2018, https://portals.broadinstitute.org/gppx/crispick/public [accessed 12/6/23]), PhytoCRISP-Ex (Rastogi *et al.*, 2016, https://www.phytocrispex.bio.ens.psl.eu/CRISP-Ex/ [accessed 12/6/23]), CRISPOR (Concordet and Haeussler, 2018, http://crispor.tefor.net/ [accessed 12/6/23]), and EuPaGDT (Peng and Tarleton, 2015, http://grna.ctegd.uga.edu/ [accessed 12/6/23]). For screening mutation and Sanger sequence analysis, DSDecode, ICE Analysis Tool and TIDE can save researchers from time-consuming manual calculations. Additionally, CRISPR/Cas9 vector pKS-diaCas9-sgRNA (addgene ID: 74923), Episome vector pPtPuc3 (addgene ID: 62863) and Episome-based CRISPR/Cas9 vector PtPuc3_diaCas9_sgRNA (addgene ID: 109219) are all available online, providing convenience for diatom researchers. All the tools and vectors mentioned above are verified by published papers of diatom genome editing researches and are reliable for genome editing designed for diatom genome. The rapid development of bioinformatics is exciting news, and diatom genome editing is to benefit from upcoming powerful tools.

5. Future Challenges

The progress of diatom genome editing is still insufficient at present. Due to its relatively complex construction process, it is difficult to popularize TALEN in diatom researches. CRISPR/Cas9 has been verified in many studies; however, CRISPR-based genome editing has not been widely used in diatom researches. The alternative CRISPR system CRISPR/Cpf1 and other advanced applications, like base editors and epigenetic modifiers, are not reported in diatoms yet. The slow growth rate of diatom cells is a huge disadvantage for genetic transformation compared with simple organisms like yeast, thus the optimization of propagation and screening should be given more prominence. Diatom species diversity is a valuable resource for diatom genome editing, and unconventional diatom species may have advantages over model diatoms in growth rate and culture conditions. Diatoms have the great potential to become leading industrial microbes. It is challenging but feasible to create a customizable genome design and improved engineering platform of diatoms as Pt-syn 1.0 project proposed (Pampuch *et al.*, 2022).

References

Aksoy, E., Yildirim, K., Kavas, M., Kayihan, C., Yerlikaya, B.A. *et al.* (2022) General guidelines for CRISPR/Cas-based genome editing in plants. *Molecular Biology Reports* 49, 12151–12164.

Ameruoso, A., Villegas Kcam, M.C., Cohen, K.P. and Chappell, J. (2022) Activating natural product synthesis using CRISPR interference and activation systems in *Streptomyces*. *Nucleic Acids Research* 50(13), 7751–7760.

Bae, S., Park, J. and Kim, J.S. (2014) Cas-OFFinder: a fast and versatile algorithm that searches for potential off-target sites of Cas9 RNA-guided endonucleases. *Bioinformatics* 30(10), 1473–1475.

Beying, N., Schmidt, C., Pacher, M., Houben, A. and Puchta, H. (2020) CRISPR-Cas9-mediated induction of heritable chromosomal translocations in *Arabidopsis*. *Nature Plants* 6(6), 638–645.

Brinkman, E.K., Chen, T., Amendola, M. and van Steensel, B. (2014) Easy quantitative assessment of genome editing by sequence trace decomposition. *Nucleic Acids Research* 42(22), e168.

Cermak, T., Doyle, E.L., Christian, M., Wang, L., Zhang, Y. *et al.* (2011) Efficient design and assembly of custom TALEN and other TAL effector-based constructs for DNA targeting. *Nucleic Acids Research* 39(12), e82.

Conant, D., Hsiau, T., Rossi, N., Oki, J., Maures, T. et al. (2022) Inference of CRISPR edits from Sanger trace data. *The CRISPR Journal* 5(1), 123–130.

Concordet, J.P. and Haeussler, M. (2018) CRISPOR: intuitive guide selection for CRISPR/Cas9 genome editing experiments and screens. *Nucleic Acids Research* 46(W1), W242–W245.

Cuozzo, C., Porcellini, A., Angrisano, T., Morano, A., Lee, B. et al. (2007) DNA damage, homology-directed repair, and DNA methylation. *PLOS Genetics* 3(7), e110.

Daboussi, F., Leduc, S., Maréchal, A., Dubois, G., Guyot, V. et al. (2014) Genome engineering empowers the diatom *Phaeodactylum tricornutum* for biotechnology. *Nature Communications* 5, 3831.

Doyle, E.L., Booher, N.J., Standage, D.S., Voytas, D.F., Brendel, V.P. et al. (2012) TAL Effector-nucleotide targeter (TALE-NT) 2.0: tools for TAL effector design and target prediction. *Nucleic Acids Research* 40(Web Server issue), W117–W122.

Gallego-Bartolomé, J., Gardiner, J., Liu, W., Papikian, A., Ghoshal, B. et al. (2018) Targeted DNA demethylation of the *Arabidopsis* genome using the human TET1 catalytic domain. *Proceedings of the National Academy of Sciences of the United States of America* 115(9), E2125–E2134.

Gaudelli, N.M., Komor, A.C., Rees, H.A., Packer, M.S., Badran, A.H. et al. (2017) Programmable base editing of A•T to G•C in genomic DNA without DNA cleavage. *Nature* 551, 464–471.

Horvath, P. and Barrangou, R. (2010) CRISPR/Cas, the immune system of bacteria and archaea. *Science* 327(5962), 167–170.

Jiang, F. and Doudna, J.A. (2017) CRISPR–Cas9 structures and mechanisms. *Annual Review of Biophysics* 46, 505–529.

Karas, B.J., Diner, R.E., Lefebvre, S.C., McQuaid, J., Phillips, A.P. et al. (2015) Designer diatom episomes delivered by bacterial conjugation. *Nature Communications* 6, 6925.

Kay, S., Hahn, S., Marois, E., Hause, G. and Bonas, U. (2007) A bacterial effector acts as a plant transcription factor and induces a cell size regulator. *Science* 318, 648–651.

Komor, A.C., Kim, Y.B., Packer, M.S., Zuris, J.A. and Liu, D.R. (2016) Programmable editing of a target base in genomic DNA without double stranded DNA cleavage. *Nature* 533, 420–424.

Li, T., Liu, B., Spalding, M.H., Weeks, D.P. and Yang, B. (2012) High-efficiency TALEN-based gene editing produces disease-resistant rice. *Nature Biotechnology* 30, 390–392.

Lieber, M.R. (2010) The mechanism of double-strand DNA break repair by the nonhomologous DNA end-joining pathway. *Annual Review of Biochemistry* 79, 181–211.

Lin, Y., Fine, E.J., Zheng, Z., Antico, C.J., Voit, R.A. et al. (2014) SAPTA: a new design tool for improving TALE nuclease activity. *Nucleic Acids Research* 42(6), e47.

Liu, W., Xie, X., Ma, X., Li, J., Chen, J. et al. (2015) DSDecode: a web-based tool for decoding of sequencing chromatograms for genotyping of targeted mutations. *Molecular Plant* 8, 1431–1433.

Miyagawa, A., Okami, T., Kira, N., Yamaguchi, H., Ohnishi, K. et al. (2009) Research note: high efficiency transformation of the diatom *Phaeodactylum tricornutum* with a promoter from the diatom *Cylindrotheca fusiformis*. *Phycological Research* 57, 142–146.

Moosburner, M.A., Gholami, P., McCarthy, J.K., Tan, M., Bielinski, V.A. et al. (2020) Multiplexed knockouts in the model diatom *Phaeodactylum* by episomal delivery of a selectable Cas9. *Frontiers in Microbiology* 11, 5.

Nymark, M., Sharma, A.K., Sparstad, T., Bones, A.M. and Winge, P. (2016) A CRISPR/Cas9 system adapted for gene editing in marine algae. *Scientific Reports* 6, 24951.

Pampuch, M., Walker, E.J.L. and Karas, B.J. (2022) Towards synthetic diatoms: the *Phaeodactylum tricornutum* Pt-syn 1.0 project. *Current Opinion in Green and Sustainable Chemistry* 35, 100611.

Papikian, A., Liu, W., Gallego-Bartolomé, J. and Jacobsen, S.E. (2019) Site-specific manipulation of *Arabidopsis* loci using CRISPR-Cas9 SunTag systems. *Nature Communications* 10(1), 729.

Park, J., Bae, S. and Kim, J.S. (2015) Cas-Designer: a web-based tool for choice of CRISPR-Cas9 target sites. *Bioinformatics* 31(24), 4014–4016.

Peng, D. and Tarleton, R. (2015) EuPaGDT: a web tool tailored to design CRISPR guide RNAs for eukaryotic pathogens. *Microbial Genomics* 1(4), e000033.

Poulsen, N., Chesley, P.M. and Kröger, N. (2006) Molecular genetic manipulation of the diatom *Thalassiosira pseudonana* (Bacillariophyceae). *Journal of Phycology* 42, 1059–1065.

Rastogi, A., Murik, O., Bowler, C. and Tirichine, L. (2016) PhytoCRISP-Ex: a web-based and stand-alone application to find specific target sequences for CRISPR/CAS editing. *BMC Bioinformatics* 17(1), 261.

Rogato, A., Richard, H., Sarazin, A., Voss, B., Cheminant Navarro, S. et al. (2014) The diversity of small non-coding RNAs in the diatom *Phaeodactylum tricornutum*. *BMC Genomics* 15, 698.

Sander, J.D. and Joung, J.K. (2014) CRISPR-Cas systems for editing, regulating and targeting genomes. *Nature Biotechnology* 32(4), 347–355.

Sanjana, N.E., Cong, L., Zhou, Y., Cunniff, M.M., Feng, G. *et al.* (2012) A transcription activator-like effector toolbox for genome engineering. *Nature Protocols* 7, 171–192.

Sanson, K.R., Hanna, R.E., Hegde, M., Donovan, K.F., Strand, C. *et al.* (2018) Optimized libraries for CRISPR-Cas9 genetic screens with multiple modalities. *Nature Communications* 9(1), 5416.

Serif, M., Dubois, G., Finoux, A.L., Teste, M.A., Jallet, D. *et al.* (2018) One-step generation of multiple gene knock-outs in the diatom *Phaeodactylum tricornutum* by DNA-free genome editing. *Nature Communications* 9(1), 3924.

Slattery, S.S., Diamond, A., Wang, H., Therrien, J.A., Lant, J.T. *et al.* (2018) An expanded plasmid-based genetic toolbox enables Cas9 genome editing and stable maintenance of synthetic pathways in *Phaeodactylum tricornutum*. *ACS Synthetic Biology* 7(2), 328–338.

Stukenberg, D., Zauner, S., Dell'Aquila, G. and Maier, U.G. (2018) Optimizing CRISPR/Cas9 for the diatom *Phaeodactylum tricornutum*. *Frontiers in Plant Science* 9, 740.

Sukegawa, S., Saika, H. and Toki, S. (2021) Plant genome editing: ever more precise and wide reaching. *The Plant Journal* 106(5), 1208–1218.

Taparia, Y., Dolui, A.K., Boussiba, S. and Khozin-Goldberg, I. (2022) Multiplexed genome editing *via* an RNA polymerase II promoter-driven sgRNA array in the diatom *Phaeodactylum tricornutum*: insights into the role of StLDP. *Frontiers in Plant Science* 12, 784780.

Weyman, P.D., Beeri, K., Lefebvre, S.C., Rivera, J., McCarthy, J.K. *et al.* (2015) Inactivation of *Phaeodactylum tricornutum* urease gene using transcription activator-like effector nuclease-based targeted mutagenesis. *Plant Biotechnology Journal* 13(4), 460–470.

Xie, X., Ma, X., Zhu, Q., Zeng, D., Li, G. *et al.* (2017) CRISPR-GE: a convenient software toolkit for CRISPR-based genome editing. *Molecular Plant* 10(9), 1246–1249.

Yin, W. and Hu, H. (2021) High-efficiency transformation of a centric diatom *Chaetoceros muelleri* by electroporation with a variety of selectable markers. *Algal Research* 55, 102274.

Zaslavskaia, L.A., Lippmeier, J.C., Kroth, P.G., Grossman, A.R. and Apt, K.E. (2000) Transformation of the diatom *Phaeodactylum tricornutum* (bacillariophyceae) with a variety of selectable marker and reporter genes. *Journal of Phycology* 36, 379–386.

Zhang, C. and Hu, H. (2014) High-efficiency nuclear transformation of the diatom *Phaeodactylum tricornutum* by electroporation. *Marine Genomics* 16, 63–66.

Zhou, H., Liu, B., Weeks, D.P., Spalding, M.H. and Yang, B. (2014) Large chromosomal deletions and heritable small genetic changes induced by CRISPR/Cas9 in rice. *Nucleic Acids Research* 42(17), 10903–10914.

Zischewski, J., Fischer, R. and Bortesi, L. (2017) Detection of on-target and off-target mutations generated by CRISPR/Cas9 and other sequence-specific nucleases. *Biotechnology Advances* 35(1), 95–104.

4 Euglena-based Synthetic Biology and Cell Factory

Chao Li[1,2,3], Zhenfan Chen[2], Zixi Chen[2], Anping Lei[2], Qiong Liu[2] and Jiangxin Wang[2]*

[1]*College of Food Engineering and Biotechnology, Hanshan Normal University, Chaozhou, China;* [2]*Shenzhen Key Laboratory of Marine Bioresource and Eco-Environmental Science, Shenzhen Engineering Laboratory for Marine Algal Biotechnology, Guangdong Provincial Key Laboratory for Plant Epigenetics, College of Life Sciences and Oceanography, Shenzhen University, Shenzhen, China;* [3]*Key Laboratory of Optoelectronic Devices and Systems of Ministry of Education and Guangdong Province, College of Optoelectronic Engineering, Shenzhen University, Shenzhen, China*

Abstract

Euglena, which flourished on Earth more than 500 million years ago, is a single-celled eukaryotic alga with dual properties of plants and animals. *E. gracilis*, an ideal model species, is of significant academic interest in studying endosymbiosis and chloroplast development. Moreover, *E. gracilis* cells produce a range of compounds, such as unsaturated fatty acids, amino acids and vitamins. Also, they can accumulate very high levels (75%+ dried weight) of the linear polysaccharide (β-1, 3-glucan, paramylon) and wax esters, which is both a high-value-added healthy food and a high-quality raw material for biofuels. In recent years, algal-based synthetic biology and cell factory have attracted much attention due to their massive potential in medical, agricultural, food and other fields. Although many researchers focused on genetic engineering tools for *E. gracilis* in recent years, little progress has been achieved because of the lack of high-quality genome information and efficient techniques for genetic operation. This chapter provides a detailed review of the progress of genetic transformation methods of *E. gracilis* and the advanced biotechnological tools for *E. gracilis*, such as CRISPR and RNA interference. In addition, the prospects and challenges of *E. gracilis* in stable transformation and genetic engineering of a 'smart' cell factory were discussed. We hope to provide a cohesive approach involving functional genomics and synthetic biology of *E. gracilis* to help in widening the horizon of the use of *E. gracilis* towards industrial biotechnology.

Keywords: *Euglena gracilis*; genetic transformation; synthetic biology; CRISPR

1. Introduction

Euglena is a single-celled eukaryotic alga without a cell wall but has flagella on the top of the cell and can swim freely. *Euglena* has multiple chloroplasts and is capable of autotrophic growth by photosynthesis under light and heterotrophic growth by using external organic matter in the absence of light (Zakryś *et al.*, 2017). Molecular phylogenetic studies have also shown that *Euglena* has close relatives to the pathogenic parasites *Trypanosoma* and *Leishmania* (O'Neill *et al.*, 2015a). The nucleus of *Euglena* has a mesokaryotic

*Address for correspondence: jxwang@szu.edu.cn

nature; the nuclear envelope and nucleolus do not disappear during cell division, chromosomes remain condensed. *Euglena* chloroplasts were proposed originated from an endosymbiotic primitive green alga (Gibbs, 1978). Unlike other algae or higher plants with chloroplasts, the chloroplasts of *Euglena* are less stable and prone to loss under stress conditions such as antibiotics, heat or pressure (Wang et al., 2002). Therefore, with its unique cellular structure and biological status, *Euglena* is a model microorgainsm for studying eukaryotic evolution, chloroplast endosymbiosis and chloroplast developments.

Euglena gracilis (*E. gracilis*), a model species, arguably represents one of the most comprehensively studied organisms within *Euglenida*. *E. gracilis* cells are ellipsoid or globular in shape and variable in length, 31–40 µm, width 9–14 µm (Torihara and Kishimoto, 2015). *E. gracilis* has been studied extensively in many fields, such as bioremediation of environmental pollution, and decreasing carbon dioxide emissions, especially for favored high-value-added products on the market (Klinthong et al., 2015; Gissibl et al., 2019; Tahira et al., 2019). *E. gracilis* cells are rich in many vitamins, amino acids and polyunsaturated fatty acids and have a high nutritional and commercial value (Matsumoto et al., 2009; Gissibl et al., 2019). During the growth of *E. gracilis*, a large amount of paramylon can be accumulated as energy storage material, accounting for 50–90% of the dry weight of *E. gracilis* (Šantek et al., 2009; Ivusic and Santek, 2015; Yasuda et al., 2018), which is much higher than that in other species such as fungi. Paramylon is a β-1, 3-glucan with enhanced immunity (Kondo et al., 1992; Yang et al., 2022), anti-allergic (Sugiyama et al., 2010), anti-viral (Koizumi et al., 1993) and anti-tumor (Quesada et al., 1976; Watanabe et al., 2013) activities. Moreover, Xie et al. (2022) reported that paramylon or sonicated and alkalized paramylon efficiently protected the liver from LPS/D-Gal-induced acute liver injury by alleviating inflammation and cell death. Under dark anaerobic conditions, *E. gracilis* cells can catabolize paramylon into wax esters, which have a wide range of industrial uses and can produce biofuels (Mahapatra et al., 2013; Grimm et al., 2015; Inui et al., 2017). Moreover, paramylon can be processed into plastics produced with succinic and lactic acids, an environmentally friendly application of *E. gracilis* under anaerobic conditions (Tomita et al., 2016). Recently, the prebiotic activity of *E. gracilis* and paramylon was reported, and the emerging prospects of *E. gracilis* as prebiotics were evaluated (Dai et al., 2022; Song et al., 2022). Therefore, *E. gracilis* is not only a health food with high-added-value but also an important raw material with great potential to produce bioenergy and is of great economic value.

At present, the biological studies of *Euglena* are mainly focused on cell physiology, enzymatic characteristics, gene cloning, and so on (Preisfeld et al., 2000; Bennett and Triemer, 2015; Ogawa et al., 2015a; Takeda et al., 2015; Watanabe et al., 2017), and the research of the genetic transformation of *Euglena* is minimal, which has limited the development of the genetics and biotechnology of *Euglena*. The reason could be the transformation efficiency and lack of complete genome information on *E. gracilis* (Doron et al., 2016). With the rapid development of synthetic biology, *E. gracilis* is one of the most promising microalgae among other industrial species, such as *Chlamydomonas reinhardtii*, *Chlorella* sp., and *Haematococcus pluvialis*. Thus, using *E. gracilis* as an example, in conjunction with our own researches, we provide a detailed review of the progress of the genetic transformation and ideas for delivery methods of exogenous materials of *E. gracilis*, and also give perspectives for future research and applications of *E. gracilis*.

2. Biological Characterization of *E. gracilis*

2.1 Genome and systemic biological of *E. gracilis*

E. gracilis contains three genomes: nucleus, chloroplast and mitochondrion. Its huge genome has a highly complex sequence structure with a high proportion of repetitive sequences, making sequencing and splicing difficult (Ebenezer et al., 2017). Ebenezer et al. (2019) had published the draft nuclear genome of E. gracilis with an estimated haploid genome size of about 500 Mb, and a total of 1,266,288 contigs were assembled with a total length of 1.43 Gb, including 1459 contigs with a length longer than 10 kb. The longest contig was 16 kb, and the N50 was only 955 bp. Compared with the previous estimation genome size of E. gracilis (1–4 Gb), the new version of the genome size was smaller, and the data can only be used for fundamental analysis. We have pioneered the completion of the draft genome of E. gracilis wild-type alga strain Z, showing that the N50 contig has reached 1000 kb, and has preliminarily confirmed that E. gracilis genome is about 2,541 M. This result has been confirmed by various methods, and is much larger than the 124 M (Merchant et al., 2007) found in the model organism Chlamydomonas reinhardtii, 27.4 M (Bowler et al., 2008) for Phaeodactylum tricornutum, 138 M (Prochnik et al., 2010) for Volvox carteri, and 46 M (Blanc et al., 2010) for Chlorella variabilis. The draft genome contains around 45% repetitive sequences, causing great difficulties in genome assembling with the combination of second- and third-generation sequencing data. In addition, using E. gracilis transcriptome and proteome obtained from our group, a huge number of non-coding RNAs were discovered to complete the identification and functional annotation of 40,128 genes. Its exceptionally complex internuclear architecture, karyotype, Hi-C, as well as genome 3D-STORM scans, and fine mapping of the genome are also underway.

Each Euglena cell contains about ten chloroplasts (Hadariová et al., 2017), while each chloroplast contains 200–600 genome copies (Rawson and Boerma, 1976), and the chloroplast genome, similar to prokaryotes, is in a circular structure, approximately 143 kb in size, encodes 87 known genes, and contains at least 149 introns within the gene (Hallick et al., 1993). The mitochondrial genome is linearly structured and contains several linear segments of 5–8 kb, encoding seven respiratory chain complex proteins, including three complex I subunits (encoded by nad1, nad4, and nad5, respectively), one complex III subunit (encoded by cob), and three complex IV subunits (encoded by cox1, cox2, and cox3, respectively). In addition, there are some small gene segments of unknown function contained within mitochondria that may be involved in DNA recombination and RNA modification (Dobáková et al., 2015).

Initial studies of the de novo transcriptome of Euglena cultured in different media have been reported (O'Neill et al., 2015b; Yoshida et al., 2016), but other omics data, including proteome, small RNA panel, and methylome have not been reported. Our research team addresses chloroplast evolution in Euglena and to nucleus chloroplast molecular communication at different molecular levels, including transcriptome, proteome, metabolome and methylome. In addition, the latest single-cell analysis techniques are being used by our group to simultaneously resolve the problem of polymorphisms in chloroplast differentiation of about 15% of cells grown in more than 4000–6000 Euglena cells supplemented with methylase inhibitors under dark-culture conditions. In summary, through various omics analyses, our group has obtained a lot of plausible key genes and non-coding small RNAs related to chloroplast differentiation, chloroplast to nucleus molecular communication, and several new hypothesis theories have also been proposed, requiring a high-throughput and efficient technical approach to gene engineering for functional validation.

2.2 Antibiotic resistance and selection markers of E. gracilis

Antibiotic resistance is the key to the screening and selection of transformants. However, studies on the sensitivity of Euglena to antibiotics are limited. Khatiwada et al. (2019) introduced vectors with the hygromycin resistance gene hptII (encoding hygromycin phosphotransferase II) into E. gracilis by electroporation, Agrobacterium-mediated transformation, and biolistic bombardment.

In our study, the antibiotic tolerance of *E. gracilis* was examined in detail and it found that 10μg/ml or more of geneticin (G418) and bleomycin (zeocin) significantly inhibited the growth of *E. gracilis* cells, whereas paromomycin, hygromycin, kanamycin, tetracycline, as well as the glufosinate, added at 30μ g/ml still had no obvious inhibitory effect on *E. gracilis* cells, which were treated with higher concentrations (> 100 μg/ml) of kanamycin and paromomycin and could also significantly inhibit the growth of *E. gracilis*, indicating that *E. gracilis* has strong tolerance to most antibiotics and herbicides, but is more sensitive to G418 and zeocin. Becker *et al.* (2021) explored the sensitivity of *E. gracilis* to a range of antibiotics in liquids and agar plates with different concentrations.

G418 is an aminoglycoside structurally similar to paromomycin, kanamycin that blocks protein synthesis by interfering with ribosome function and is toxic to both bacterial and eukaryotic cells. In cell transfection, the aminoglycoside phosphotransferase gene (*neo*) is commonly used as a screening marker, which catalyzes the phosphorylation inactivation of G418, making cells acquire resistance.

E. gracilis screening with G418 has not been reported. Still, the use of paromomycin for *E. gracilis* mutant strain screening with the neomycin phosphotransferase gene (*NPTII*) as a resistance marker has been documented (Ogawa *et al.*, 2015c).

Zeocin, a glycoprotein antibiotic competing for the mithramycin family, degrades intracellular DNA and has broad inhibitory effects on prokaryotic and eukaryotic cells. In the transformation of *Chlamydomonas reinhardtii*, the adoption of a *ble* gene derived from *Streptoalloteichus hindustanus*, which encodes a 13.5 kD protein capable of binding to zeocin to inhibit its activity (Jain *et al.*, 1992), was shown to confer zeocin resistance to the algal cells (Ogawa *et al.*, 2015b). Within *E. gracilis* cells, transformation using *ble* gene and zeocin screening has also been reported (Krajcovic *et al.*, 2011). In addition, the literature also reports the transformation of *E. gracilis* with *aadA* genes and the screening of positive transformants with streptomycin and spectinomycin (Doetsch *et al.*, 2001), which encode aminoglycoside adenylyltransferases capable of transferring the adenine group of ATP onto antibiotic molecules and thereby inactivate it.

In our experiments using the method of biolistic bombardment to transform linearized plasmids containing *ble* and *NPTII*, we have generated single clones of resistant *E. gracilis* cells capable of harboring the plasmid in the presence of 80 μg/ml zeocin and 250 μg/ml paromomycin in liquid CM medium and were also able to grow and passage stably on solid plates at the same antibiotic concentration. In contrast, wild-type *E. gracilis* cells were non-viable and passaged in the medium at this antibiotic concentration.

3. The Delivery Methods of Exogenous Materials for *E. gracilis*

In addition to *Euglena*, the genetic transformation system of eukaryotic microalgae such as *Chlamydomonas reinhardtii* has been studied systematically, and various transformation methods have been established, including biolistic, electroporation, *Agrobacterium*-mediated transformation, and glass beads. Krajcovic *et al.* (2011) first reported the successful nuclear genome transformation of *E. gracilis* with a zeocin resistance transformation cassette by electroporation. Khatiwada *et al.* (2019) explored the genetic engineering tools to obtain stable *E. gracilis* nuclear transformants via constructing plasmid vectors with the selection marker of hygromycin by *Agrobacterium*-mediated transformation, biolistic bombardment, and electroporation techniques. Moreover, with the rise of single-cell-analysis technologies, it is proposed that single-cell microinjection and electroporation could be the efficient molecule delivery ways for *E. gracilis*.

3.1 Biolistic bombardment

Biolistic bombardment, which bombards DNA-coated microparticles into cells by a high-pressure gas flow, is currently one of the most efficient transformation methods, which is not limited by cell type and is widely used in transgenic research. This method has been used for the transformation of *Chlamydomonas reinhardtii* (Mayfield and Kindle, 1990), *Phaeodactylum tricornutum* (Kira et al., 2016), *Dunaliella salina* (Tan et al., 2005), *Nannochloropsis* (Anley, 2015), *Synechocystis* (Hallmann and Rappel, 1999), and *Trypanosoma* (Vainstein et al., 1994), a close relative of *Euglena*. In addition, this method approach can introduce exogenous genes into organelles with high efficiency. For example, Boynton et al. (1988) bombarded the *atpB* gene of wild-type *Chlamydomonas* into three chloroplast *atpB* mutant strains of *Chlamydomonas* to restore its photosynthetic capacity. Hu et al. (2011) bombarded the resistance gene *ble* to *Chlamydomonas* mitochondria and stably expressed it.

The earlier reports of the genetic transformation of *E. gracilis* were focused on the chloroplast genome. Biolistic bombardments and selection biomarkers have been used for plastid transformation (Doetsch et al., 2001; Ogawa et al., 2015b). Doetsch et al. (2001) first modified the Biolistic Method established by Sanford et al. (1993), and used for transformation of *E. gracilis* with the following main steps: *E. gracilis* cells grown to the end of log phase under heterotrophic conditions were collected, flattened onto filters by low-pressure vacuum suction filtration (diameter 45 mm, pore-diamete 0.22µm); a thin layer of cells was formed; filters were transferred onto heterotrophic plates and transformation was completed within 2 h; 2.5 µg of plasmid DNA was coated with tungsten powder particles of M5 type; DuPont pPDS-100/He-type gene gun and ruptured membrane of 7600 kN/m², setting a vacuum of 3.4 kPa and a targeting distance of 12.3 cm, after one bombardment of the *E. gracilis* cells, the filters were transferred to screening plates. Transformants were picked approximately six weeks later.

Ogawa et al. (2015b) further improved the process of biolistic transformation in order to improve the transformation efficiency of *Euglena*, and the main steps were as follows: centrifugation to collect the *Euglena* cells in the logarithmic growth period, washed with sterile water and re-suspended, the cells were spread onto filter membrane with low-pressure vacuum suction filter, and the filter membrane was placed on CM medium plate for dark overnight. When 0.25 µm of gold nanoparticles were suspended in 100 µL sterile water and added 6 µg of DNA, 100 µL 2.5 mol/L CaCl2, 40 µL 0.1 mol/L spermidine, and mixed for 20 min at 4°C to allow sufficient binding of DNA to gold particles. After one bombardment of *E. gracilis* cells with a Bio-Rad PDS-1000/He gene gun and ruptured membrane at 900 psi, a targeting distance of 9 cm was set, the algal cells were washed off with 2 mL CM medium, re-suspended in light overnight, and then spread onto screening plates and incubated until transformants were obtained.

3.2 Electroporation

Electroporation is the breakdown of the cell membrane using a high-voltage electric field that transiently increases the membrane's permeability, and exogenous nucleic acids or protein molecules diffuse into the cell under the action of an electric field. Electroporation is the preferred transformation method when converting cells due to its easy operation, low cost and little damage to cells. In microalgal studies, the electroporation technique has a higher transformation efficiency than the biolistic method, but fewer studies have focused on the transformation of organelles. Stable electroporation transformation methods have been developed for *Chlamydomonas reinhardtii* (Shimogawara et al., 1998), *Chlorella vulgaris* (Chow and Tung, 1999), *Phaeodactylum tricornutum* (Niu et al., 2012), *salina donovani* (Sun et al., 2005), *Bacteroides thetaiotaomicron* (Li et al., 2014), and the *Chlamydomonas* orthologs *Trypanosoma* and *Leishmania* (Beverley and Clayton, 1993).

Krajcovic et al. (2011) reported introducing the *ble* gene into *E. gracilis* cells using the electroporation technique, but did not describe the condition of electric shock. Iseki et al. (2002), regarding the electroporation conditions of *Trypanosoma brucei*, modified it to successfully introduce double-stranded RNA (dsRNA) molecules into *Euglena* cells (Ngô et al., 1998; Iseki et al.,

2002). The main steps are: the *Euglena* cells cultured in the dark for two days were collected, re-suspended with inorganic salt medium, and taken 90 μL algal fluid (containing 1×10⁶ cells) was transferred to a 0.4 cm gap electric shock cup, the RNA solution was added to mix, and the electric shock was set to 1.2 kV and 25 μF, using Bio-Rad Gene PulserII electroporator, and for 30 min at room temperature before seeding into fresh medium. Ishikawa *et al.* (2010) were still able to introduce dsRNA into *Euglena* cells after lowering the voltage to 0.5 kV, while Nakazawa *et al.* (2015a) introduced 400 μL algal fluid (containing 4×10⁶ cells) that were transferred into a 0.2 cm gap electric shock cup using a BTX-ECM60 type electroporator with the electric shock parameters set to 0.5 kV and 200 μF.

Another efficient way to deliver exogenous materials for *E. gracilis* cells is through the technique of single-cell electroporation (Ohmachi *et al.*, 2016), that is, using a fine-caliber micropipette connected to a syringe to fix a single algal cell by negative pressure adsorption, in the pipette built into the electrode and a fluorescent probe such as green fluorescent protein (GFP), application of square wave electric field successfully injected the fluorescent probe into a single algal cell. The method used a smaller voltage of 5–20 V, a pulse time of 30–500 ms, a frequency of 1–5 Hz and duration of 1–5 s.

3.3 *Agrobacterium*-mediated transformation

Agrobacterium tumefaciens can naturally infect plant cells and insert a piece of T-DNA from its Ti plasmid into the plant genome. Researchers have used this principle to establish an *Agrobacterium*-mediated transformation method (Bechtold, 1993), and it is widely used in genetic engineering studies of higher plants. Currently, *Agrobacterium*-mediated transgenesis methods have also been established in microalgae such as *Chlamydomonas reinhardtii* (Kumar *et al.*, 2004), *Chlorella vulgaris* (Cha *et al.*, 2012), *Haematococcus pluvialis* (Kathiresan *et al.*, 2009), *Salina duchenna* (Fang *et al.*, 2012), and *Bacteroides thetaiotaomicron* (Cha *et al.*, 2001). Japanese investigators have used *Agrobacterium*-mediated introduction of resistance genes for zeocin, G418, as well as hygromycin into *Euglena* cells and have detected the corresponding resistance genes in the genome and transcriptome of the resistant transformants, and the method has been patented (Nakazawa *et al.*, 2015b). Khatiwada *et al.* (2019) explored the genetic engineering tools to obtain stable *E. gracilis* nuclear transformants via constructing plasmid vectors with the selection marker of hygromycin by *Agrobacterium*-mediated transformation, biolistic bombardment, and electroporation techniques. However, the results showed that only *Agrobacterium*-mediated transformation could produce stable nuclear transformants, while the others lost their property after repeated rounds of cultivation. Recently, Becker *et al.* (2021) have successfully carried out another *Agrobacterium*-mediated nuclear transformation of *E. gracilis* using the selective markers of hygromycin and zeocin, and the transformation efficiency reached 8.26 ± 4.9% after optimizing co-cultivation parameters.

3.4 Glass beads

High transformation efficiency was reported once when the *ble* gene was introduced into *Chlamydomonas reinhardtii* with cell wall defects by grinding with glass beads (Kindle, 1990). Since *Euglena* cells have no cell walls, the authors attempted to introduce the pRI101-AN plasmid containing the NPT II gene into *E. gracilis* by grinding with glass beads and screening with paromomycin. Multiple repeated experiments also did not obtain positive transformants, presumably because of the tough structure of the *Euglena* protoplast membrane, which makes it difficult to form efficient perforations by grinding.

3.5 Single-cell microinjection

Single-cell microinjection, as a mechanical delivery tool, has been successfully used to deliver exogenous proteins, cDNA constructs, peptides, drugs and particles into transfection-challenged cells. With precisely controlled delivery dosage and timing, microinjection has been used in many studies of primary cultured cells, transgenic animal production, in vitro fertilization and RNA inference (Zhang and Yu, 2008). The microinjection method has been successfully used in mice, zebrafish, rabbits and other animals as a result of its high efficiency and low cell-death rates (Gordon et al., 1980; Hammer et al., 1985; Yuan and Sun, 2009). However, it is rarely conducted in microalgae cells. In 1986, Neuhaus et al. (1986) delivered SV40 DNA and pSV2neo into the nuclei of *Acetabularia* sp. by using the microinjection technique. Meindl et al. (1994) had injected phalloidin into the cells of the green alga *Micrasterias denticulata* (the cell size is larger than 100 μm). However, it is difficult to perform microalgae injection when the size of the cells is less than 20 μm, such as the cells of *E. gracilis* in sphere form. In 1978, Nichols and Rikmenspoel (1978) injected EDTA, EGTA, Zn^{2+} and Mn^{2+} into *E. gracilis* and *C. reinhardtii*, respectively, and explored the relationship between microalgae flagellar movement and divalent cations. Since then, no delivery of active molecules was reported by microinjection into *Euglena* and the other microalgae with a cell size of less than 100 μm. The reason why single-celled eukaryotic alga cells are less amenable to the gene editing or genetic transformation than animal cells may be due to the fact that exogenous substances are not easily delivered into the cell (electrotransduction), and mutant strains would be difficult to obtain without employing resistance screening. Therefore, using microinjection-based single-cell microinjection, our team successfully achieved, for the first time, the delivery of macromolecular exo-somes such as CRISRP/Cas9 into *E. gracilis* cells and achieved gene editing (Chen et al., 2022). Previously, in the field of single-cell microalgal biotechnology, the research in the directions of single-cell gene editing, single-cell genetic transformation, and single-cell synthetic biology has been almost blank.

4. The Genetic Transformation and Editing Tools of *E. gracilis*

4.1 Transformation of nuclear genome of *E. gracilis*

Krajcovic et al. (2011) first reported the method for nuclear transformation of *E. gracilis* cells by cloning the gene *ble* between the 5' and 3' ends of the Light Harvesting Pigment complex II gene (*lhcp*II) to make it subject to regulated expression from the *lhcp*II promoter, and the expressed sequence was transferred into *E. gracilis* cells by electroporation or biolistic and screened on zeocin plates, and the resulting transformants were able to detect the *ble* gene and could be stably stored for more than one year.

Cyanobacterial FBP/SBPase is a bifunctional enzyme possessing both fructose-1,6-bisphosphatase and sedoheptulose-1,7-bisphosphatase activities. Ogawa et al. (2015c) constructed a fusion protein of FBP/SBPase with the small subunit of tomato ribulose-1,5-bisphosphate carboxylase (RBCs), which was used for chloroplast localization, and cloned in frame into the plant expression vector pRI101-AN under the control of the CaMV 35S promoter and NOS Terminator. The recombinant plasmid was transformed into algal cells using biolistic. The FBP/SBPase gene sequence could be detected in the positive transformants obtained on paromomycin plates, as well as in the chloroplasts of the transformants. The photosynthetic efficiency of the transformants under high light and high CO_2 conditions, as well as the accumulation of paramylon by *E. gracilis*, were significantly improved (Ogawa et al., 2015c). The authors adopted the method to transform *Euglena* cells with pRI101-AN linearized plasmid alone, and the resulting transformants were able to maintain paromomycin resistance for a long time, but the resistance gene *NPT*II was not detected in both genome and transcriptome. The analysis reason may be that *E. gracilis* itself can develop stronger antibiotic tolerance. Therefore, the positive transformants were identified with multiple methods.

4.2 Transformation of chloroplast genome of *E. gracilis*

Doetsch et al. (2001) cloned the *aad*A gene between the 5′ and 3′ ends of the photosystem II core protein gene *psb*A so that it was under the regulation of the *psb*A promoter, and introduced the vector containing this expression frame into *E. gracilis* chloroplasts using biolistic, screened the transformants on plates containing high concentrations of streptomycin and spectinomycin. Only transformants introduced with the *aad*A gene and efficiently expressed were able to maintain a green growth state, as streptomycin and spectinomycin inhibit chloroplast protein synthesis, leading to retarded chloroplast development and algal cell-fading apoptosis. In accordance with this phenotype, Doetsch et al. (2001) obtained several positive transformants in which the *aad*A gene could be detected in their chloroplasts. Still, this DNA fragment was not integrated into the chloroplast genome but was present in the chloroplast stroma as a free element and maintained a low level of expression. Similar results were obtained by Doetsch et al. (2001), who also introduced vectors containing chloroplast homologous recombination fragments into *E. gracilis* cells. Moreover, the presence of this exogenous carrier in the form of free elements has also been found in the transformation of *Trypanosoma* (Vainstein et al., 1994). Because the genetic background of *E. gracilis* is not known, the phenomenon that exogenous vectors replicate within chloroplasts and are stably inherited remains unexplained.

4.3 Application of CRISPR-Cas9 of gene editing in *E. gracilis*

In 2013, Zhang's and Church's team successfully used the CRISPR-Cas9 system derived from *Streptococcus* to successfully perform genome editing in mammalian cells (Cong et al., 2013; Mali et al., 2013). Since then, the technology has been extensively researched and applied for rapid gene editing and transcriptional regulation, and many Cas9-based application tools have been developed.

CRISPR/Cas9 can be used in many genetic engineering organisms, including animal cells, plant cells and microalgae cells. Shin et al. (2016) introduced RNPs into *C. reinhardtii* cells by electroporation and successfully obtained CRISPR-/Cas9-induced mutations at the targeted sites of *MAA7*, *CpSRP43* and *ChlM* genes. Compared with the mutation induced by transgenic Cas9 reported for the first time, the efficiency of Cas9 RNP delivered is 100 times higher, significantly improving the gene knockout efficiency (Jiang et al., 2014). In the past ten years, researchers have successfully applied CRISPR/Cas9 gene editing technology in microalgae, such as *Nannochloropsis* spp., *Phaeodactylum tricornutum* and *Thalassiosira pseudonana* (Hopes et al., 2016; Nymark et al., 2016; Wang et al., 2016).

Our team studied genome editing in *E. gracilis* since early 2014. After a long eight-year reticent exploration, both commonly used transformation methods and different gene editing enzymes (Cpf1, Cas9, etc.) had failed. Nomura's research team reported successful CRISPR/Cas9-mediated genome editing in *E. gracilis* using electroporation of Cas9 ribonucleoproteins (RNPs) targeting *EgGSL2* and obtained mutant rates of 70–90% based on morphology and amplicon-sequencing (Nomura et al., 2019). The protocol for implementing the *E. gracilis* CRISPR experiments was also described in further detail and published (Nomura et al., 2020; Ishikawa et al., 2022). However, since their publication, no other successful report has been published by other groups, including ours. Recently, Chen et al. (2022) of our team, revealed through the high-throughput sequencing results that it is difficult to achieve CRISPR high-efficiency gene editing in *E. gracilis* by electroconversion method, with less than 1% editing efficiency and inability to obtain mutant strain, which is quite different from the results of the ultra-high-efficiency article with Japanese scholars approaching 80% in 2019 (Nomura et al., 2019). In the article, we revealed that a 10–20 μm microinjection-based single-cell microinjection method could directly deliver drugs into *E. gracilis* cells, and we conducted CRISPR/Cas9-mediated genome editing of *E.gracilis* and successfully knocked out the *crtP*1 gene with relatively high efficiency (16.7%) (Chen et al., 2022). To the best of our knowledge, this is the first application of microinjection to genome editing in microalgae.

4.4 The research progress of RNA interference technology in *Euglena* cells

RNA interference (RNAi) is a particular degradation phenomenon of homologous mRNA induced by dsRNA, which is a regulation mode of gene expression in organisms. Using this mechanism, we can artificially construct the homologous dsRNA of the target gene and then transfect it into cells to achieve the expression inhibition of specific genes. Iseki et al. (2002) first used RNAi in *E. gracilis* to study the function of nuclear genes, and using electroporation to introduce the homologous dsRNA of a blue light receptor, photosensitized adenylyl cyclase (PAC), into the cells of *Euglena*, which significantly reduced the amount of PAC in the cells and remained effective after several passages of cell division. Subsequently, the RNAi technique was widely used for gene function validation in *E. gracilis*, and many studies about gene function have been carried out in *E. gracilis* by RNAi technology (Table 4.1), which are involved in various aspects of the primary metabolism, light response, metabolite synthesis, and so on. The methods of DNA introduction and gene cloning used in these documents can be optimized for the genetic transformation studies of *E. gracilis*.

5. Discussion

The unique biological characteristics of *E. gracilis* make it very important for scientific research and economic value. Still, mature genetic engineering methods are indispensable regardless of genetic studies or industrial applications. Only a few laboratories in the world have conducted studies on the genetic transformation of *E. gracilis*, and the number of officially published articles is minimal. Presently, algae-based synthetic biology has followed the workflow of analyzing the omics data, designing the transformation systems, delivering exogenous molecules, and screening, identification, and obtaining mutants with a predictable outcome. To some extent, the draft genome of *E. gracilis* has adequately improved the subsequent design of novel biological tools. For the gene function studies in *E. gracilis*, we recommend using RNAi tools that can silence genes with high efficiency but will not cause permanent phenotypes. The delivery methods of exogenous materials for *E. gracilis* have been established, including biolistic, electroporation, *Agrobacterium*-mediated transformation, and glass beads. With the development of new technology, it is proposed that single-cell microinjection and electroporation could be efficient molecule delivery ways for *E. gracilis*. The rapid development of CRISPR technology provides the opportunity to permanently obtain mutants with predictable results. However, at present, only Nomura et al. (2019, 2020) have reported the research articles of CRISPR/Cas9 for *Euglena*, using electroporation without a knock-in selected marker gene, with an efficiency rate of approximately 70–90%, which has hardly been reproducible by other teams, including ours, since 2020. The reason may be that electroporation is challenging to optimize parameters, or there is a lack of the knock-in selected marker for mutation screening. For example, Shamoto et al. (2018) have obtained a 55% gene editing efficiency rate of *fap70* in *C. reinhardtii* by using a knock-in paromomycin-resistance gene as a selected marker, which was much higher than that in the report of CRISPR/Cas9 in *C. reinhardtii* without selected markers by delivering RNPs using electroporation (Baek et al., 2016). It is reported that *Agrobacterium*-mediated transformation may be the most efficient way among electroporation and biolistic bombardment for obtaining stable mutants in bulk cells of *E. gracilis* (Khatiwada et al., 2019; Becker et al., 2021). However, with the rise of single-cell analysis

Table 4.1. Studies on the gene function of *E. gracilis* by RNAi technique. (Adapted from Chen *et al.*, 2022)

Gene	Full name	Main function	Reference
ALase	Aldonolactonase	Catalyzes the synthesis of ascorbic acid	Ishikawa *et al.* (2008)
APXs	Ascorbate peroxidase	Participates in the metabolism of reactive oxygen species	Ishikawa *et al.* (2010)
Cals, CaMs	Calmodulins	Transient receptor potential channel, related to the gravity axis	Häder *et al.* (2005), Häder *et al.* (2009), Daiker *et al.* (2010), Nasir *et al.* (2018)
CRT B	Phytoene synthase	Key enzymes for synthesizing phytoene	Kato *et al.* (2017), Tamaki *et al.* (2020)
CYP97H1	Carotene hydroxylase	Involved in the hydroxylation of β-carotene	Tamaki *et al.* (2019)
FBPase	Fructose-1,6-bisphosphatase	Key enzymes of Calvin cycle	Ogawa *et al.* (2015a), Ogawa *et al.* (2015b)
GSL2	Glucan synthase–like 2	Key enzymes in the synthesis of paramylon	Tanaka *et al.* (2017)
KAT	3-ketoacyl-CoA Thiolase	*Euglena* mitochondrial fatty-acid-condensing enzyme	Nakazawa *et al.* (2015a)
NTRs	NADPH-dependent thioredoxin reductase, NADPH	Redox reaction regulator	Tamaki *et al.* (2015), Ozasa *et al.* (2017)
Prxs	Peroxiredoxins	Participates in the metabolism of reactive oxygen species	Tamaki *et al.* (2014), Ozasa *et al.* (2017)
PAC	Photoactivated adenylyl cyclase	*Euglena* blue light receptor, related to phototaxis	Iseki *et al.* (2002), Ntefidou *et al.* (2003), Ozasa *et al.* (2017)
PKA	Protein kinase A	Participates in signal transduction of *Euglena* phototaxis and gravity axis	Daiker *et al.* (2011)
PNO	Pyruvate: NADP+ oxidoreductase	Participates in the fermentation of *Euglena* wax fat	Nakazawa *et al.* (2017)
2-ogdc	2-oxoglutarate decarboxylase	Key enzymes of the tricarboxylic acid cycle	Nakazawa *et al.* (2017)
RGA IU	UDP-glucose pyrophosphorylase	Involved in the synthesis of paramylon	Muchut *et al.* (2021)
STDs	Starch degradation	*Euglena* growth and proliferation-related protein	Kimura and Ishikawa (2018)
TERs	Trans-2-enoyl-CoA reductases	Plays the role of transferring hydrogen in the metabolic process	Tomiyama *et al.* (2019)
WSDs	Wax ester synthase	Key enzymes of *Euglena* wax fermentation	Tomiyama *et al.* (2017)

technology, microinjection and single-cell electroporation will become powerful delivery tools to introduce exogenous molecules into microalgal cells with high transformation efficiency and high cell viability. Using microinjection-based single-cell microinjection, our team successfully delivered macromolecular exosomes (such as CRISPR/Cas9) to *E. gracilis* cells for the first time. It successfully knocked out the *crtP*1 gene with relatively high gene editing efficiency (16.7%) (Chen *et al.*, 2022). Our team have carried out a long study on the genetic transformation system of *E. gracilis* and found that there were the following problems in the stable transformation of *E. gracilis*: (i) *E. gracilis* cells are prone to develop strong tolerance under the action of antibiotics, leading to screening failure; (ii) the insertion mutation of the genome may not be achieved when the exogenous DNA is introduced into the cell, which may exist in an episomal form. The replication mechanism of the exogenous DNA inside the cell is also unclear. Based on the summary and analysis of the literature data, the authors suggest that there are several research directions for the genetic transformation of *E. gracilis*: (i) after the sequencing of *E. gracilis* genome is completed, the replication mechanism of DNA inside and outside the nucleus remains to be elucidated; (ii) more efficient *E. gracilis* screening markers have been constructed in various ways, such as high-expression promoters, codon optimization, and intron embedding, as well as versatile tool vectors; (iii) to establish a mitochondrial transformation system of *Euglen* and carry out studies on the genetic characteristics of mitochondria; (iv) after optimization, gene editing tools, such as CRISPR/cas9 and cpf1 (which have not been reported in *E. gracilis*), should be used for genetic modification in *E. gracilis*.

In summary, *E. gracilis*, with its special biological characteristics, has excellent potential for academic and economic value. In the future, with the improvement of the genetic transformation system, algae-based synthetic biology will make steady progress, and genetic engineering will also greatly promote the industrial application and the scale-up production of *E. gracilis*.

References

Anley, K.A. (2015) *Developing Molecular Tools to Genetically Engineer the Microalga Nannochloropsis*. Norwegian University of Science and Technology, Trondheim, Norway.

Baek, K., Kim, D.H., Jeong, J., Sim, S.J., Melis, A. *et al.* (2016) DNA-free two-gene knockout in *Chlamydomonas reinhardtii* via CRISPR-Cas9 ribonucleoproteins. *Scientific Reports* 6, 30620.

Bechtold, N. (1993) In planta *Agrobacterium* mediated gene transfer by infiltration of adult *Arabidopsis thaliana* plants. *Methods in Molecular Biology* 82(10), 259.

Becker, I., Prasad, B., Ntefidou, M., Daiker, V., Richter, P. *et al.* (2021) *Agrobacterium tumefaciens*-mediated nuclear transformation of a biotechnologically important microalga-*Euglena gracilis*. *International Journal of Molecular Sciences* 22(12), 6299.

Bennett, M.S. and Triemer, R.E. (2015) Chloroplast genome evolution in the Euglenaceae. *Journal of Eukaryotic Microbiology* 62(6), 773–785.

Beverley, S.M. and Clayton, C.E. (1993) Transfection of *Leishmania* and *Trypanosoma brucei* by electroporation. *Protocols in Molecular Parasitology* 21, 333–348.

Blanc, G., Duncan, G., Agarkova, I., Borodovsky, M., Gurnon, J. *et al.* (2010) The *Chlorella variabilis* NC64A genome reveals adaptation to photosymbiosis, coevolution with viruses, and cryptic sex. *Plant Cell* 22(9), 2943–2955.

Bowler, C., Allen, A.E., Badger, J.H., Grimwood, J., Jabbari, K. *et al.* (2008) The *Phaeodactylum* genome reveals the evolutionary history of diatom genomes. *Nature* 456(7219), 239–244.

Boynton, J.E., Gillham, N.W., Harris, E.H., Hosler, J.P., Johnson, A.M. *et al.* (1988) Chloroplast transformation in *Chlamydomonas* with high velocity microprojectiles. *Science* 240(4858), 1534–1538.

Cha, T.S., Chen, C.F., Yee, W., Aziz, A. and Loh, S.H. (2001) Cinnamic acid, coumarin and vanillin: alternative phenolic compounds for efficient *Agrobacterium*-mediated transformation of the unicellular green alga, *Nannochloropsis* sp. *Journal of Microbiological Methods* 84(3), 430–434.

Cha, T.S., Yee, W. and Aziz, A. (2012) Assessment of factors affecting *Agrobacterium*-mediated genetic transformation of the unicellular green alga, *Chlorella vulgaris*. *World Journal of Microbiology and Biotechnology* 28(4), 1771–1779.

Chen, Z., Zhu, J., Chen, Z., Du, M., Yao, R. *et al.* (2022) High-throughput sequencing revealed low-efficacy genome editing using Cas9 RNPs electroporation and single-celled microinjection provided an alternative to deliver CRISPR reagents into *Euglena gracilis*. *Plant Biotechnology Journal* 20(11), 2048–2050.

Chow, K.-C. and Tung, W. (1999) Electrotransformation of *Chlorella vulgaris*. *Plant Cell Reports* 18(9), 778–780.

Cong, L., Ran, F.A., Cox, D., Lin, S., Barretto, R. et al. (2013) Multiplex genome engineering using CRISPR/Cas systems. *Science (New York, N.Y.)* 339(6121), 819–823.

Dai, J., He, J., Chen, Z., Qin, H., Du, M. et al (2022) *Euglena gracilis* promotes lactobacillus growth and antioxidants accumulation as a potential next-generation prebiotic. *Frontiers in Nutrition* 9, 864565.

Daiker, V., Lebert, M., Richter, P. and Häder, D.P. (2010) Molecular characterization of a calmodulin involved in the signal transduction chain of gravitaxis in *Euglena gracilis*. *Planta* 231(5), 1229–1236.

Daiker, V., Häder, D.P., Richter, P.R. and Lebert, M. (2011) The involvement of a protein kinase in phototaxis and gravitaxis of *Euglena gracilis*. *Planta* 233(5), 1055–1062.

Dobáková, E., Flegontov, P., Skalický, T. and Lukeš, J. (2015) Unexpectedly streamlined mitochondrial genome of the euglenozoan *Euglena gracilis*. *Genome Biology and Evolution* 7(12), 3358–3367.

Doetsch, N.A., Favreau, M.R., Kuscuoglu, N., Thompson, M.D. and Hallick, R.B. (2001) Chloroplast transformation in *Euglena gracilis*: splicing of a group III twintron transcribed from a transgenic *psbK* operon. *Current Genetics* 39(1), 49–60.

Doron, L., Segal, N. and Shapira, M. (2016) Transgene expression in microalgae-from tools to applications. *Frontiers in Plant Science* 7, 505.

Ebenezer, T.E., Carrington, M., Lebert, M., Kelly, S. and Field, M.C. (2017) *Euglena gracilis* genome and transcriptome: organelles, nuclear genome assembly strategies and initial features. *Advances in Experimental Medicine and Biology* 979, 125–140.

Ebenezer, T.E., Zoltner, M., Burrell, A., Nenarokova, A., Novák Vanclová, A.M. et al. (2019) Transcriptome, proteome and draft genome of *Euglena gracilis*. *BMC Biology* 17(1), 11.

Fang, L., Lin, H.X., Low, C.S., Wu, M.H., Chow, Y. et al. (2012) Expression of the *Chlamydomonas reinhardtii* sedoheptulose-1,7-bisphosphatase in *Dunaliella bardawil* leads to enhanced photosynthesis and increased glycerol production. *Plant Biotechnology Journal* 10(9), 1129–1135.

Gibbs, S.P. (1978) The chloroplasts of *Euglena* may have evolved from symbiotic green algae. *Canadian Journal of Botany* 56(22), 2883–2889.

Gissibl, A., Sun, A., Care, A., Nevalainen, H. and Sunna, A. (2019) Bioproducts from *Euglena gracilis*: synthesis and applications. *Frontiers in Bioengineering and Biotechnology* 7, 108.

Gordon, J.W., Scangos, G.A., Plotkin, D.J., Barbosa, J.A. and Ruddle, F.H. (1980) Genetic transformation of mouse embryos by microinjection of purified DNA. *Proceedings of the National Academy of Sciences of the United States of America* 77(12), 7380–7384.

Grimm, P., Risse, J.M., Cholewa, D., Müller, J.M., Beshay, U. et al. (2015) Applicability of *Euglena gracilis* for biorefineries demonstrated by the production of α-tocopherol and paramylon followed by anaerobic digestion. *Journal of Biotechnology* 215, 72–79.

Hadariová, L., Vesteg, M., Birčák, E., Schwartzbach, S.D. and Krajčovič, J. (2017) An intact plastid genome is essential for the survival of colorless *Euglena longa* but not *Euglena gracilis*. *Current Genetics* 63(2), 331–341.

Häder, D.P., Richter, P., Ntefidou, M. and Lebert, M. (2005) Gravitational sensory transduction chain in flagellates. *Advances in Space Research* 36(7), 1182–1188.

Häder D.-P., Richter P.R., Schuster M., Daiker V. and Lebert M. (2009) Molecular analysis of the graviperception signal transduction in the flagellate *Euglena gracilis*: involvement of a transient receptor potential-like channel and a calmodulin. *Advances in Space Research* 43(8), 1179–1184.

Hallick, R.B., Hong, L., Drager, R.G., Favreau, M.R., Monfort, A. et al. (1993) Complete sequence of *Euglena gracilis* chloroplast DNA. *Nucleic Acids Research* 21(15), 3537–3544.

Hallmann, A. and Rappel, A. (1999) Genetic engineering of the multicellular green alga *Volvox*: a modified and multiplied bacterial antibiotic resistance gene as a dominant selectable marker. *The Plant Journal* 17(1), 99–109.

Hammer, R.E., Pursel, V.G., Rexroad, C.E. Jr, Wall, R.J., Bolt, D.J. et al. (1985) Production of transgenic rabbits, sheep and pigs by microinjection. *Nature* 315(6021), 680–683.

Hopes, A., Nekrasov, V., Kamoun, S. and Mock, T. (2016) Editing of the urease gene by CRISPR-Cas in the diatom *Thalassiosira pseudonana*. *Plant Methods* 12, 49.

Hu, Z., Zhao, Z., Wu, Z., Fan, Z., Chen, J. et al. (2011) Successful expression of heterologous *egfp* gene in the mitochondria of a photosynthetic eukaryote *Chlamydomonas reinhardtii*. *Mitochondrion* 11(5), 716–721.

Inui, H., Ishikawa, T. and Tamoi, M. (2017) Wax ester fermentation and its application for biofuel production. *Advances in Experimental Medicine and Biology* 979, 269–283.

Iseki, M., Matsunaga, S., Murakami, A., Ohno, K., Shiga, K. et al. (2002) A blue-light activated adenylyl cyclase mediates photoavoidance in *Euglena gracilis*. *Nature* 415(6875), 1047–1051.

Ishikawa, T., Nishikawa, H., Gao, Y., Sawa, Y., Shibata, H. et al. (2008) The pathway via D-galacturonate/L-galactonate is significant for ascorbate biosynthesis in Euglena gracilis identification and functional characterization of aldonolactonase. Journal of Biological Chemistry 283(45), 31133–31141.

Ishikawa, T., Tajima, N., Nishikawa, H., Gao, Y., Rapolu, M. et al. (2010) Euglena gracilis ascorbate peroxidase forms an intramolecular dimeric structure: its unique molecular characterization. Biochemical Journal 426(2), 125–134.

Ishikawa, M., Nomura, T., Tamaki, S., Ozasa, K., Suzuki, T. et al. (2022) CRISPR/Cas9-mediated generation of non-motile mutants to improve the harvesting efficiency of mass-cultivated Euglena gracilis. Plant Biotechnology Journal 20(11), 2042–2044.

Ivusic, F. and Santek, B. (2015) Optimization of complex medium composition for heterotrophic cultivation of Euglena gracilis and paramylon production. Bioprocess and Biosystems Engineering 38(6), 1103–1112.

Jain, S., Durand, H. and Tiraby, G. (1992) Development of a transformation system for the thermophilic fungus Talaromyces sp. CL240 based on the use of phleomycin resistance as a dominant selectable marker. Molecular Genetics and Genomics 234(3), 489–493.

Jiang, W., Brueggeman, A.J., Horken, K.M., Plucinak, T.M. and Weeks, D.P. (2014) Successful transient expression of Cas9 and single guide RNA genes in Chlamydomonas reinhardtii. Eukaryotic Cell 13(11), 1465–1469.

Kathiresan, S., Chandrashekar, A., Ravishankar, G.A. and Sarada, R. (2009) Agrobacterium-mediated transformation in the green alga Haematococcus pluvalis (chlorophyceae, volvocales). Journal of Phycology 45(3), 642–649.

Kato, S., Soshino, M., Takaichi, S., Ishikawa, T., Nagata, N. et al. (2017) Suppression of the phytoene synthase gene (EgcrtB) alters carotenoid content and intracellular structure of Euglena gracilis. BMC Plant Biology 17(1), 125.

Khatiwada, B., Kautto, L., Sunna, A., Sun, A. and Nevalainen, H. (2019) Nuclear transformation of the versatile microalga Euglena gracilis. Algal Research 37, 178–185.

Kimura, M. and Ishikawa, T. (2018) Suppression of DYRK ortholog expression affects wax ester fermentation in Euglena gracilis. Journal of Applied Phycology 30(1), 367–373.

Kindle, K.L. (1990) High-frequency nuclear transformation of Chlamydomonas reinhardtii. Proceedings of the National Academy of Sciences 87(3), 1228–1232.

Kira, N., Ohnishi, K., Miyagawa-Yamaguchi, A., Kadono, T. and Adachi, M. (2016) Nuclear transformation of the diatom Phaeodactylum tricornutum using PCR-amplified DNA fragments by micro-particle bombardment. Marine Genomics 25, 49–56.

Klinthong, W., Yang, Y.H., Huang, C.H. and Tan, C.S. (2015) A review: microalgae and their applications in CO_2 capture and renewable energy. Aerosol and Air Quality Research 15(2), 712–742.

Koizumi, N., Sakagami, H., Utsumi, A., Fujinaga, S., Takeda, M. et al. (1993) Anti-hiv (human immunodeficiency virus) activity of sulfated paramylon. Antiviral Research 21(1), 1–14.

Kondo, Y., Kato, A., Hojo, H., Nozoe, S., Takeuchi, M. et al. (1992) Cytokine-related immunopotentiating activities of paramylon, a β-(1→3)-D-glucan from Euglena gracilis. Journal of Pharmacobio-Dynamics 15(11), 617–662.

Krajcovic J., Vejerla V.K., Vacula R., Dobáková E., Gavurníková G. et al. (2011) Development of an effective transformation system for the nuclear genome of the flagellate Euglena gracilis. Current Opinion in Biotechnology 22(5), S45.

Kumar, S.V., Misquitta, R.W., Reddy, V.S., Rao, B.J. and Rajam, M.V. (2004) Genetic transformation of the greenalga-Chlamydomonas reinhardtii by Agrobacterium tumefaciens. Plant Science 166(3), 731–738.

Li, F., Gao, D. and Hu, H. (2014) High-efficiency nuclear transformation of the oleaginous marine Nannochloropsis species using PCR product. Bioscience, Biotechnology, and Biochemistry 78(5), 812–817.

Mahapatra, D.M., Chanakya, H. and Ramachandra, T. (2013) Euglena sp. as a suitable source of lipids for potential use as biofuel and sustainable wastewater treatment. Journal of Applied Phycology 25(3), 855–865.

Mali, P., Yang, L., Esvelt, K.M., Aach, J., Guell, M. et al. (2013) RNA-guided human genome engineering via Cas9. Science (New York, N.Y.) 339(6121), 823–826.

Matsumoto, T., Inui, H., Miyatake, K., Nakano, Y. and Murakami, K. (2009) Comparison of nutrients in Euglena with those in other representative food sources. Ecological Engineering 21(2), 81–86.

Mayfield, S.P. and Kindle, K.L. (1990) Stable nuclear transformation of Chlamydomonas reinhardtii by using a C. reinhardtii gene as the selectable marker. Proceedings of the National Academy of Sciences 87(6), 2087–2091.

Meindl, U., Zhang, D. and Hepler, P.K. (1994) Actin microfilaments are associated with the migrating nucleus and the cell cortex in the green alga Micrasterias. Studies on living cells. *Journal of Cell Science* 107(Pt 7), 1929–1934.

Merchant, S.S., Prochnik, S.E., Vallon, O., Harris, E.H., Karpowicz, S.J. et al. (2007) The *Chlamydomonas* genome reveals the evolution of key animal and plant functions. *Science* 318(5848), 245–250.

Muchut, R.J., Calloni, R.D., Arias, D.G., Arce, A.L., Iglesias, A.A. et al. (2021) Elucidating carbohydrate metabolism in *Euglena gracilis*: reverse genetics-based evaluation of genes coding for enzymes linked to paramylon accumulation. *Biochimie* 184, 125–131.

Nakazawa, M., Andoh, H., Koyama, K., Watanabe, Y., Nakai, T. et al. (2015a) Alteration of wax ester content and composition in *Euglena gracilis* with gene silencing of 3-ketoacyl-CoA thiolase isozymes. *Lipids* 50(5), 483–492.

Nakazawa, M., Haruguchi, D., Ueda, M. and Miyatake, K. (2015b) Transformed *Euglena* and process for producing same. United States Application, US(2015)0368655 A1.

Nakazawa, M., Hayashi, R., Takenaka, S., Inui, H., Ishikawa, T. et al. (2017) Physiological functions of pyruvate: NADP$^+$ oxidoreductase and 2-oxoglutarate decarboxylase in *Euglena gracilis* under aerobic and anaerobic conditions. *Bioscience, Biotechnology, and Biochemistry* 81(7), 1386–1393.

Nasir, A., Le Bail, A., Daiker, V., Klima, J., Richter, P. et al. (2018) Identification of a flagellar protein implicated in the gravitaxis in the flagellate *Euglena gracilis*. *Scientific Reports* 8(1), 7605.

Neuhaus, G., Neuhaus-Url, G., de Groot, E.J. and Schweiger, H.G. (1986) High yield and stable transformation of the unicellular green alga *Acetabularia* by microinjection of SV40 DNA and pSV2neo. *The EMBO Journal* 5(7), 1437–1444.

Ngô, H., Tschudi, C., Gull, K. and Ullu, E. (1998) Double-stranded RNA induces mRNA degradation in *Trypanosoma brucei*. *Proceedings of the National Academy of Sciences of the United States of America* 95(25), 14687–14692.

Nichols, K.M. and Rikmenspoel, R. (1978) Control of flagellar motility in *Euglena* and *Chlamydomonas*. Microinjection of EDTA, EGTA, Mn(2)+, and Zn(2)+. *Experimental Cell Research* 116(2), 333–340.

Niu, Y.F., Yang, Z.K., Zhang, M.H., Zhu, C.C., Yang, W.D. et al. (2012) Transformation of diatom *Phaeodactylum tricornutum* by electroporation and establishment of inducible selection marker. *Biotechniques* 52(6), 1–3.

Nomura, T., Inoue, K., Uehara-Yamaguchi, Y., Yamada, K., Iwata, O. et al. (2019) Highly efficient transgene-free targeted mutagenesis and single-stranded oligodeoxynucleotide-mediated precise knock-in in the industrial microalga *Euglena gracilis* using Cas9 ribonucleoproteins. *Plant Biotechnology Journal* 17(11), 2032–2034.

Nomura, T., Yoshikawa, M., Suzuki, K. and Mochida, K. (2020) Highly efficient CRISPR-associated protein 9 ribonucleoprotein-based genome editing in *Euglena gracilis*. *STAR Protocols* 1(1), 100023.

Ntefidou, M., Iseki, M., Watanabe, M., Lebert, M. and Häder, D.P. (2003) Photoactivated adenylyl cyclase controls phototaxis in the flagellate *Euglena gracilis*. *Plant Physiology* 133(4), 1517–1521.

Nymark, M., Sharma, A.K., Sparstad, T., Bones, A.M. and Winge, P. (2016) A CRISPR/Cas9 system adapted for gene editing in marine algae. *Scientific Reports* 6, 24951.

Ogawa, T., Kimura, A., Sakuyama, H., Tamoi, M., Ishikawa, T. et al. (2015a) Characterization and physiological role of two types of chloroplastic fructose-1,6-bisphosphatases in *Euglena gracilis*. *Archives of Biochemistry and Biophysics* 575, 61–68.

Ogawa, T., Kimura, A., Sakuyama, H., Tamoi, M., Ishikawa, T. et al. (2015b) Identification and characterization of cytosolic fructose-1,6-bisphosphatase in *Euglena gracilis*. *Bioscience Biotechnology and Biochemistry* 79(12), 1957–1964.

Ogawa, T., Tamoi, M., Kimura, A., Mine, A., Sakuyama, H. et al. (2015c) Enhancement of photosynthetic capacity in *Euglena gracilis* by expression of cyanobacterial fructose-1,6-/sedoheptulose-1,7-bisphosphatase leads to increases in biomass and wax ester production. *Biotechnology for Biofuels* 8, 80.

Ohmachi, M., Fujiwara, Y., Muramatsu, S., Yamada, K., Iwata, O. et al. (2016) A modified single-cell electroporation method for molecule delivery into a motile protist, *Euglena gracilis*. *Journal of Microbiological Methods* 130, 106–111.

O'Neill, E.C., Trick, M., Henrissat, B. and Field, R.A. (2015a) Euglena in time: evolution, control of central metabolic processes and multi-domain proteins in carbohydrate and natural product biochemistry. *Perspectives in Science* 6, 84–93.

O'Neill, E.C., Trick, M., Hill, L., Rejzek, M., Dusi, R.G. et al. (2015b) The transcriptome of *Euglena gracilis* reveals unexpected metabolic capabilities for carbohydrate and natural product biochemistry. *Molecular Biosystems* 11(10), 2808–2820.

Ozasa, K., Won, J., Song, S., Tamaki, S., Ishikawa, T. et al. (2017) Temporal change of photophobic step-up responses of Euglena gracilis investigated through motion analysis. PLOS ONE 12(2), e0172813.

Preisfeld, A., Berger, S., Busse, I., Liller, S. and Ruppel, H.G. (2000) Phylogenetic analyses of various euglenoid taxa (euglenozoa) based on 18s rDNA sequence data. Journal of Phycology 36(1), 220–226.

Prochnik, S.E., Umen, J., Nedelcu, A.M., Hallmann, A., Miller, S.M. et al. (2010) Genomic analysis of organismal complexity in the multicellular green alga Volvox carteri. Science 329(5988), 223–226.

Quesada, L.A., de Lustig, E.S., Marechal, L.R. and Belocopitow, E. (1976) Antitumor activity of Paramylon on sarcoma-180 in mice. Gan 67(3), 455–459.

Rawson, J.R. and Boerma, C. (1976) Influence of growth conditions upon the number of chloroplast DNA molecules in Euglena gracilis. Proceedings of the National Academy of Sciences of the United States of America 73(7), 2401–2404.

Sanford, J.C., Smith, F.D. and Russell, J. (1993) Optimizing the biolistic process for different biological applications. Methods in Enzymology 217, 483–509.

Šantek, B., Felski, M., Friehs, K., Lotz, M. and Flaschel, E. (2009) Production of Paramylon, a β-1,3-glucan, by heterotrophic cultivation of Euglena gracilis on a synthetic medium. Engineering in Life Sciences 9(1), 23–28.

Shamoto, N., Narita, K., Kubo, T., Oda, T. and Takeda, S. (2018) CFAP70 is a novel axoneme-binding protein that localizes at the base of the outer dynein arm and regulates ciliary motility. Cells 7(9), 124.

Shimogawara, K., Fujiwara, S., Grossman, A. and Usuda, H. (1998) High-efficiency transformation of Chlamydomonas reinhardtii by electroporation. Genetics 148(4), 1821–1828.

Shin, S.E., Lim, J.M., Koh, H.G., Kim, E.K., Kang, N.K. et al. (2016) CRISPR/Cas9-induced knockout and knock-in mutations in Chlamydomonas reinhardtii. Scientific Reports 6, 27810.

Song, Y., Shin, H., Sianipar, H.G.J., Park, J.Y., Lee, M., Hah, J. et al. (2022) Oral administration of Euglena gracilis paramylon ameliorates chemotherapy-induced leukocytopenia and gut dysbiosis in mice. International Journal of Biological Macromolecules 211, 47–56.

Sugiyama, A., Hata, S., Suzuki, K., Yoshida, E., Nakano, R. et al. (2010) Oral administration of paramylon, a β-1,3-D-glucan isolated from Euglena gracilis z inhibits development of atopic dermatitis-like skin lesions in NC/Nga mice. Journal of Veterinary Medical Science 72(6), 755–763.

Sun, Y., Yang, Z., Gao, X., Li, Q., Zhang, Q. et al. (2005) Expression of foreign genes in Dunaliella by electroporation. Molecular Biotechnology 30(3), 185–192.

Tahira, S., Khan, S., Samrana, S., Shahi, L., Ali, I. et al. (2019) Bio-assessment and remediation of arsenic (arsenite As-III) in water by Euglena gracilis. Journal of Applied Phycology 31(1), 423–433.

Takeda, T., Nakano, Y., Takahashi, M., Konno, N., Sakamoto, Y. et al. (2015) Identification and enzymatic characterization of an endo-1,3-β-glucanase from Euglena gracilis. Phytochemistry 116, 21–27.

Tamaki, S., Maruta, T., Sawa, Y., Shigeoka, S. and Ishikawa, T. (2014) Identification and functional analysis of peroxiredoxin isoforms in Euglena gracilis. Bioscience, Biotechnology, and Biochemistry 78(4), 593–601.

Tamaki, S., Maruta, T., Sawa, Y., Shigeoka, S. and Ishikawa, T. (2015) Biochemical and physiological analyses of NADPH-dependent thioredoxin reductase isozymes in Euglena gracilis. Plant Science 236, 29–36.

Tamaki, S., Kato, S., Shinomura, T., Ishikawa, T. and Imaishi, H. (2019) Physiological role of β-carotene monohydroxylase (CYP97H1) in carotenoid biosynthesis in Euglena gracilis. Plant Science: An International Journal of Experimental Plant Biology 278, 80–87.

Tamaki, S., Tanno, Y., Kato, S., Ozasa, K., Wakazaki, M. et al. (2020) Carotenoid accumulation in the eyespot apparatus required for phototaxis is independent of chloroplast development in Euglena gracilis. Plant Science: An International Journal of Experimental Plant Biology 298, 110564.

Tan, C., Qin, S., Zhang, Q., Jiang, P. and Zhao, F. (2005) Establishment of a microparticle bombardment transformation system for Dunaliella salina. Journal of Microbiology 43(4), 361.

Tanaka, Y., Ogawa, T., Maruta, T., Yoshida, Y., Arakawa, K. et al. (2017) Glucan synthaselike 2 is indispensable for paramylon synthesis in Euglena gracilis. FEBS Letters 591(10), 1360–1370.

Tomita, Y., Yoshioka, K., Iijima, H., Nakashima, A., Iwata, O. et al. (2016) Succinate and lactate production from Euglena gracilis during dark, anaerobic conditions. Frontiers in Microbiology 7, 2050.

Tomiyama, T., Kurihara, K., Ogawa, T., Maruta, T., Ogawa, T. et al. (2017) Wax ester sSynthase/diacylglycerol acyltransferase isoenzymes play a pivotal role in wax ester biosynthesis in Euglena gracilis. Scientific Reports 7(1), 13504.

Tomiyama, T., Goto, K., Tanaka, Y., Maruta, T., Ogawa, T. et al (2019) A major isoform of mitochondrial trans-2-enoyl-CoA reductase is dispensable for wax ester production in Euglena gracilis under anaerobic conditions. PLOS ONE 14(1), e0210755.

Torihara K. and Kishimoto N. (2015) Evaluation of growth characteristics of *Euglena gracilis* for microalgal biomass production using wastewater. *Journal of Water and Environment Technology* 13(3), 195–205.

Vainstein, M.H., Alves, S.A., de Lima, B.D., Aragao, F.J. and Rech, E.L. (1994) Stable DNA transfection in a flagellate trypanosomatid by microparticle bombardment. *Nucleic Acids Research* 22(15), 3263–3264.

Wang, J.X., Shi, Z.X. and Xu, X.D. (2002) Chloroplast-less mutants of two species of *Euglena*. *Acta Hydrobiologica Sinica* 26(2), 175–179.

Wang, Q., Lu, Y., Xin, Y., Wei, L., Huang, S. *et al.* (2016) Genome editing of model oleaginous microalgae Nannochloropsis spp. by CRISPR/Cas9. *The Plant Journal: for Cell and Molecular Biology* 88(6), 1071–1081.

Watanabe, T., Shimada, R., Matsuyama, A., Yuasa, M., Sawamura, H. *et al.* (2013) Anti-tumor activity of the β-glucan Paramylon from *Euglena* against preneoplastic colonic aberrant crypt foci in mice. *Food Function* 4(11), 1685–1690.

Watanabe, F., Yoshimura, K. and Shigeoka, S. (2017) Biochemistry and physiology of vitamins in *Euglena*. *Advances in Experimental Medicine and Biology* 979, 68–90.

Xie, Y., Li, J., Qin, H., Wang, Q., Chen, Z. *et al.* (2022) Paramylon from *Euglena gracilis* prevents lipopolysaccharide-induced acute liver injury. *Frontiers in Immunology* 12, 797096.

Yang, H., Choi, K., Kim, K.J., Park, S.Y., Jeon, J.Y. *et al.* (2022) Immunoenhancing effects of *Euglena gracilis* on a cyclophosphamide-induced immunosuppressive mouse model. *Journal of Microbiology and Biotechnology* 32(2), 228–237.

Yasuda, K., Ogushi, M., Nakashima, A., Nakano, Y. and Suzuki, K. (2018) Accelerated wound healing on the skin using a film dressing with β-glucan paramylon. *In vivo (Athens, Greece)* 32(4), 799–805.

Yoshida, Y., Tomiyama, T., Maruta, T., Tomita, M., Ishikawa, T. *et al.* (2016) De novo assembly and comparative transcriptome analysis of *Euglena gracilis* in response to anaerobic conditions. *BMC Genomics* 17(1), 1–10.

Yuan, S. and Sun, Z. (2009) Microinjection of mRNA and morpholino antisense oligonucleotides in zebrafish embryos. *Journal of Visualized Experiments: JoVE* 27, 1113.

Zakryś, B., Milanowski, R. and Karnkowska, A. (2017) Evolutionary origin of *Euglena*. *Advances in Experimental Medicine and Biology* 979, 3–17.

Zhang, Y. and Yu, L.C. (2008) Single-cell microinjection technology in cell biology. *BioEssays* 30(6), 606–610.

5 Visualizing Native Supramolecular Architectures of Photosynthetic Membranes in Cyanobacteria and Red Algae Using Atomic Force Microscopy

Long-Sheng Zhao[1,2,3], Yu-Zhong Zhang[2,3] and Lu-Ning Liu[2,4]

[1]Marine Biotechnology Research Center, State Key Laboratory of Microbial Technology, Shandong University, Qingdao, 266237 China; [2]MOE Key Laboratory of Evolution and Marine Biodiversity, Frontiers Science Center for Deep Ocean Multispheres and Earth System & College of Marine Life Sciences, Ocean University of China, Qingdao, 266003 China; [3]Laboratory for Marine Biology and Biotechnology, Laoshan Laboratory, Qingdao, 266237 China; [4]Institute of Systems, Molecular and Integrative Biology, University of Liverpool, L69 7ZB United Kingdom

Abstract

Sunlight can be efficiently captured and converted into chemical energy by phototrophic organisms through photosynthesis to support most of life on Earth. Thylakoid membranes are specialized intracellular membrane systems in algae, cyanobacteria (also called blue-green algae), and plants to accommodate photosynthetic complexes for oxygenic photosynthesis. Despite substantial information on the structures of individual photosynthetic complexes, we need a better understanding of how the photosynthetic complexes are spatially organized and functionally coordinate with each other in thylakoid membranes to perform efficient energy and electron transfer and dynamically adapt to environmental changes. In this chapter, we summarize the recent advances of atomic force microscopy (AFM) in deciphering the structural landscapes and dynamic organization of thylakoid membranes in cyanobacteria and red algae, which are fundamental for efficient photosynthesis and photoadaptation to cope with environmental variations. Advanced understanding of the molecular details of natural thylakoid membrane architectures and regulation provides essential information required for rational design and engineering of artificial photosynthetic systems to underpin the development and production of sustainable energy.

Keywords: photosynthesis; thylakoid membrane; atomic force microscopy; cyanobacteria; red algae; membrane structure

1. Photosynthetic Membrane

The photosynthetic light-dependent reactions, performed by plants, algae, cyanobacteria (also referred to as blue-green algae), and some anoxygenic phototrophic bacteria, take place in the specialized membrane – photosynthetic membranes. The photosynthetic membrane

*Address for correspondence: luning.liu@liverpool.ac.uk

© CAB International 2023. *Algal Biotechnology* (Qiang Wang)
DOI: 10.1079/9781800621954.0005

houses multiple pigment–protein macromolecular complexes, which execute efficient absorption and conversion of light energy to chemical energy to sustain most of the living organisms on Earth (Tavano and Donohue, 2006; Scholes et al., 2011; Mullineaux and Liu, 2020). Despite the similar photochemical functions, the photosynthetic membranes exhibit notable architectural variability, implicating their functional adaption to specific ecological niches.

Anoxygenic phototrophic purple bacteria are widely distributed in nature, primarily in aquatic habitats. Purple bacteria carry out light-dependent photosynthetic reactions and energy conversion in specialized intracytoplasmic membranes (ICMs, also termed chromatophores). The purple bacterial photosynthetic apparatus involves a series of multicomponent pigment–protein complexes, including light-harvesting complexes 2 (LH2), light-harvesting complexes 1 (LH1), reaction centers (RC), cytochrome (cyt) bc_1 complexes and ATP synthases (ATPases) (Cogdell et al., 2006; Miller et al., 2020a). LH1 form closed or open ring-like structures surrounding the RC to construct an RC–LH1 core complex (Qian et al., 2018; Yu et al., 2018; Bracun et al., 2021; Swainsbury et al., 2021; Cao et al., 2022). LH1 and LH2 coordinate bacteriochlorophylls (BChls) and carotenoids, and capture solar light and transfer to the RC to execute the primary energy conversion reactions of photosynthesis.

Unlike purple bacteria, other anoxygenic phototrophic bacteria such as green sulfur bacteria, heliobacteria, halophilic archaea employing bacteriorhodopsin, proteobacteria employing proteorhodopsin, *Chloroflexus* and its relatives, and *Acidobacteria* accomplish photosynthetic activities in the plasma membrane (Beja et al., 2001; Saer and Blankenship, 2017). These organisms evolve the large supramolecular antenna system, chlorosomes, which are composed of lipids and BChls and are located on the cytoplasmic side of the plasma membrane via a protein baseplate, harvesting light energy and transferring to the RCs in the plasma membrane (Nielsen et al., 2016; Saer and Blankenship, 2017).

In green plants, algae and cyanobacteria, photosynthetic light reactions take place in thylakoid membranes mainly through four membrane-spanning macromolecular complexes – photosystem I (PSI), photosystem II (PSII), cytochrome b_6f complex (Cyt b_6f) and ATPase (Liu, 2016; Mullineaux and Liu, 2020). The plant chloroplast thylakoid constitutes cylindrical stacked regions (grana) and unstacked regions (stromal lamellae). The grana are composed of stacked membrane discs closely appressed at their stromal faces and interconnected by non-appressed stromal lamellae (Shimoni et al., 2005; Daum et al., 2010; Austin and Staehelin, 2011; Nevo et al., 2012; Ruban and Johnson, 2015). The distribution of photosynthetic complexes in the plant chloroplast thylakoid possesses lateral heterogeneity. PSI and ATPase are located primarily in the stromal lamellae as well as the non-appressed grana membranes at both ends of each stack and the attachment area of the grana and stromal lamellae. In contrast, PSII complexes appear mostly in appressed grana membranes. In addition, Cyt b_6f and light-harvesting complexes (LHCs) are located in both membrane regions (Anderson and Andersson, 1982; Albertsson, 2001; Dekker and Boekema, 2005).

Cyanobacterial thylakoid membranes, a superior example of bacterial intracytoplasmic membranes, are situated between the plasma membrane and central cytoplasm, forming an array of extended, flattened sacs instead of tightly appressed grana stacks in plant chloroplasts (Sun et al., 2019). Cyanobacterial thylakoid membranes accommodate membrane protein complexes that are involved in photosynthetic and respiratory electron transport, representing one of the most important and unique bioenergetic membrane systems in nature (Mullineaux, 2014; Liu, 2016; Nagarajan and Pakrasi, 2016; Mullineaux and Liu, 2020; Huokko et al., 2021). In addition to photosynthetic complexes, the respiratory electron transport complexes in cyanobacterial thylakoid membranes include type-I NADH dehydrogenase-like complex (NDH-1), succinate dehydrogenase (SDH) and cytochrome oxidases (Cooley and Vermaas,

2001; Liu et al., 2012; Lea-Smith et al., 2013; Ermakova et al., 2016). The thylakoid membranes and membrane-spanning components perform a stepwise biosynthesis process in cyanobacterial cells (Mahbub et al., 2020; Huokko et al., 2021). The cytoplasmic surface of the cyanobacterial thylakoid membrane is covered by phycobilisomes, which serve as the major extrinsic antenna supercomplex for both photosystems (Adir, 2005; Watanabe and Ikeuchi, 2013; Zheng et al., 2021; Dominguez-Martin et al., 2022; Kawakami et al., 2022). Phycobilisomes are self-assembled supercomplexes made of stacks of chromophore-containing phycobiliproteins and colorless linker polypeptides (Liu et al., 2005), capturing light at the wavelength of 500–670 nm, greatly extending the absorbance range of chlorophyll a. The association of phycobilisomes has also been observed in the photosynthetic membranes in red algae (Arteni et al., 2008; Liu et al., 2008; Zhao et al., 2016; Li et al., 2021). Moreover, a membrane-spanning protein associated with PSI, known as IsiA, is expressed in cyanobacteria under stress conditions (Vinnemeier et al., 1998; Bibby et al., 2001; Boekema et al., 2001; Yousef et al., 2003; Havaux et al., 2005; Chen et al., 2018), acting as an alternative light-harvesting antenna to increase the effective absorption cross-section of PSI (Zhao et al., 2020).

Efficient light capture, electron transport and energy conversion rely on not only the activities of individual photosynthetic complexes but also on the lateral organization and inter-complex association of photosynthetic complexes in the thylakoid membrane (Liu, 2016; Mullineaux and Liu, 2020). The elaborate photosynthetic apparatus assembles in a coordinated way to improve the efficiency of photosynthetic electron transport. On the other hand, the photosynthetic membrane organization is plastic, allowing photosynthetic complexes to be dynamically remodeled. This feature is essential for protein repair and photoprotection processes of cells, enabling rapid responses and cell fitness to changes in the environment. Therefore, advanced knowledge of the overall organization and dynamics of thylakoid membranes is required for a comprehensive understanding of the molecular mechanisms of efficient light harvesting and energy conversion, and platform development for green biotechnologies.

2. Atomic Force Microscopy (AFM) Studies on Thylakoid Membrane Architectures

Multiple techniques have been applied to study the organization and dynamics of photosynthetic membranes, such as electron microscopy (EM), atomic force microscopy (AFM), fluorescence microscopy and neutron scattering (Mullineaux and Liu, 2020). These techniques complement each other, and each technique has its specific advantages. Among these techniques, AFM is a unique and powerful tool for directly visualizing the supramolecular structures of functional biological samples (e.g. membranes, multi-protein complex assemblies, DNA and living cells) at submolecular resolution, as well as their molecular forces and nanomechanics in the near-physiological context (Liu et al., 2011; Liu and Scheuring, 2013; Faulkner et al., 2017; Miller et al., 2020b). AFM has exhibited extraordinary features in gaining structural information on biological samples, including simplicity of sample preparation (Scheuring and Sturgis, 2009), near-physiological imaging conditions (in buffer, ambient temperature and pressure), high signal-to-noise ratio (Engel and Gaub, 2008; Muller, 2008; Frederix et al., 2009), direct visualization of multicomponent supercomplexes, manipulating samples and detecting the intra- and intermolecular forces (Scheuring et al., 2003; Liu et al., 2011; Muller and Dufrene, 2011; Miller et al., 2020b), real-time visualization of dynamic biological processes by high-speed AFM (Sutter et al., 2016; Ando, 2018; Uchihashi and Scheuring, 2018; Faulkner et al., 2019) and mapping of specific proteins via recognition imaging (Chtcheglova and Hinterdorfer, 2018). Taking advantage of these characteristics, AFM has been widely used in studying the supramolecular architecture and protein organization of purple photosynthetic ICMs, photosynthetic plasma membranes, and thylakoid membranes (Zhu et al., 1995; Martinez-Planells et al., 2002; Montano et al., 2003; Liu and Scheuring, 2013; Johnson et al., 2014; Onoa et al., 2014; Phuthong et al., 2015; Zhao et al., 2016; Liu, 2016; Casella et al., 2017; Faulkner et al., 2017; Kumar et al., 2017; MacGregor-Chatwin et al., 2017, 2019; Miller et al., 2020b). These studies provide detailed insight into the organizational underpinnings of the adaptation of photosynthetic machinery under diverse environmental conditions.

3. AFM Imaging to Study Cyanobacterial Thylakoid Membrane Organizations

The initial study of cyanobacterial photosynthesis by AFM is the visualization of 2D crystals of individual integral photosynthetic complexes reconstituted in artificial lipid bilayers, such as PSI of *Synechococcus* (Fotiadis et al., 1998) and the C ring of cyanobacterial ATPase (Pogoryelov et al., 2005, 2007). These studies revealed the size, shape, precise surfaces, oligomerization and sidedness of photosynthetic complexes with a lateral resolution of approximately 1 nm and a vertical resolution of 0.1 nm. In recent years, the organization and adaptation of photosynthetic complexes in cyanobacterial thylakoid membranes have been delineated by AFM imaging (Fig. 5.1A). To date, AFM topographs have been recorded to characterize the structural landscapes of thylakoid membranes from *Synechococcus elongatus* PCC 7942 (Syn7942), *Synechococcus elongatus* UTEX 2973 (Syn2973), *Thermosynechococcus elongatus*, *Synechococcus* sp. PCC 7002 (Syn7002), *Synechocystis* sp. PCC 6803 (Syn6803), *Prochlorococcus marinus* MED4, MIT9313 and SS120, and cyanobacterial species capable of far-red light-induced photoacclimation (Casella et al., 2017; MacGregor-Chatwin et al., 2017, 2019; Ho et al., 2020; Zhao et al., 2020; MacGregor-Chatwin et al., 2022; Zhao et al., 2022). The AFM studies provide insight into the structural variability and plasticity of the cyanobacterial photosynthetic apparatus.

Cyanobacterial thylakoids are densely packed with membrane-integral protein complexes, as shown by AFM and cryo-electron tomography (Casella et al., 2017; MacGregor-Chatwin et al., 2017, 2019; Rast et al., 2019; Zhao et al., 2020; Ho et al., 2020; MacGregor-Chatwin et al., 2022; Zhao et al., 2022). The self-aggregation and lateral segregation of photosynthetic complexes in plant thylakoid membranes are critical for the stable and efficient photosynthetic electron transport and state transition (Dekker and Boekema, 2005; Kouril et al., 2012; Johnson et al., 2014). In cyanobacterial thylakoid membrane, PSI and PSII also self-aggregate (Casella et al., 2017; MacGregor-Chatwin et al., 2017, 2019; Straskova et al., 2019; Zhao et al., 2020; Ho et al., 2020; MacGregor-Chatwin et al., 2022; Zhao et al., 2022) (Fig. 5.1). The lateral segregation of photosystems in cyanobacterial thylakoid membranes may provide favorable micro-environments for photosynthetic linear electron flow (Busch et al., 2013). In addition, the oligomeric state of cyanobacterial photosystems and their organization in thylakoid membranes vary according to the ecophysiological environments to fulfill the needs for energy transduction (MacGregor-Chatwin et al., 2017, 2019; Zhao et al., 2020; MacGregor-Chatwin et al., 2022).

3.1 Thylakoid membrane architecture of *Synechococcus*, *Synechocystis* and *Thermosynechococcus*

Photosynthetic complexes, such as PSI, PSII, Cyt b_6f, NDH-1 and ATPase, have been identified in high-resolution AFM topographs of cyanobacterial thylakoid membranes (Casella et al., 2017; MacGregor-Chatwin et al., 2017; Zhao et al., 2020) (Fig. 5.1A). The orientation and long-range distribution of photosynthetic complexes were determined from both cytoplasmic and lumenal surfaces (Casella et al., 2017; MacGregor-Chatwin et al., 2017, 2019; Ho et al., 2020; Zhao et al., 2020; MacGregor-Chatwin et al., 2022; Zhao et al., 2022). AFM imaging on native thylakoid membranes from Syn7942 reveals that over 75% of the thylakoid membrane surface area is occupied by membrane proteins

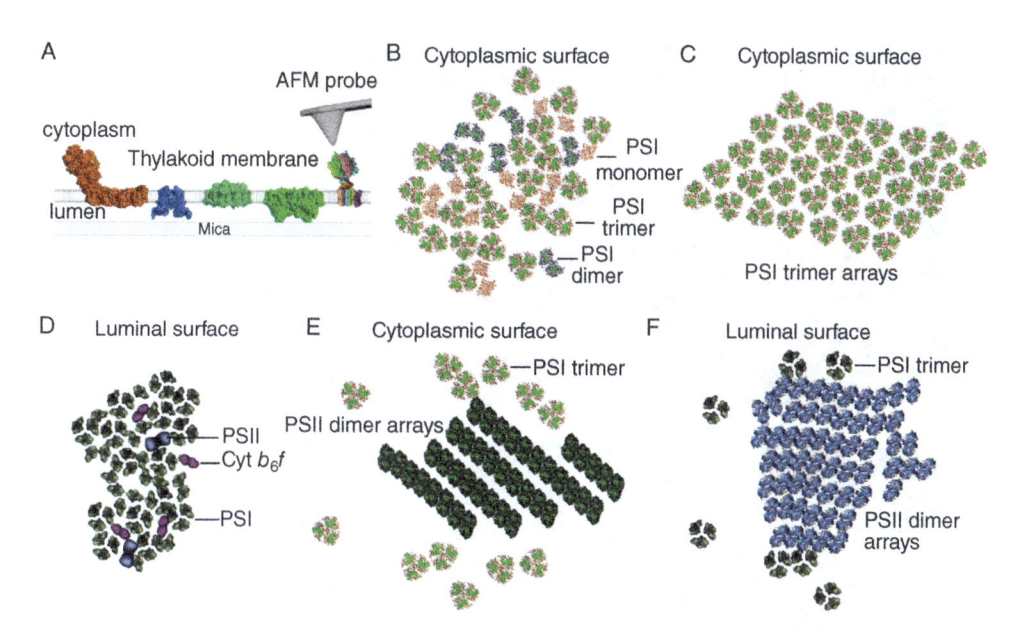

Fig. 5.1. Organization of photosynthetic apparatus in cyanobacterial thylakoid membranes. **A**, Schematic representation of AFM imaging on photosynthetic membranes immobilized on the mica substrate surface. **B**, Structural model of disordered PSI arrangement in PSI-rich membrane region based on AFM topograph viewed from cytoplasmic surface (Zhao et al., 2020). **C**, Structural model of regular PSI arrangement in PSI-rich membrane region based on AFM topograph viewed from cytoplasmic surface (MacGregor-Chatwin et al., 2017). **D**, Structural model of PSI, PSII and Cyt b_6f within the thylakoid membrane based on AFM topograph viewed from lumenal surface. Reproduced from Reference (Zhao et al., 2020) with permission. **E**, Structural model of the organization of PSII arrays and surrounding PSI trimers based on AFM topograph viewed from cytoplasmic surface. Reproduced from Reference (Zhao et al., 2022) with permission. **F**, Structural model of the organization of PSII arrays and surrounding PSI trimers based on AFM topograph viewed from lumenal surface. Reproduced from Reference (Zhao et al., 2022) with permission. PDB ID: PSI, 1JB0; PSII, 3WU2; Cyt b_6f, 2E74.

(Casella et al., 2017). PSI-enriched domains were observed to be intermixed with PSII in the thylakoid membranes from Syn7942, Syn2973, Syn7002, Syn6803 and *Thermosynechococcus* (Fig. 5.1B, 5.1D). The integration of PSI and PSII provides the structural basis for the dynamic movement of phycobilisomes between PSI and PSII to balance the energy captured by phycobilisomes to PSI or PSII in response to environmental changes (Joshua and Mullineaux, 2004; Mullineaux and Emlyn-Jones, 2005). In contrast, PSI-only membrane patches in thylakoid membranes from Syn7002 and *Thermosynechococcus* showed hexagonally packed PSI trimers (MacGregor-Chatwin et al., 2017) (Fig. 5.1C). Most of the PSII were randomly distributed in cyanobacterial thylakoid membranes (Fig. 5.1D), whereas parallel arrays of PSII dimers were visualized in Syn7942 and Syn2973 thylakoid membranes by AFM (Fig. 5.1E, 5.1F) (Zhao et al., 2020, 2022). This specific PSII organization may facilitate the physical attachment of phycobilisomes.

High-resolution AFM imaging of native thylakoid membranes permits the visualization of photosynthetic supercomplexes formed by flexible or weak interactions. In cyanobacterial thylakoid membranes, PSI complexes could form direct contacts with PSII, Cyt b_6f, ATPase and NDH-1 (Zhao et al., 2020, 2022) (Fig. 5.1D–5.1F, Fig. 5.2). The close association of PSI, Cyt b_6f and PSII could facilitate the photosynthetic linear electron flow, whereas the close associations between PSI and NDH-1 could increase the rate of cyclic electron transport. Diverse forms of PSI–NDH-1 supercomplexes

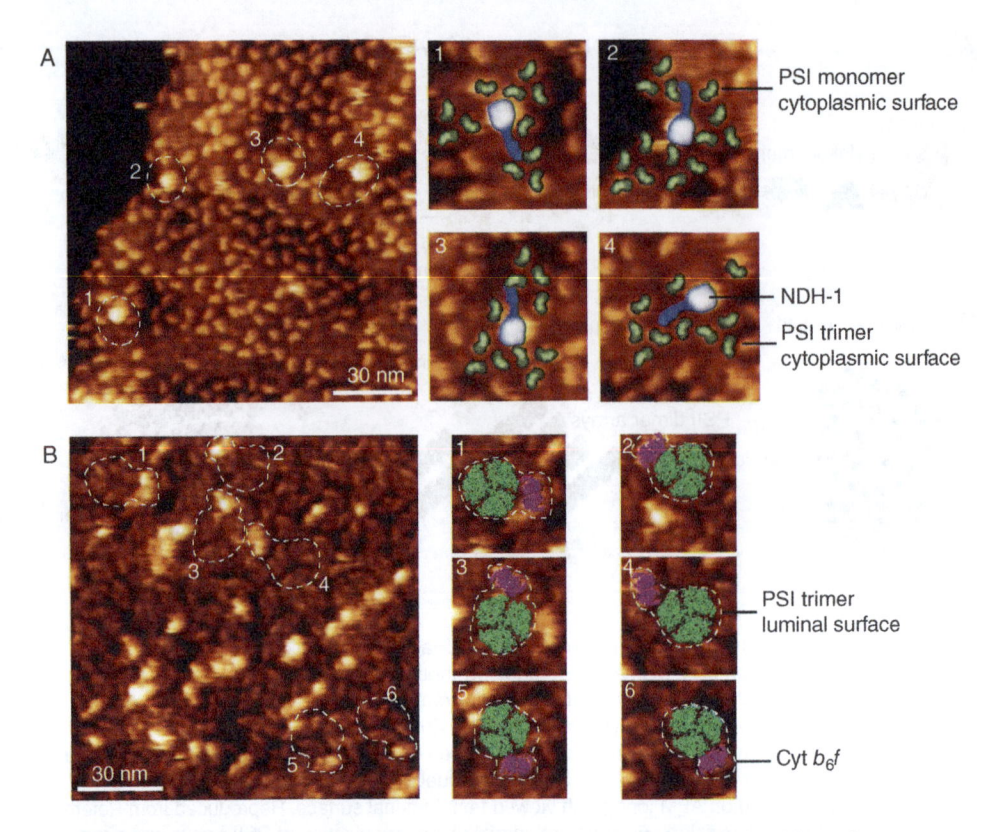

Fig. 5.2. Arrangement of PSI, NDH-1 complex and Cyt b_6f in cyanobacterial thylakoid membrane. A (*left panel*): association of PSI and NDH-1 in thylakoid membrane viewed from cytoplasmic surface; (*right panel*): structural model of the PSI–NDH-1 supercomplex based on the AFM topograph. B (*left panel*): association of PSI and Cyt b_6f in thylakoid membrane viewed from lumenal surface; (*right panel*): structural model of the PSI–Cyt b_6f supercomplex based on the AFM topograph. (Reproduced from Zhao et al., 2020). PDB ID: PSI, 1JB0; NDH-1, 6HUM; Cyt b_6f, 2E74

(Fig. 5.2A) and PSI–Cyt b_6f (Fig. 5.2B) have been imaged by AFM, indicating their flexible associations (Zhao et al., 2020, 2022).

High-resolution AFM has also been exploited to study the supramolecular architecture of high light (HL)-adapted and iron-stressed (Fe–) cyanobacterial thylakoid membranes (Zhao et al., 2020, 2022). The chlorophyll-binding proteins IsiA are highly expressed in HL-adapted Syn7942, Fe–Syn7942 and Fe–Syn2973, and the produced IsiA proteins associate with PSI to form IsiA–PSI supercomplexes to increase the absorption cross-section of PSI. AFM imaging characterized explicitly the highly variable structures of the IsiA–PSI supercomplexes in thylakoid membranes (Fig. 5.3A, B): IsiA single, double, triple or multimeric rings can surround the PSI trimers, dimers or monomers, forming diverse IsiA–PSI assemblies varying in dimension and structure (Zhao et al., 2020, 2022). Different from Syn7942, HL-adapted thylakoid membranes from the fast-growing cyanobacterium Syn2973 accommodate a large amount of PSI complexes, without the incorporation of IsiA assemblies (Fig. 5.3C), facilitating the photosynthetic efficiency (Zhao et al., 2022).

3.2 Thylakoid membrane architecture of *Prochlorococcus* from marine ecological niches

In addition to the freshwater cyanobacteria discussed above, AFM has also been applied

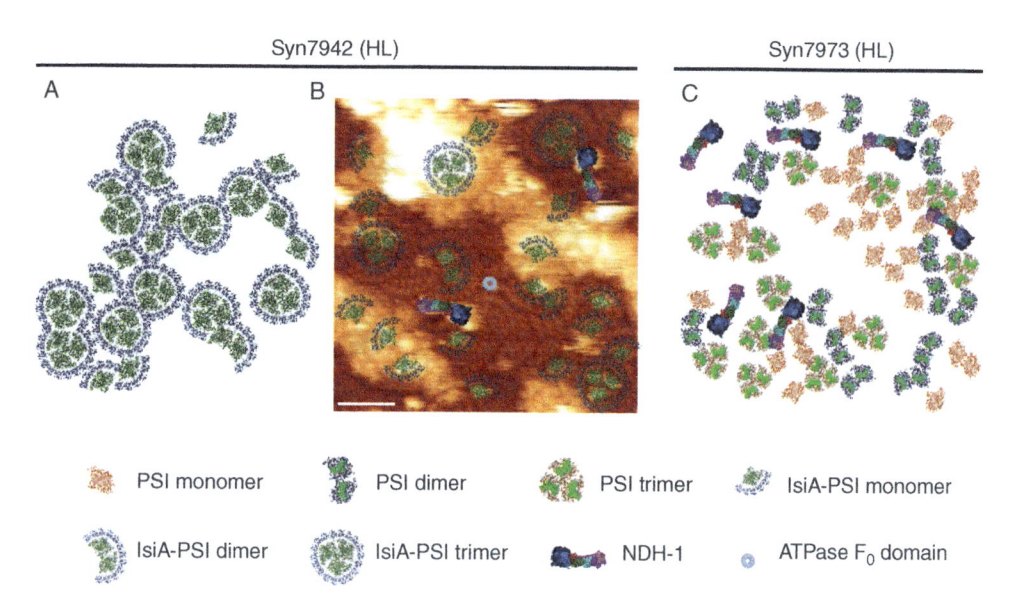

Fig. 5.3. High light (HL) acclimation of cyanobacterial thylakoid membrane. A. Structural model of densely packed IsiA-PSI supercomplexes based on AFM topograph of thylakoid membrane from HL-adapted Syn7942 viewed from cytoplasmic surface. B. AFM topograph of thylakoid membrane from HL-adapted Syn7942 viewed from cytoplasmic surface. Structural model of the arrangement of photosynthetic apparatus based on the AFM topograph shows the sparse distribution of IsiA-PSI. (Reproduced from Zhao et al., 2020). C. Structural model of the organization of photosynthetic apparatus based on AFM topograph of thylakoid membrane from HL-adapted Syn2973 viewed from cytoplasmic surface (Zhao et al., 2022). PDB ID: IsiA–PSI supercomplex, 6NWA; PSI, 1JB0; NDH-1, 6HUM; ATPase, 6FKF

to study the supramolecular architectures of thylakoid membranes in different ecotypes of *Prochlorococcus marinus*, a genus of picocyanobacterial in surface ocean, grown under different light intensities (MacGregor-Chatwin et al., 2019). The variability of the PSI organization in thylakoid membranes was revealed in the high-light ecotype (*Prochlorococcus marinus* MED4) grown under HL and low-light (LL) conditions and the low-light ecotype (*Prochlorococcus marinus* MIT9313 and SS120) grown under LL conditions. PSI-only membrane patches were observed in thylakoid membranes from LL-adapted MED4, MIT9313 and SS120, while thylakoid membranes from MIT9313 and SS120 showed paracrystalline PSI organization. PSI complexes in MIT9313 thylakoids are organized into a tightly packed pseudo-hexagonal lattice to maximize the capturing of light, as found in *Thermosynechococcus* thylakoid membranes (MacGregor-Chatwin et al., 2017). PSI complexes of LL-adapted SS120 associate with 18 Pcb proteins (prochlorophyte Chl b-binding proteins) to form the Pcb–PSI supercomplexes, resembling the IsiA-PSI supercomplexes (Zhao et al., 2020, 2022). The closely packed Pcb–PSI supercomplexes could be a different tactic to cope with low-light levels in nutrient-limiting environment. The PSI densities of membrane patches from HL-adapted MED4 is lower than their LL-adapted counterparts.

3.3 Thylakoid membrane architecture of cyanobacteria capable of far-red light photoacclimation

AFM has recently been used to unveil the remodeling of photosynthetic apparatus in thylakoid membranes of *Synechococcus* sp. PCC 7335 (Syn7335), *Chlorogloeopsis fritschii* PCC 9212 (*C. fritschii* 9212) and *Chroococcidiopsis thermalis* PCC 7203 (*C. thermalis* 7203) during Far-Red Light Photoacclimation

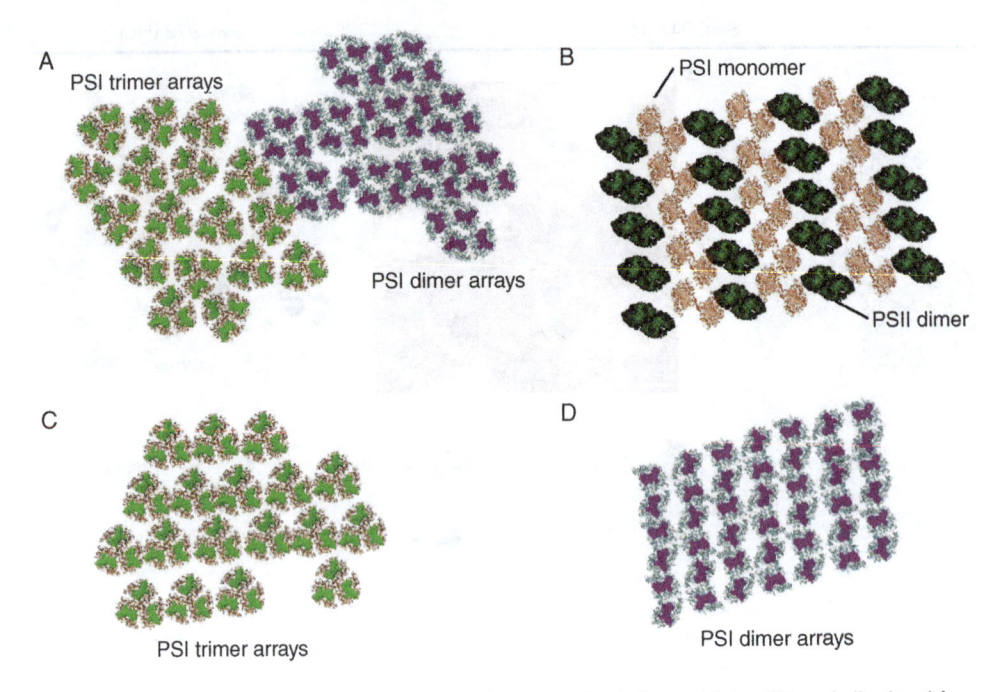

Fig. 5.4. Photoacclimation of thylakoid membrane from cyanobacteria containing chlorophylls *d* and *f*. A. Structural model of regular PSI arrangement based on AFM topograph of red light-adapted thylakoid membrane viewed from cytoplasmic surface (Ho *et al.*, 2020). PSI dimer arrays and PSI trimer arrays are neighboring. B. Structural model of regular PSI and PSII arrangement based on AFM topograph of red light-adapted thylakoid membrane viewed from cytoplasmic surface (Ho *et al.*, 2020). C. Structural model of regular PSI arrangement based on AFM topograph of far-red light-adapted thylakoid membrane viewed from cytoplasmic surface (MacGregor-Chatwin *et al.*, 2022). D. Structural model of regular PSI arrangement based on AFM topograph of white-light-adapted thylakoid membrane viewed from cytoplasmic surface (MacGregor-Chatwin *et al.*, 2022). PDB ID: PSI trimer, 1JB0; PSI dimer, 6K61; PSI monomer, 1JB0; PSII, 3WU2.

(FaRLiP) (Ho *et al.*, 2020; MacGregor-Chatwin *et al.*, 2022). These cyanobacterial species could synthesize chlorophylls *d* and *f* to acclimate to far-red light (FRL). Supramolecular organization of thylakoid membranes from red light (RL)-adapted Syn7335 showed three structural arrangements of RL-PSI: very closely packed quasi-regular arrays of PSI trimers (Fig. 5.4A), tightly packed and highly regular arrays of likely PSI dimers (Fig. 5.4A), and alternating rows of closely packed dimeric PSI and dimeric PSII (Fig. 4B) (Ho *et al.*, 2020). In contrast, only regular arrays of PSI trimers were observed in the thylakoid membranes from FRL-adapted Syn7335 (Fig. 5.4C). In white light (WL)-adapted Syn7335, non-trimeric PSI were densely packed in thylakoid membranes (Fig. 5.4D). The supramolecular assembly of PSI was affected by the substitution of PSI subunits with paralogous proteins during FaRLiP. The change in PSI from a non-trimeric state in WL to a trimeric state in FRL was also identified in *C. fritschii* 9212 and *C. thermalis* 7203 by AFM imaging (MacGregor-Chatwin *et al.*, 2022).

4. Thylakoid Membrane Architecture of Red Algae

AFM has also been applied to the studies of red algal thylakoid membranes (Liu *et al.*, 2008; Zhao *et al.*, 2016). It was reported that

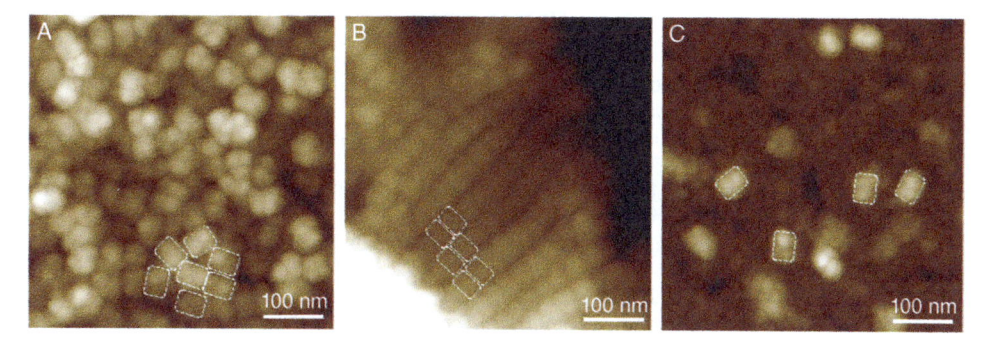

Fig. 5.5. Architecture and acclimation of photosynthetic membranes in red algae. A. AFM imaging of moderate light-adapted thylakoid membrane from *Porphyridium cruentum*, showing the disordered arrangement of phycobilisomes. B. AFM imaging of low light-adapted thylakoid membrane from *Porphyridium cruentum*, showing the regular arrays of phycobilisomes. C. AFM imaging of thylakoid membrane from nitrogen-starved *Porphyridium cruentum*, showing the decreased copy number of phycobilisomes.

the arrangement of phycobilisomes on the thylakoid membrane surface of red alga *Porphyridium cruentum* was regulated by light intensity and nitrogen availability (Liu *et al.*, 2008; Zhao *et al.*, 2016) (Fig. 5.5). Under moderate light, phycobilisomes were densely packed on the thylakoid membrane surface with a relatively random and clustered distribution (Liu *et al.*, 2008) (Fig. 5.5A). In contrast, parallel arrays of phycobilisomes appeared in low light-adapted thylakoid membranes (Fig. 5.5B), increasing the density of phycobilisomes per membrane surface to maximize the light-harvesting efficiency. AFM imaging also showed that nitrogen starvation resulted in the degradation of phycobilisomes and decreases in phycobilisome density and size (Fig. 5.5C), implicating the role of phycobilisomes in nitrogen storage (Zhao *et al.*, 2016).

5. Conclusions and Perspectives

Thylakoid membranes form a complicated intracellular membrane system in cyanobacteria and red algae to conduct efficient light harvesting and energy transfer. The thylakoid membrane is organized to form a dynamic and variable pigment-protein network and the architecture of the thylakoid membrane varies among species. AFM represents a powerful tool and has opened new vistas for visualizing, in unprecedented detail, the structural landscapes of thylakoid membranes and the organization of photosynthetic apparatus in various phototrophic organisms. The recent AFM observations of thylakoid membranes have provided detailed insight into the structural variability and plasticity of cyanobacterial thylakoid membranes, the organizational underpinnings of the adaptation of cyanobacterial and red algal photosynthetic machinery under diverse environmental conditions with variable light wavelengths and quantities as well as limiting nutrients. In addition to its unique imaging capability, AFM-based techniques such as real-time molecule tracking, force spectroscopy and correlation with fluorescence will facilitate further evaluation of thylakoid membrane protein folding, assembly, interactions and functional correlations. The valuable architectural information of photosynthetic machinery achieved by AFM has implications for fine-tuning natural photosynthetic systems as well as rational design and bioengineering of artificial photosynthetic systems for green energy production.

References

Adir, N. (2005) Elucidation of the molecular structures of components of the phycobilisome: reconstructing a giant. *Photosynthesis Research* 85, 15–32.

Albertsson, P. (2001) A quantitative model of the domain structure of the photosynthetic membrane. *Trends in Plant Science* 6, 349–358.

Anderson, J.M. and Andersson, B. (1982) The architecture of photosynthetic membranes: lateral and transverse organization. *Trends in Biochemical Sciences* 7, 288–292.

Ando, T. (2018) High-speed atomic force microscopy and its future prospects. *Biophysical Reviews* 10, 285–292.

Arteni, A.A., Liu, L.N., Aartsma, T.J., Zhang, Y.Z., Zhou, B.C. et al. (2008) Structure and organization of phycobilisomes on membranes of the red alga *Porphyridium cruentum*. *Photosynthesis Research* 95, 169–174.

Austin, J.R., 2nd and Staehelin, L.A. (2011) Three-dimensional architecture of grana and stroma thylakoids of higher plants as determined by electron tomography. *Plant physiology* 155, 1601–1611.

Beja, O., Spudich, E.N., Spudich, J.L., Leclerc, M. and Delong, E.F. (2001) Proteorhodopsin phototrophy in the ocean. *Nature* 411, 786–789.

Bibby, T.S., Nield, J. and Barber, J. (2001) Iron deficiency induces the formation of an antenna ring around trimeric photosystem I in cyanobacteria. *Nature* 412, 743–745.

Boekema, E.J., Hifney, A., Yakushevska, A.E., Piotrowski, M., Keegstra, W. et al. (2001). A giant chlorophyll-protein complex induced by iron deficiency in cyanobacteria. *Nature* 412, 745–748.

Bracun, L., Yamagata, A., Christianson, B.M., Terada, T., Canniffe, D.P. et al. (2021) Cryo-EM structure of the photosynthetic RC-LH1-PufX supercomplex at 2.8-Å resolution. *Science Advances* 7, eabf8864.

Busch, K.B., Deckers-Hebestreit, G., Hanke, G.T. and Mulkidjanian, A.Y. (2013) Dynamics of bioenergetic microcompartments. *Biological Chemistry* 394, 163–188.

Cao, P., Bracun, L., Yamagata, A., Christianson, B.M., Negami, T. et al. (2022) Structural basis for the assembly and quinone transport mechanisms of the dimeric photosynthetic RC–LH1 supercomplex. *Nature Communications* 13, 1977.

Casella, S., Huang, F., Mason, D., Zhao, G.Y., Johnson, G.N. et al. (2017) Dissecting the native architecture and dynamics of cyanobacterial photosynthetic machinery. *Molecular Plant* 10, 1434–1448.

Chen, H.Y.S., Bandyopadhyay, A. and Pakrasi, H.B. (2018) Function, regulation and distribution of IsiA, a membrane-bound chlorophyll a-antenna protein in cyanobacteria. *Photosynthetica* 56, 322–333.

Chtcheglova, L.A. and Hinterdorfer, P. (2018) Simultaneous AFM topography and recognition imaging at the plasma membrane of mammalian cells. *Seminars in Cell & Developmental Biology* 73, 45–56.

Cogdell, R.J., Gall, A. and Kohler, J. (2006) The architecture and function of the light-harvesting apparatus of purple bacteria: from single molecules to in vivo membranes. *Quarterly Reviews of Biophysics* 39, 227–324.

Cooley, J.W. and Vermaas, W.F. (2001) Succinate dehydrogenase and other respiratory pathways in thylakoid membranes of Synechocystis sp. strain PCC 6803: capacity comparisons and physiological function. *Journal of Bacteriology* 183, 4251–4258.

Daum, B., Nicastro, D., Austin, J. 2nd, Mcintosh, J.R. and Kuhlbrandt, W. (2010) Arrangement of photosystem II and ATP synthase in chloroplast membranes of spinach and pea. *Plant Cell* 22, 1299–1312.

Dekker, J.P. and Boekema, E.J. (2005) Supramolecular organization of thylakoid membrane proteins in green plants. *Biochimica et Biophysica Acta (BBA) – Bioenergetics* 1706, 12–39.

Dominguez-Martin, M.A., Sauer, P.V., Kirst, H., Sutter, M., Bina, D. et al. (2022) Structures of a phycobilisome in light-harvesting and photoprotected states. *Nature* 609, 835–845.

Engel, A. and Gaub, H.E. (2008) Structure and mechanics of membrane proteins. *Annual Review of Biochemistry* 77, 127–148.

Ermakova, M., Huokko, T., Richaud, P., Bersanini, L., Howe, C.J. et al. (2016) Distinguishing the roles of thylakoid respiratory terminal oxidases in the cyanobacterium Synechocystis sp. PCC 6803. *Plant Physiology* 171, 1307–1319.

Faulkner, M., Rodriguez-Ramos, J., Dykes, G.F., Owen, S.V., Casella, S. et al. (2017) Direct characterization of the native structure and mechanics of cyanobacterial carboxysomes. *Nanoscale* 9, 10662–10673.

Faulkner, M., Zhao, L.S., Barrett, S. and Liu, L.N. (2019) Self-assembly stability and variability of bacterial microcompartment shell proteins in response to the environmental change. *Nanoscale Research Letters* 14, 54.

Fotiadis, D., Muller, D.J., Tsiotis, G., Hasler, L., Tittmann, P. et al. (1998) Surface analysis of the photosystem I complex by electron and atomic force microscopy. *Journal of Molecular Biology* 283, 83–94.

Frederix, P.L., Bosshart, P.D. and Engel, A. (2009) Atomic force microscopy of biological membranes. *Biophysical Journal* 96, 329–338.

Havaux, M., Guedeney, G., Hagemann, M., Yeremenko, N., Matthijs, H.C. et al. (2005) The chlorophyll-binding protein IsiA is inducible by high light and protects the cyanobacterium Synechocystis PCC6803 from photooxidative stress. *FEBS Letters* 579, 2289–2293.

Ho, M.Y., Niedzwiedzki, D.M., Macgregor-Chatwin, C., Gerstenecker, G., Hunter, C.N. et al. (2020) Extensive remodeling of the photosynthetic apparatus alters energy transfer among photosynthetic complexes when cyanobacteria acclimate to far-red light. *Biochimica et Biophysica Acta (BBA) – Bioenergetics* 1861, 148064.

Huokko, T., Ni, T., Dykes, G.F., Simpson, D.M., Brownridge, P. et al. (2021) Probing the biogenesis pathway and dynamics of thylakoid membranes. *Nature Communications* 12, 3475.

Johnson, M.P., Vasilev, C., Olsen, J.D. and Hunter, C.N. (2014) Nanodomains of cytochrome b_6f and photosystem II complexes in spinach grana thylakoid membranes. *Plant Cell* 26, 3051–3061.

Joshua, S. and Mullineaux, C.W. (2004) Phycobilisome diffusion is required for light-state transitions in cyanobacteria. *Plant Physiology* 135, 2112–2119.

Kawakami, K., Hamaguchi, T., Hirose, Y., Kosumi, D., Miyata, M. et al. (2022) Core and rod structures of a thermophilic cyanobacterial light-harvesting phycobilisome. *Nature Communications* 13, 3389.

Kouril, R., Dekker, J.P. and Boekema, E.J. (2012) Supramolecular organization of photosystem II in green plants. *Biochimica et Biophysica Acta (BBA) – Bioenergetics* 1817, 2–12.

Kumar, S., Cartron, M.L., Mullin, N., Qian, P., Leggett, G.J. et al. (2017) Direct imaging of protein organization in an intact bacterial organelle using high-resolution atomic force microscopy. *ACS Nano* 11, 126–133.

Lea-Smith, D.J., Ross, N., Zori, M., Bendall, D.S., Dennis, J.S. et al. (2013) Thylakoid terminal oxidases are essential for the cyanobacterium Synechocystis sp. PCC 6803 to survive rapidly changing light intensities. *Plant Physiology* 162, 484–495.

Li, M., Ma, J., Li, X. and Sui, S.F. (2021) In situ cryo-ET structure of phycobilisome-photosystem II supercomplex from red alga. *Elife,* 10.

Liu, L.N. (2016) Distribution and dynamics of electron transport complexes in cyanobacterial thylakoid membranes. *Biochim Biophys Acta* 1857, 256–265.

Liu, L.N. and Scheuring, S. (2013) Investigation of photosynthetic membrane structure using atomic force microscopy. *Trends in Plant Science* 18, 277–286.

Liu, L.N., Chen, X.L., Zhang, Y.Z. and Zhou, B.C. (2005) Characterization, structure and function of linker polypeptides in phycobilisomes of cyanobacteria and red algae: an overview. *Biochimica et Biophysica Acta (BBA) – Bioenergetics* 1708, 133–142.

Liu, L.N., Aartsma, T.J., Thomas, J.C., Lamers, G.E., Zhou, B.C. et al. (2008) Watching the native supramolecular architecture of photosynthetic membrane in red algae: topography of phycobilisomes and their crowding, diverse distribution patterns. *The Journal of Biological Chemistry* 283, 34946–34953.

Liu, L.N., Duquesne, K., Oesterhelt, F., Sturgis, J.N. and Scheuring, S. (2011) Forces guiding assembly of light-harvesting complex 2 in native membranes. *Proceedings of the National Academy of Sciences of the United States of America* 108, 9455–9459.

Liu, L.N., Bryan, S.J., Huang, F., Yu, J.F., Nixon, P.J. et al. (2012) Control of electron transport routes through redox-regulated redistribution of respiratory complexes. *Proceedings of the National Academy of Sciences of the United States of America* 109, 11431–11436.

Macgregor-Chatwin, C., Sener, M., Barnett, S.F.H., Hitchcock, A., Barnhart-Dailey, M.C. et al. (2017) Lateral segregation of photosystem I in cyanobacterial thylakoids. *Plant Cell* 29, 1119–1136.

Macgregor-Chatwin, C., Jackson, P.J., Sener, M., Chidgey, J.W., Hitchcock, A. et al. (2019) Membrane organization of photosystem I complexes in the most abundant phototroph on Earth. *Nature Plants* 5, 879–889.

Macgregor-Chatwin, C., Nurnberg, D.J., Jackson, P.J., Vasilev, C., Hitchcock, A. et al. (2022) Changes in supramolecular organization of cyanobacterial thylakoid membrane complexes in response to far-red light photoacclimation. *Science Advances* 8, eabj4437.

Mahbub, M., Hemm, L., Yang, Y., Kaur, R., Carmen, H. et al. (2020) mRNA localisation, reaction centre biogenesis and thylakoid membrane targeting in cyanobacteria. *Nature Plants* 6, 1179–1191.

Martinez-Planells, A., Arellano, J.B., Borrego, C.M., Lopez-Iglesias, C., Gich, F. et al. (2002) Determination of the topography and biometry of chlorosomes by atomic force microscopy. *Photosynthesis Research* 71, 83–90.

Miller, L.C., Martin, D.S., Liu, L.-N. and Canniffe, D.P. (2020a) Composition, organisation and function of purple photosynthetic machinery. In: Wang, Q. (ed.) *Microbial Photosynthesis*. Springer Singapore, pp. 73–114.

Miller, L.C., Zhao, L., Canniffe, D.P., Martin, D. and Liu, L.N. (2020b) Unfolding pathway and intermolecular interactions of the cytochrome subunit in the bacterial photosynthetic reaction center. *Biochimica et Biophysica Acta (BBA) – Bioenergetics* 1861, 148204.

Montano, G.A., Bowen, B.P., Labelle, J.T., Woodbury, N.W., Pizziconi, V.B. *et al.* (2003) Characterization of *Chlorobium tepidum* chlorosomes: a calculation of bacteriochlorophyll c per chlorosome and oligomer modeling. *Biophysical Journal* 85, 2560–2565.

Muller, D.J. (2008) AFM: a nanotool in membrane biology. *Biochemistry* 47, 7986–7998.

Muller, D.J. and Dufrene, Y.F. (2011) Atomic force microscopy: a nanoscopic window on the cell surface. *Trends in Cell Biology* 21, 461–469.

Mullineaux, C.W. (2014) Co-existence of photosynthetic and respiratory activities in cyanobacterial thylakoid membranes. *Biochimica et Biophysica Acta (BBA) – Bioenergetics* 1837, 503–511.

Mullineaux, C.W. and Emlyn-Jones, D. (2005) State transitions: an example of acclimation to low-light stress. *Journal of Experimental Botany* 56, 389–393.

Mullineaux, C.W. and Liu, L.N. (2020) Membrane dynamics in phototrophic bacteria. *Annual Review of Microbiology* 74, 633–654.

Nagarajan, A. and Pakrasi, H.B. (2016) Membrane-bound protein complexes for photosynthesis and respiration in cyanobacteria. *eLS*, 1–8.

Nevo, R., Charuvi, D., Tsabari, O. and Reich, Z. (2012) Composition, architecture and dynamics of the photosynthetic apparatus in higher plants. *Plant Journal* 70, 157–176.

Nielsen, J.T., Kulminskaya, N.V., Bjerring, M., Linnanto, J.M., Ratsep, M. *et al.* (2016) In situ high-resolution structure of the baseplate antenna complex in Chlorobaculum tepidum. *Nature Communications* 7, 12454.

Onoa, B., Schneider, A.R., Brooks, M.D., Grob, P., Nogales, E. *et al.* (2014) Atomic force microscopy of photosystem II and its unit cell clustering quantitatively delineate the mesoscale variability in Arabidopsis thylakoids. *PLOS ONE* 9, e101470.

Phuthong, W., Huang, Z., Wittkopp, T.M., Sznee, K., Heinnickel, M.L. *et al.* (2015) The use of contact mode atomic force microscopy in aqueous medium for structural analysis of spinach photosynthetic complexes. *Plant Physiology* 169, 1318–1332.

Pogoryelov, D., Yu, J., Meier, T., Vonck, J., Dimroth, P. *et al.* (2005) The c15 ring of the Spirulina platensis F-ATP synthase: F1/F0 symmetry mismatch is not obligatory. *EMBO Reports* 6, 1040–1044.

Pogoryelov, D., Reichen, C., Klyszejko, A.L., Brunisholz, R., Müller, D.J. *et al.* (2007) The oligomeric state of c rings from cyanobacterial F-ATP synthases varies from 13 to 15. *Journal of Bacteriology* 189, 5895–5902.

Qian, P., Siebert, C.A., Wang, P., Canniffe, D.P. and Hunter, C.N. (2018) Cryo-EM structure of the *Blastochloris viridis* LH1-RC complex at 2.9 Å. *Nature* 556, 203–208.

Rast, A., Schaffer, M., Albert, S., Wan, W., Pfeffer, S. *et al.* (2019) Biogenic regions of cyanobacterial thylakoids form contact sites with the plasma membrane. *Nature Plants* 5, 436–446.

Ruban, A.V. and Johnson, M.P. (2015) Visualizing the dynamic structure of the plant photosynthetic membrane. *Nature Plants* 1, 15161.

Saer, R.G. and Blankenship, R.E. (2017) Light harvesting in phototrophic bacteria: structure and function. *The Biochemical Journal* 474, 2107–2131.

Scheuring, S. and Sturgis, J.N. (2009) Atomic force microscopy of the bacterial photosynthetic apparatus: plain pictures of an elaborate machinery. *Photosynthesis Research* 102, 197–211.

Scheuring, S., Seguin, J., Marco, S., Levy, D., Robert, B. *et al.* (2003) Nanodissection and high-resolution imaging of the Rhodopseudomonas viridis photosynthetic core complex in native membranes by AFM. *Proceedings of the National Academy of Sciences of the United States of America* 100, 1690–1693.

Scholes, G.D., Fleming, G.R., Olaya-Castro, A. and Van Grondelle, R. (2011) Lessons from nature about solar light harvesting. *Nature Chemistry* 3, 763–774.

Shimoni, E., Rav-Hon, O., Ohad, I., Brumfeld, V. and Reich, Z. (2005) Three-dimensional organization of higher-plant chloroplast thylakoid membranes revealed by electron tomography. *Plant Cell* 17, 2580–2586.

Straskova, A., Steinbach, G., Konert, G., Kotabova, E., Komenda, J. *et al.* (2019) Pigment-protein complexes are organized into stable microdomains in cyanobacterial thylakoids. *Biochim Biophys Acta Bioenerg* 1860, 148053.

Sun, Y., Wollman, A.J.M., Huang, F., Leake, M.C. and Liu, L.N. (2019) Single-organelle quantification reveals stoichiometric and structural variability of carboxysomes dependent on the environment. *Plant Cell* 31, 1648–1664.

Sutter, M., Faulkner, M., Aussignargues, C., Paasch, B.C., Barrett, S. *et al.* (2016) Visualization of bacterial microcompartment facet assembly using high-speed atomic force microscopy. *Nano Letters* 16, 1590–1595.

Swainsbury, D.J.K., Qian, P., Jackson, P.J., Faries, K.M., Niedzwiedzki, D.M. *et al.* (2021) Structures of *Rhodopseudomonas palustris* RC-LH1 complexes with open or closed quinone channels. *Science Advances* 7, eabe2631.

Tavano, C.L. and Donohue, T.J. (2006) Development of the bacterial photosynthetic apparatus. *Current Opinion in Microbiology* 9, 625–631.

Uchihashi, T. and Scheuring, S. (2018) Applications of high-speed atomic force microscopy to real-time visualization of dynamic biomolecular processes. *Biochim Biophys Acta Gen Subj* 1862, 229–240.

Vinnemeier, J., Kunert, A. and Hagemann, M. (1998) Transcriptional analysis of the isiAB operon in salt-stressed cells of the cyanobacterium Synechocystis sp. PCC 6803. *Fems Microbiology Letters* 169, 323–330.

Watanabe, M. and Ikeuchi, M. (2013) Phycobilisome: architecture of a light-harvesting supercomplex. *Photosynthesis Research* 116, 265–276.

Yousef, N., Pistorius, E.K. and Michel, K.P. (2003) Comparative analysis of idiA and isiA transcription under iron starvation and oxidative stress in Synechococcus elongatus PCC 7942 wild-type and selected mutants. *Archives of Microbiology* 180, 471–483.

Yu, L.J., Suga, M., Wang-Otomo, Z.Y. and Shen, J.R. (2018) Structure of photosynthetic LH1-RC supercomplex at 1.9 Å resolution. *Nature* 556, 209–213.

Zhao, L.S., Su, H.N., Li, K., Xie, B.B., Liu, L.N. *et al.* (2016) Supramolecular architecture of photosynthetic membrane in red algae in response to nitrogen starvation. *Biochimica et Biophysica Acta (BBA) – Bioenergetics* 1857, 1751–1758.

Zhao, L.S., Huokko, T., Wilson, S., Simpson, D.M., Wang, Q. *et al.* (2020) Structural variability, coordination and adaptation of a native photosynthetic machinery. *Nature Plants* 6, 869–882.

Zhao, L.S., Li, C.Y., Chen, X.L., Wang, Q., Zhang, Y.Z. *et al.* (2022) Native architecture and acclimation of photosynthetic membranes in a fast-growing cyanobacterium. *Plant Physiology* 190, 1883–1895.

Zheng, L., Zheng, Z., Li, X., Wang, G., Zhang, K. *et al.* (2021) Structural insight into the mechanism of energy transfer in cyanobacterial phycobilisomes. *Nature Communications* 12, 5497.

Zhu, Y., Ramakrishna, B.L., Van Noort, P.I. and Blankenship, R.E. (1995) Microscopic and spectroscopic studies of untreated and hexanol-treated chlorosomes from Chloroflexus aurantiacus. *Biochimica et Biophysica Acta (BBA)-Bioenergetics* 1232, 197–207.

6 Heterotrophic High-cell-density Cultivation for Effective Production of Microalgae Biomass and Metabolites of Commercial Interest

Hu Jin[1,2] and Feng Ge[1,2]*

[1]*State Key Laboratory of Freshwater Ecology and Biotechnology, Institute of Hydrobiology, Chinese Academy of Sciences, Wuhan, China;* [2]*Key Laboratory of Algal Biology, Institute of Hydrobiology, Chinese Academy of Sciences, Wuhan, China*

Abstract

Microalgae of numerous genera are capable of heterotrophic growth and exhibit considerable metabolic versatility and flexibility. Heterotrophic cultivation of microalgae eliminates the dependence on light and thereby both biomass concentration and productivity could be markedly improved. By implementing effective optimization and fermentation process control, many microalgae have achieved high-cell-density cultivation and the biomass concentrations of some green algae have even reached an ultra-high level of above 200 g L^{-1} of dry cell weight (DCW), making possible the applications of microalgae from small volume, high-value products to bulk low-value commodities. This chapter summarizes the major factors affecting heterotrophic growth of microalgae and provides a review of recent progress in high-cell-density cultivation for enhanced biomass or target product production.

Keywords: microalgae; heterotrophic cultivation; high-cell-density; fermentation; fermentor; optimization

1. Introduction

Microalgae are the most primitive photosynthetic organisms on Earth and their photosynthetic mechanisms are similar to land plants. Most microalgae are photoautotrophs by harvesting light as their energy source and using carbon dioxide as their carbon source. As one of the oldest culture systems, open outdoor pond cultivation is still currently the main culture mode for industrial microalgal biomass production (Perez-Garcia *et al.*, 2011; Morales-Sanchez *et al.*, 2017). Although open culture systems have the advantage of minimal construction and operation costs, they also have several inherent drawbacks: (i) low biomass concentration/productivity and high harvesting cost; (ii) subject to contamination by external microbial predators and eukaryotic grazers; (iii) uncontrolled environmental parameters for algal growth; (iv) large land occupation and water requirement (Xu *et al.*, 2009; Perez-Garcia *et al.*, 2011; Morales-Sanchez *et al.*, 2017). To date, large-scale open culture systems are feasible only for a few extreme or fast-growth microalgae like *Spirulina*, *Dunaliella* and *Chlorella* (Fernandez *et al.*, 2013).

To overcome the inherent disadvantages of open culture systems, numerous enclosed

*Address for correspondence: gefeng@ihb.ac.cn

photobioreactor (PBR) systems with various volumes and shapes have been designed. Due to the improved control of culture environment and gas diffusion, enclosed PBRs could achieve higher biomass concentration and also increase the protection from environmental contamination (Posten, 2009). However, large-scale PBRs have three major disadvantages that make them uneconomical for low-cost products: (i) low light dispersion efficiency at large operational volumes; (ii) light penetration limitation by algal biofilm formed in PBR surfaces; and (iii) high initial investment in infrastructure and requirement of continuous maintenance.

Heterotrophic cultivation of microalgae is very similar to heterotrophic bacterial cultivation and can be carried out in a conventional bacterial fermentor without any illumination under dark conditions. Compared to photoautotrophic cultivation, heterotrophic cultivation in well-controlled fermentor eliminates the requirement on light and is performed under strict axenic conditions. Thus, the major problem faced in photoautotrophic cultivation is overcome and the biomass concentration and productivities are significantly increased, and the harvesting cost could be reduced greatly (Nagarajan et al., 2018). However, heterotrophic cultures also have some drawbacks as follows: (i) few algae species capable of heterotrophic growth; (ii) high costs of nutrition (e.g. glucose) and power consumption; (iii) inability to produce light-induced metabolites; and (iv) release of large amount of CO_2. In this work, we review the latest developments in the heterotrophic cultivation of microalgae, focusing on the introduction of reported microalgal species capable of heterotrophic cultivation, key nutritional and environmental factors affecting microalgae growth, culture modes applied in high-cell-density cultivation, and strategies for enhanced target metabolites production.

2. Microalgal Species Capable of Heterotrophic Growth

It has been estimated that there are about 200,000–800,000 species of microalgae, of which over 40,000 species have already been identified and classified in multiple major groupings as follows: Cyanophyceae, Chlorophyceae, Bacillariophyceae, Xanthophyceae, Chrysophyceae, Rhodophyceae, Phaeophyceae, dinoflagellates Dinophyceae, Prasinophyceae and Eustigmatophyceae (Hu et al., 2008; Han et al., 2015). Photoautotroph and heterotrophy are two major nutrition forms for algal growth. By far, most microalgae belong to photoautotrophs and they only require inorganic mineral ions and minimal quantities of essential organic compounds (e.g. vitamins) for cell growth (Grobbelaar, 2003). Heterotrophic microalgae obtain their substrate and energy from organic compounds produced by other organisms. A number of microalgal species from five groupings have been reported to be capable of growing heterotrophically, and most of them belong to the grouping of green algae (Chlorophyceae) (Table 6.1). Only a very few heterotrophic microalgae (e.g. *Nitzschia alba*) are obligate heterotrophs, and most of them are capable of both autotrophic and heterotrophic growth (Hu et al., 2018).

3. Factors Affecting Heterotrophic Growth of Microalgae

3.1. Nutritional factors

3.1.1. Carbon source

Glucose and acetate are the two most commonly used carbon sources for heterotrophic culture of microalgae. It appears that glucose might be considered a preferred substrate and microalgae will require a lag period to develop the specific transport systems for the uptake of other substrates (Perez-Garcia et al., 2011). *Chlorella* species can grow on various carbon substrates including sodium acetate, fructose, glucose, glycerol, sucrose and acetate (Qiao and Wang, 2009; Gao et al., 2010; Heredia-Arroyo et al., 2010; Yeh and Chang, 2012). The optimal type of carbon source was usually different among different algal species. For instance, sucrose was found to be the optimal carbon source for *Chlorella* sp. Y8-1 growth among four different carbon

Table 6.1. Representative heterotrophic microalgal species and their maximum biomass concentrations under high-cell-density cultivation.

Algal group	Species	Target product	Maximum biomass conc. (g L^{-1})	References
Chlorophyceae	Chlorella sorokiniana	Protein	232	Xu et al. (2021)
	Chlorella sorokiniana	Biomass	271	Jin et al. (2021)
	Chlorella regularis	Phytochemicals	90	Sansawa and Endo (2004)
	Chlorella vulgaris	Biomass	117.2	Doucha and Livansky (2012)
	Chlorella protothecoides	Lipid	64	Ceron-Garcia et al. (2013)
	Chlorella pyrenoidosa	Biomass	116.2	Wu and Shi (2007)
	Scenedesmus acuminatus	Lipid	287	Jin et al. (2020)
	Chlamydomonas reinhardtii	Biomass	25.44	Zhang et al. (2019)
	Euglena gracilis	α-tocopherol	39.5	Ogbonna et al. (1998)
	Haematococcus pluvialis	Astaxanthin	26	Wan et al. (2015)
	Chromochloris zofingiensis	Astaxanthin	235.4	Chen et al. (2022)
	Chlorococcum sp.	Ketocarotenoid	18	Zhang and Lee (2001)
	Neochloris oleoabundans	Lipid	14.2	Morales-Sanchez et al. (2013)
Bacillariophyceae	Nitzschia laevis	EPA	40	Wen and Chen (2002)
	Nitzschia alba	EPA	48	Kyle David and Gladue (1992)
Phodophyceae	Galdieria sulphuraria	Phycocyanin	113.2	Graverholt and Eriksen (2007)
Chrysophyceae	Poterioochromonas malhamensis	β-1,3-glucan	32.8	Ma et al. (2021)
Dinophyceae	Cryptecodinium cohnii	DHA	83	de Swaaf et al. (2003a)

sources (fructose, glucose, glycerol and sucrose) (Lin and Wu, 2015). The algal strains of *Nannochloropsis* sp., *Rhodomonas reticulate* and *Cyclotella cryptica* prefer glycerol over glucose or acetate (Wood et al., 1999). In addition, ethanol could also be well utilized by some algal species. For instance, *Euglena* could use ethanol as its sole carbon source and a high biomass concentration of 39.5 g L^{-1} was achieved by a fed-batch culture with ethanol (Ogbonna et al., 1998).

Beside the type of carbon source, the substrate concentration also should be well controlled during fermentation due to their inhibitory effect on cell growth under high concentration level. Maintaining a low glucose concentration is essential for sustaining a higher specific growth rate. For example, a glucose concentration of 2.5 g L^{-1} is optimum for the growth of *C. saccharophila*, whereas inhibition occurred at the concentration >25 g L^{-1} (Tan and Johns, 1991). For an eicosapentaenoic acid-producing diatom *Nitschia laevis*, the growth rate decreased with the increase of glucose concentration from 1 to 40 g L^{-1} (Wen and Chen, 2000). The highest average biomass productivity (1.21 g·L^{-1}·h^{-1}) of *Scenedesmus acuminatus* and glucose conversion efficiency (60.09%) were achieved by maintaining the glucose at a relatively low constant concentration below 5 g L^{-1} (Jin et al., 2020). Similarly, several microalgae can use acetate as their sole carbon source as long as the acetate concentration is maintained at a low level (Perez-Garcia et al., 2011). Although *G. sulphuraria* could tolerate glucose and fructose up to 166 g L^{-1} (0.9 M) without negative effects on the specific growth rate, growth was completely inhibited when the total substrate concentration exceeded 1-2 M (Schmidt et al., 2005).

3.1.2. Nitrogen source

After carbon, nitrogen is the most important nutrient contributing to the biomass produced. Nitrogen, which generally accounts for about 7-10% of cell dry weight, is an essential constituent of all structural and functional proteins in algal cells. A wide variety of nitrogen sources, such as nitrate, ammonia and urea, can be used as a sole source of nitrogen for sustaining algal growth and reproduction. Ammonium is the most preferred and most energetically efficient nitrogen source (Perez-Garcia et al., 2011). It was found that among three common N sources (ammonium chloride, potassium nitrate and urea), ammonium chloride was the optimal nitrogen source for *C. sorokiniana* CMBB276 growth when the same pH was maintained (Xu et al., 2021). Besides cellular growth, it was found that the cellular protein content of *C. sorokiniana* CMBB276 was also affected by the type of nitrogen source. Under the same C:N ratio, higher protein content could be obtained when culturing *C. sorokiniana* CMBB276 with ammonium as nitrogen source (Xu et al., 2021). However, *C. protothecoides*, *N. laevis* and *P. tricornutum* exhibit a preference for nitrate or urea over ammonium (Shi et al., 2000; Lee and Lee, 2002; Wen and Chen, 2003).

3.1.3. Silicon

Silicon is a major limiting nutrient for diatom growth and hence is a controlling factor in primary productivity (Martin-Jezequel et al., 2000). To most diatoms, silicon is essential for cell growth because the siliceous cell wall formation was tightly coupled with cell cycle (Martin-Jezequel et al., 2000). Depending on the types of diatoms, silicate limitation could inhibit cell growth, distort cell morphology and change lipid production (Jiang et al., 2015). It was found that high silicate addition stimulated biomass growth and accumulation of fucoxanthin and EPA in the marine diatom *Nitzschia laevis* (Mao et al., 2020). Moreover, the combination of different glucose and silicate concentration affects the growth of *N. laevis* and eicosapentaenoic acid (EPA) production. The highest cell dry weight was obtained at a medium concentration of glucose (20 g L^{-1}) and silicate (32 mg L^{-1}), while the highest specific growth rate was obtained at a relatively low glucose concentration (5 g L^{-1}) and high silicate concentration (32-64 mg L^{-1}). Higher EPA content was obtained at low silicate concentration below 16 mg L^{-1} or high glucose concentration above 20 g L^{-1} (Wen and Chen, 2000).

3.1.4. Vitamins

Vitamins are essential for all organisms because they provide the precursors to enzyme cofactors important for metabolism. Animals must obtain these organic micronutrients in their diet, but plants and microorganisms generally synthesize *de novo* the cofactors they need (Smith et al., 2007). Many algae exhibit vitamin auxotrophy, which is the inability to synthesize an organic nutrient essential for growth. Over half of all microalgal species require an exogenous supply of vitamin B12, while just over 20% require vitamin B1 and a smaller proportion (5%) require biotin (Croft et al., 2006). Land plants and fungi neither synthesize nor require vitamin B12 because they do not contain methylmalonyl-CoA mutase, and have an alternative B12-independent methionine synthase (METE). Many studies described algal species that required different combinations of three B vitamins: vitamin B12, vitamin B1 and vitamin B7. It was found that the *P. malhamensis* cells grew slowly in the vitamin-free growth medium, and the addition of VB1 or VB12 alone improved the growth rate of *P. malhamensis* cells. A better growth-promoting effect was observed when both vitamins were added into the inorganic-nitrogen culture of *P. malhamensi* (Ma et al., 2021).

3.1.5. C:N ratio

The C:N ratio of a culture medium is thought to be one of the most critical nutritional factors affecting microbial growth (Huang et al., 2010). The C:N ratio of 12 sustained the highest growth rate of *S. acuminatus* GT-2 and maximum final dry biomass concentration of 220 g L^{-1} (Jin et al., 2020). The effects of different mole C:N ratios (6, 12, 18 and 24) on the growth of *C. sorokiniana* CMBB276 were investigated. Among this C:N range, a low C:N ratio of 6 did not benefit to the growth of *C. sorokiniana* CMBB276, and both its cellular growth rate and maximum biomass concentration were the lowest. It was found that culturing *C. sorokiniana* CMBB276 under the C:N ratio of 18 could achieve the fastest growth rate and the highest biomass concentration (Xu et al., 2021).

3.2. Environmental factors

3.2.1. Temperature

Temperature is an important environmental parameter affecting microalgal growth and the optimal temperature differs between microalgal species. The optimal growth temperature for most industrial microalgal strains is commonly between 20°C and 30°C (Barten et al., 2020). Chlorophyta could grow at a wide range of temperatures from 15°C to 35°C, with the optimum temperature of 20–30°C, mainly around 25°C; whereas cyanobacteria had a higher tolerance to high temperature than chlorophyta, the optimum temperature for cyanobacteria growth was between 30°C and 35°C (Yu et al., 2014). The red alga *Galdieria sulphuraria* could survive under a high temperature of 56°C and its optimal growth temperature under heterotrophic cultivation was 42°C (Schmidt et al., 2005). When the culture temperature was increased from 25°C to 30°C, *S. acuminatus* GT-2 grew more rapidly, reaching the highest cell density of 223 g L^{-1} at 30°C. Further increase in growth occurred when raising the temperature to 35°C and 37°C during the first 168 h, after which the cellular growth leveled off and then declined. When the temperature was increased to 40°C, little growth was observed. Although it took less time to achieve the highest biomass yield at 35°C or 37°C as compared to 30°C, a longer period of time of imposing the maximum aeration rate (600 rpm) was required to maintain the DO level at 40%, which indicated more energy input (Jin et al., 2020).

3.2.2. pH

Microalgae have an optimum pH for their best growth. Changes in pH can lead to denaturing of enzymes and other proteins and can interfere with pumping of ions at the cell membrane. The green alga *S. acuminatus* GT-2 favored a weak acidic and neutral pH environment (Jin et al., 2020). Under the optimum pH of 6, the highest biomass concentration (205.4 g L^{-1}) was achieved. *C. sorokiniana* GT–1 cells also prefer a weak acidic environment and can even grow well at pH 5, better than that at a weak alkaline pH 7.5. Culturing *C. sorokiniana* GT-1 cells at pH 6 resulted in the fastest biomass growth (Jin et al., 2021). *Euglena gracilis* could grow under a wide pH range from 2 to 7. Under heterotrophic cultivation, optimal pH for paramylon synthesis in the range of 6.5–8.0 has been reported, whereas an optimal pH for biomass production should be found in the range of 3.0–5.0 (Santek et al., 2009). *G. sulphuraria* could grow under extreme low pH between 0.05 and 3, and the optimal growth pH was found at pH 1–2 (Schmidt et al., 2005).

3.2.3. Dissolved oxygen (DO)

Oxygen supply is a key factor in heterotrophic cultivation of microalgae. It appears that the high-cell-density cultivation of microalgae have different oxygen requirement and growth metabolic characteristics compared with other industrial microorganisms. For example, it was extremely difficult to maintain the DO level when the cell density of *E. coli* or yeast reached above 50 g L^{-1} due to their fast oxygen consumption (Jin et al., 2021). The single pure oxygen or mixed pure oxygen/air supply mode were commonly applied to improve the oxygen supply in their fermentation processes. By contrast, the requirement of oxygen by green alga *C. sorokiniana* GT-1 cells was not so intensive as compared to that of the industrial *E. coli* or yeasts. Additionally, it seemed that *C. sorokiniana* GT-1 cells were less sensitive to the low oxygen environment. In response to low oxygen stress, *E. coli* and yeast are easy to produce and accumulate inhibitory by-products such as ethanol or acetate by quickly shifting their metabolic pathway from TCA cycle to fermentation. However, these metabolic by-products of ethanol or acetate were undetectable during the cultivation of *C. sorokiniana* GT-1, even though the long-term maintenance of low DO (< 5%) for six days (Jin et al., 2021). Even though the DO decreased to the extremely low level (~0), continuous glucose consumption and biomass growth could also be observed. It was found that maintaining the DO above 20% could achieve a high *Chlorella* biomass concentration (> 250 g L^{-1}) and productivity (> 70 g L^{-1} d^{-1}) simultaneously (Jin et al., 2021). However, for the heterotrophic cultivation of *Euglena*, adverse effects on both cell growth and α-tocopherol production were found at high DO level, and the optimum DO was below 0.8 ppm (Ogbonna et al., 1998).

4. Culture Strategies Applied in High-cell-density Cultivation of Microalgae

4.1. Fed-batch cultivation

Fed-batch culture by controlling the nutrient feeding is one of the most popular methods to achieve high cell density, which is often necessary for high yield and productivity of the desired product (Kim et al., 2004). To achieve the high-cell-density cultivation of microalgae, various feeding modes have been applied based on their differences in substrate utilization.

4.1.1. Pulsed feeding mode

Pulsed feeding strategy is a traditional and simple feeding method by repeated addition of concentrated carbon source containing medium when the carbon source (e.g. glucose) dropped below a defined concentration. The substrate concentration under pulsed feeding could be maintained within a rough range below inhibitory level. Therefore, pulsed feeding strategy is appropriate to the cultivation of algal species whose growth is inhibited by very high substrate concentrations. This feeding strategy has been widely employed in the high-cell-density cultivation of various microalgae such as *Chlorella* sp., *E. gracilis* and *G. sulphuraria* and many microalgae have achieved high biomass concentrations by using this control strategy (Bumbak et al., 2011). Recently, ultra-high *Chlorella* biomass concentrations of 271 and 247 g L^{-1} were achieved in 7.5 L bench-scale and 1000 L pilot-scale fermenters, respectively, by adopting the pulsed feeding strategy and controlling glucose in the range of 0–30 g L^{-1} (Jin et al., 2021).

4.1.2. Stepwise constant feeding mode

Different from pulsed feeding, the substrate (e.g. glucose) concentration under stepwise constant feeding could be finely controlled at a relatively stable level by more frequently sampling, measuring and adjusting feeding rate. Jin et al. (2020) compared the difference of traditional pulsed feeding and stepwise constant feeding strategies on the high-cell-density cultivation of *S. acuminatus* GT-2. Under stepwise constant feeding, the glucose could be well controlled within different range concentrations with less fluctuation (0–5, 5–10 and 15–20 g L^{-1}). As a result, the highest biomass concentration reached 273.5 g L^{-1} when the glucose concentrations were controlled at 0–5 g L^{-1}, which was 1.2-fold higher than that with the pulsed feeding strategy (226 g L^{-1}). Moreover, the biomass yields on glucose obtained by using the stepwise constant feeding strategy were higher than that of the pulsed feeding mode, indicating that utilization of carbon source by *S. acuminatus* cells was more efficient when constant glucose concentrations were well controlled (Jin et al., 2020).

4.1.3. pH-stat feeding mode

The pH-stat is a simple indirect feedback control scheme that couples nutrient feeding with measurement of pH and it is based on the fact that pH rises when the principal carbon source is depleted (Kim et al., 2004). Feeding is started when pH rises above a preset upper limit caused by the depletion of nutrient substrate. Generally, the substrate concentration under pH-stat feeding could be controlled at an extremely low level without inhibitory metabolic by-product production. The pH-stat feeding strategy has been widely applied in the fed-batch process of acetate-utilizing microalgae. By using this feeding strategy, high biomass concentrations of 109, 25.44 and 26 g L^{-1} have been achieved by *Crypthecodinium cohnii*, *Chlamydomonas reinhardtii* and *Haematococcus pluvialis*, respectively (de Swaaf et al., 2003b; Wan et al., 2015; Zhang et al., 2019).

4.1.4. DO-stat feeding mode

Similar to pH-stat feeding strategy, DO-stat fed-batch cultivation is another feedback control strategy which is based on the change of DO signal. A rapid increase of DO occurs when the culture respiration rate decreases due to the lack of substrates. Thus, the rapid rise in DO is used as an indicator for substrate feeding and the feeding is started once the DO rises above the set value. A high biomass concentration of 83 g L^{-1} was obtained by adopting DO-stat fed-batch cultivation of the docosahexaenoic-acid-producing marine alga *C. cohnii* on ethanol (de Swaaf et al., 2003a). DO-stat cultivation has also been successfully applied in the high-cell-density culture of the phycocyanin-producing red alga *G. sulphuraria* and achieved high biomass dry weight of 80–120 g L^{-1} (Schmidt et al., 2005).

4.2. Repeated-batch cultivation

Repeated-batch cultivation involves the inoculation of a part or all of microbial cells into the next batch of fresh medium. Compared with fed-batch cultivation, repeated-batch cultivation lasts longer, requires less labor, and could reduce the inhibition of the process by metabolites (Li et al., 2020). Repeated-batch operation was thought to be a simple and economic way and particularly suitable for long-term heterotrophic cultivation of *E. gracilis* for paramylon production, leading to biomass concentrations in excess of 20 g L^{-1} with a consistent paramylon mass fraction of about 75% (Santek et al., 2010).

4.3. Continuous cultivation

Continuous culture systems have been widely used to culture microbes for industrial and research purposes. In a continuous culture system, the nutrients are supplied to the culture at a constant rate and an equal volume of culture is removed simultaneously. Continuous cultivation could produce microalgae with more predictable quality, increase the reliability of the system and reduce the need for labor. A novel two-stage continuous cultivation strategy was proposed to increase the lipid productivity by *Chlorella* sp. FC2 IITG. By using this culture strategy, the obtained maximum biomass productivity of 92.7 g L^{-1} d^{-1} and lipid productivity of 9.76 g L^{-1} d^{-1} were significantly higher than other reported bioprocesses for microalgal biodiesel production (Palabhanvi et al., 2017). Further, the continuous production of lipid-rich biomass offers additional advantages including reducing unproductive time and costs of harvesting and biomass processing (Palabhanvi et al., 2017).

4.4. Perfusion cultivation

Perfusion culture systems are continuous culture systems that are modeled after *in vivo* blood-flow systems. In perfusion culture systems, the filtered medium is pumped to a product reservoir and fresh medium is pumped into the culture vessel, where the rate of addition and removal can be regulated depending on the cell concentration in the vessel. Perfusion cultivation could help alleviate the accumulation of inhibitory metabolic by-products during high-cell-density cultivation. A large number of extracellular metabolic products have been identified in the cultures of microalgae (Eppley, 1977; Srivastava et al., 1999). For the production of EPA by the diatom *Nitzschia laevis*, the perfusion culture was applied to eliminate glucose limitation and by-product inhibition simultaneously, and a high biomass concentration of 40 g L^{-1} and an ultra-high EPA yield of 1.11 g L^{-1} were achieved (Wen and Chen, 2002).

5. Combined Culture Strategies Based on Heterotrophic High-cell-density Cultivation

5.1. Sequential heterotrophic-autotrophic cultivation for enhanced target metabolites production

Sequential heterotrophic-autotrophic cultivation mode has been widely applied by combining the advantage of heterotrophic cultivation on achieving high biomass concentration and autotrophic cultivation on inducing the accumulation of light-dependent metabolites. For example, a high α-tocopherol productivity of 100 mg L^{-1} h^{-1} was achieved by sequential heterotrophic/photoautotrophic cultivation of *Euglena gracilis*, which was 9.5 and 4.6 times higher than those obtained under photoautotrophic and heterotrophic cultivation, respectively (Ogbonna et al., 1999). Moreover, the exhaust gas from the heterotrophic phase was used for aeration of the autotrophic phase in order to reduce the CO$_2$ emission into the atmosphere and about 27% of the CO$_2$ produced in the heterotrophic phase was re-utilized in the autotrophic phase (Ogbonna et al., 1997). Recently, a two-stage hetero-photoautotrophic cultivation process was established for production of fucoxanthin by the marine diatom *Nitzschia laevis*, and a superior fucoxanthin productivity of 16.5 mg L^{-1} d^{-1} was achieved (Lu et al., 2018).

5.2. Two-stage high-cell-density cultivation and stress stimulation for enhanced target metabolites production

Besides light, various inducers or stress factors could also be applied to improve the cellular accumulation of certain metabolites in microalgae under dark heterotrophic cultivation. For example, the application of peroxynitrite as an inducer increased astaxanthin accumulation from 9.9 to 11.8 mg L^{-1} (Ip and Chen, 2005b). Hydroxyl radicals generated by hydrogen peroxide were also effective in inducing carotenogenesis in *C. zofingensis*, where astaxanthin accumulation increased from 9.9 to 12.58 mg L^{-1} (Ip and Chen, 2005a). However, the addition of inducers or the stress condition has adverse effect on cell growth. Therefore, a two-stage heterotrophic process was usually applied to improve metabolites production. Recently, a two-stage heterotrophic cultivation process was developed for *C. zofingiensis* with fist optimal growth and subsequent induction by combining the addition of phytochrome GA3 with high C:N and high salinity, which enabled achieving the highest biomass yield of 235.4 g L^{-1} and astaxanthin content 0.144% of DCW (Chen et al., 2022).

6. Technoeconomic (TE) Analysis for High-cell-density Cultivation in Fermentor

Currently, heterotrophic cultivation is economically feasible only for a few high-value products including polyunsaturated fatty acids (PUFAs), pigments, antioxidants, polysaccharides, proteins (Morales-Sanchez et al., 2017). TE analysis has recently been conducted for the high-cell-density cultivation of *S. acuminatus* GT-2 and *C. sorokiniana* GT-1 for biomass production (Jin et al., 2020, 2021). Ultra-high-productivity fermentation is thought to be a prerequisite for cost-effective production of microalgae biomass. When the biomass concentration of *C. sorokiniana* GT-1 reached 200 g L^{-1}, the fermentation cost was $1601.27 and it was economically feasible for food and feed applications (Jin et al., 2021). It was found that the costs of both equipment investment and power consumption could be significantly reduced by elevated biomass concentration (Jin et al., 2020). The cost of nutrient accounts for more than 60% of the total fermentation cost. Thus, the fermentation cost could be further reduced by increasing the biomass yield on glucose and/or replacing glucose with other low-cost carbon sources.

7. Challenges and Future Perspectives

Heterotrophic cultivation of microalgae has many advantages for large-scale cultivation, but this culture mode is also confronted with many challenges. First, the metabolic potential of microalgae to produce products of commercial interest in heterotrophic cultivation relies greatly on extensive screening and evaluation of microalgae capable of utilizing organic carbon sources. Second, another major problem of heterotrophic cultivation is the difficulty in the maintenance of axenic cultures of microalgae especially for long-term fermentation cultivation, as bacteria have much slower doubling times and they can easily grow and outnumber microalgae. Other concerns regarding heterotrophic growth is the inability to produce light-induced metabolites like pigments (Perez-Garcia et al., 2011). With current developments in genetic tools and techniques, some obligate photoautotrophs have been transformed to heterotrophy to achieve high-cell-density cultivation. Meanwhile, like *E. coli* or yeast, some excellent microalgal species that have achieved ultra-high-cell-density cultivation (e.g. *Chlorella* and *Scenedesmus*) might also be developed as chassis cells for effective production of therapeutic and industrially relevant recombinant proteins.

In conclusion, cultivation of microalgae that are primarily photosynthetic under heterotrophic dark conditions for production of economically useful metabolites or technological processes is a tempting option, leading to significant reductions in production cost. Heterotrophic cultivation of microalgae is a niche of microalgae research field and its potential is limitless.

Conflicts of interest

The authors declare that they have no conflicts of interest.

Acknowledgments

This work was supported by the Strategic Priority Research Program of the Chinese Academy of Sciences (Grant No. XDB31040103), the National Natural Science Foundation of China (Grant No. 31870756) and the Chinese Academy of Sciences (Grant QYZDY-SSW-SMC00).

References

Barten, R.J.P., Wijffels, R.H. and Barbosa, M.J. (2020) Bioprospecting and characterization of temperature tolerant microalgae from Bonaire. *Algal Research* 50, 102008.

Bumbak, F., Cook, S., Zachleder, V., Hauser, S. and Kovar, K. (2011) Best practices in heterotrophic high-cell-density microalgal processes: achievements, potential and possible limitations. *Applied Microbiology and Biotechnology* 91(1), 31–46.

Ceron-Garcia, M.C., Macias-Sanchez, M.D., Sanchez-Miron, A., Garcia-Camacho, F. and Molina-Grima, E. (2013) A process for biodiesel production involving the heterotrophic fermentation of *Chlorella protothecoides* with glycerol as the carbon source. *Applied Energy* 103, 341–349.

Chen, Q.H., Chen, Y., Xu, Q., Jin, H., Hu, Q. *et al.* (2022) Effective two-stage heterotrophic cultivation of the unicellular green microalga *Chromochloris zofingiensis* enabled ultrahigh biomass and astaxanthin production. *Frontiers in Bioengineering and Biotechnology* 10, 834230.

Croft, M.T., Warren, M.J. and Smith, A.G. (2006) Algae need their vitamins. *Eukaryotic Cell* 5(8), 1175–1183.

de Swaaf, M.E., Pronk, J.T. and Sijtsma, L. (2003a) Fed-batch cultivation of the docosahexaenoic-acid-producing marine alga *Crypthecodinium cohnii* on ethanol. *Applied Microbiology and Biotechnology* 61(1), 40–43.

de Swaaf, M.E., Sijtsma, L. and Pronk, J.T. (2003b) High-cell-density fed-batch cultivation of the docosahexaenoic acid producing marine alga *Crypthecodinium cohnii*. *Biotechnology and Bioengineering* 81(6), 666–672.

Doucha, J. and Livansky, K. (2012) Production of high-density Chlorella culture grown in fermenters. *Journal of Applied Phycology*, 24(1), 35–43.

Eppley, R.W. (1977) The growth and culture of diatoms. In: Werner, D. (ed.) *The Biology of Diatoms*. University of California Press, Berkeley and Los Angeles, pp. 24–64.

Fernandez, F.G.A., Sevilla, J.M.F. and Grima, E.M. (2013) Photobioreactors for the production of microalgae. *Reviews in Environmental Science and Bio-Technology* 12(2), 131–151.

Gao, C.F., Zhai, Y., Ding, Y. and Wu, Q.Y. (2010) Application of sweet sorghum for biodiesel production by heterotrophic microalga *Chlorella protothecoides*. *Applied Energy* 87(3), 756–761.

Graverholt, O.S. and Eriksen, N.T. (2007) Heterotrophic high-cell-density fed-batch and continuous-flow cultures of *Galdieria sulphuraria* and production of phycocyanin. *Applied Microbiology and Biotechnology* 77(1), 69–75.

Grobbelaar, J.U. (2003) Algal nutrition – mineral nutrition. In: Richmond, A. (ed.) *Handbook of Microalgal Culture*. Wiley online, pp. 95–115.

Han, S.F., Jin, W.B., Tu, R.J. and Wu, W.M. (2015) Biofuel production from microalgae as feedstock: current status and potential. *Critical Reviews in Biotechnology* 35(2), 255–268.

Heredia-Arroyo, T., Wei, W. and Hu, B. (2010) Oil accumulation via heterotrophic/mixotrophic *Chlorella protothecoides*. *Applied Biochemistry and Biotechnology* 162(7), 1978–1995.

Hu, J., Nagarajan, D., Zhang, Q., Chang, J.-S. and Lee, D.-J. (2018) Heterotrophic cultivation of microalgae for pigment production: a review. *Biotechnology Advances* 36(1), 54–67.

Hu, Q., Sommerfeld, M., Jarvis, E., Ghirardi, M., Posewitz, M. *et al.* (2008) Microalgal triacylglycerols as feedstocks for biofuel production: perspectives and advances. *Plant Journal* 54(4), 621–639.

Huang, G., Chen, F., Wei, D., Zhang, X. and Chen, G. (2010) Biodiesel production by microalgal biotechnology. *Applied Energy* 87(1), 38–46.

Ip, P.F. and Chen, F. (2005a) Employment of reactive oxygen species to enhance astaxanthin formation in *Chlorella zofingiensis* in heterotrophic culture. *Process Biochemistry* 40(11), 3491–3496.

Ip, P.F. and Chen, F. (2005b) Peroxynitrite and nitryl chloride enhance astaxanthin production by the green microalga *Chlorella zofingiensis* in heterotrophic culture. *Process Biochemistry* 40(11), 3595–3599.

Jiang, Y., Laverty, K.S., Brown, J., Brown, L., Chagoya, J. *et al.* (2015) Effect of silicate limitation on growth, cell composition, and lipid production of three native diatoms to Southwest Texas desert. *Journal of Applied Phycology* 27(4), 1433–1442.

Jin, H., Zhang, H., Zhou, Z., Li, K., Hou, G. *et al.* (2020) Ultrahigh-cell-density heterotrophic cultivation of the unicellular green microalga *Scenedesmus acuminatus* and application of the cells to photoautotrophic culture enhance biomass and lipid production. *Biotechnology and Bioengineering* 117(1), 96–108.

Jin, H., Chuai, W., Li, K., Hou, G., Wu, M. *et al.* (2021) Ultrahigh-cell-density heterotrophic cultivation of the unicellular green alga *Chlorella sorokiniana* for biomass production. *Biotechnology and Bioengineering* 118(10), 4138–4151.

Kim, B.S., Lee, S.C., Lee, S.Y., Chang, Y.K. and Chang, H.N. (2004) High cell density fed-batch cultivation of *Escherichia coli* using exponential feeding combined with pH-stat. *Bioprocess and Biosystems Engineering* 26(3), 147–150.

Kyle David, J. and Gladue, R. (1992) *Eicosapentaenoic Acid-containing Oil and Methods for Its Production*. Martek Biosciences Corp., Columbia, Maryland.

Lee, K. and Lee, C.G. (2002) Nitrogen removal from wastewaters by microalgae without consuming organic carbon sources. *Journal of Microbiology and Biotechnology* 12(6), 979–985.

Li, G., Chen, Z., Chen, N. and Xu, Q. (2020) Enhancing the efficiency of L-tyrosine by repeated batch fermentation. *Bioengineered* 11(1), 852–861.

Lin, T.S. and Wu, J.Y. (2015) Effect of carbon sources on growth and lipid accumulation of newly isolated microalgae cultured under mixotrophic condition. *Bioresource Technology* 184, 100–107.

Lu, X., Sun, H., Zhao, W., Cheng, K.-W., Chen, F. et al. (2018) A hetero-photoautotrophic two-stage cultivation process for production of fucoxanthin by the marine diatom *Nitzschia laevis*. *Marine Drugs* 16(7), 219.

Ma, M., Li, Y., Chen, J., Wang, F., Yuan, L. et al. (2021) High-cell-density cultivation of the flagellate alga *Poterioochromonas malhamensis* for biomanufacturing the water-soluble β-1,3-glucan with multiple biological activities. *Bioresource Technology* 337, 125447.

Mao, X., Chen, S.H.Y., Lu, X., Yu, J. and Liu, B. (2020) High silicate concentration facilitates fucoxanthin and eicosapentaenoic acid (EPA) production under heterotrophic condition in the marine diatom *Nitzschia laevis*. *Algal Research* 52, 102086.

Martin-Jezequel, V., Hildebrand, M. and Brzezinski, M.A. (2000) Silicon metabolism in diatoms: implications for growth. *Journal of Phycology* 36(5), 821–840.

Morales-Sanchez, D., Tinoco-Valencia, R., Kyndt, J. and Martinez, A. (2013) Heterotrophic growth of *Neochloris oleoabundans* using glucose as a carbon source. *Biotechnology for Biofuels* 6, 100.

Morales-Sanchez, D., Martinez-Rodriguez, O.A. and Martinez, A. (2017) Heterotrophic cultivation of microalgae: production of metabolites of commercial interest. *Journal of Chemical Technology and Biotechnology* 92(5), 925–936.

Nagarajan, D., Lee, D.-J. and Chang, J.-S. (2018) Heterotrophic microalgal cultivation. In: Liao, Q., Chang, J.-S., Herrmann, C. and Xia, A. (eds) *Bioreactors for Microbial Biomass and Energy Conversion*. Springer Singapore, pp. 117–160.

Ogbonna, J.C., Masui, H. and Tanaka, H. (1997) Sequential heterotrophic/autotrophic cultivation – an efficient method of producing *Chlorella* biomass for health food and animal feed. *Journal of Applied Phycology* 9(4), 359–366.

Ogbonna, J.C., Tomiyama, S. and Tanaka, H. (1998) Heterotrophic cultivation of *Euglena gracilis* Z for efficient production of alpha-tocopherol. *Journal of Applied Phycology* 10(1), 67–74.

Ogbonna, J.C., Tomiyama, S. and Tanaka, H. (1999) Production of α-tocopherol by sequential heterotrophic-photoautotrophic cultivation of *Euglena gracilis*. In: Osinga, R., Tramper, J., Burgess, J.G. and Wijffels, R.H. (eds) *Progress in Industrial Microbiology*, Vol. 35. Elsevier, pp. 213–221.

Palabhanvi, B., Muthuraj, M., Kumar, V., Mukherjee, M., Ahlawat, S. et al. (2017) Continuous cultivation of lipid rich microalga *Chlorella* sp. FC2 IITG for improved biodiesel productivity via control variable optimization and substrate driven pH control. *Bioresource Technology* 224, 481–489.

Perez-Garcia, O., Escalante, F.M.E., de-Bashan, L.E. and Bashan, Y. (2011) Heterotrophic cultures of microalgae: metabolism and potential products. *Water Research* 45(1), 11–36.

Posten, C. (2009) Design principles of photo-bioreactors for cultivation of microalgae. *Engineering in Life Sciences* 9(3), 165–177.

Qiao, H.J. and Wang, G. (2009) Effect of carbon source on growth and lipid accumulation in *Chlorella sorokiniana* GXNN01. *Chinese Journal of Oceanology and Limnology* 27(4), 762–768.

Sansawa, H. and Endo, H. (2004) Production of intracellular phytochemicals in *Chlorella* under heterotrophic conditions. *Journal of Bioscience and Bioengineering* 98(6), 437–444.

Santek, B., Felski, M., Friehs, K., Lotz, M. and Flaschel, E. (2009) Production of paramylon, a beta-1,3-glucan, by heterotrophic cultivation of *Euglena gracilis* on a synthetic medium. *Engineering in Life Sciences* 9(1), 23–28.

Santek, B., Felski, M., Friehs, K., Lotz, M. and Flaschel, E. (2010) Production of paramylon, a beta-1,3-glucan, by heterotrophic cultivation of *Euglena gracilis* on potato liquor. *Engineering in Life Sciences* 10(2), 165–170.

Schmidt, R.A., Wiebe, M.G. and Eriksen, N.T. (2005) Heterotrophic high cell-density fed-batch cultures of the phycocyanin-producing red alga *Galdieria sulphuraria*. *Biotechnology and Bioengineering* 90(1), 77–84.

Shi, X.M., Zhang, X.W. and Chen, F. (2000) Heterotrophic production of biomass and lutein by *Chlorella protothecoides* on various nitrogen sources. *Enzyme and Microbial Technology* 27(3–5), 312–318.

Smith, A.G., Croft, M.T., Moulin, M. and Webb, M.E. (2007) Plants need their vitamins too. *Current Opinion in Plant Biology* 10(3), 266–275.

Srivastava, V.C., Manderson, G.J. and Bhamidimarri, R. (1999) Inhibitory metabolites production by the cyanobacterium *Fischerella muscicola*. *Microbiological Research* 153(4), 309–317.

Tan, C.K. and Johns, M.R. (1991) Fatty-acid production by heterotrophic *Chlorella saccharophila*. *Hydrobiologia* 215(1), 13–19.

Wan, M.X., Zhang, Z., Wang, J., Huang, J.K., Fan, J.H. et al. (2015) Sequential heterotrophy-dilution-photoinduction cultivation of *Haematococcus pluvialis* for efficient production of astaxanthin. *Bioresource Technology* 198, 557–563.

Wen, Z.Y. and Chen, F. (2000) Heterotrophic production of eicosapentaenoid acid by the diatom *Nitzschia laevis*: effects of silicate and glucose. *Journal of Industrial Microbiology & Biotechnology* 25(4), 218–224.

Wen, Z.Y. and Chen, F. (2002) Perfusion culture of the diatom *Nitzschia laevis* for ultra-high yield of eicosapentaenoic acid. *Process Biochemistry* 38(4), 523–529.

Wen, Z.Y. and Chen, F. (2003) Heterotrophic production of eicosapentaenoic acid by microalgae. *Biotechnology Advances* 21(4), 273–294.

Wood, B.J.B., Grimson, P.H.K., German, J.B. and Turner, M. (1999) Photoheterotrophy in the production of phytoplankton organisms. *Journal of Biotechnology* 70(1–3), 175–183.

Wu, Z. and Shi, X. (2007) Optimization for high-density cultivation of heterotrophic *Chlorella* based on a hybrid neural network model. *Letters in Applied Microbiology* 44(1), 13–18.

Xu, L., Weathers, P.J., Xiong, X.R. and Liu, C.Z. (2009) Microalgal bioreactors: challenges and opportunities. *Engineering in Life Sciences* 9(3), 178–189.

Xu, Q., Hou, G., Chen, J., Wang, H., Yuan, L. et al. (2021) Heterotrophically ultrahigh-cell-density cultivation of a high protein-yielding unicellular alga *Chlorella* with a novel nitrogen-supply strategy. *Frontiers in Bioengineering and Biotechnology* 9, 774854.

Yeh, K.L. and Chang, J.S. (2012) Effects of cultivation conditions and media composition on cell growth and lipid productivity of indigenous microalga *Chlorella vulgaris* ESP-31. *Bioresource Technology* 105, 120–127.

Yu, J., Wang, C., Su, Z.Y., Xiong, P. and Liu, J.Q. (2014) Response of microalgae growth and cell characteristics to various temperatures. *Asian Journal of Chemistry* 26(11), 3366–3370.

Zhang, D.H. and Lee, Y.K. (2001) Two-step process for ketocarotenoid production by a green alga, *Chlorococcum* sp strain MA-1. *Applied Microbiology and Biotechnology* 55(5), 537–540.

Zhang, Z., Tan, Y., Wang, W., Bai, W., Fan, J. et al. (2019) Efficient heterotrophic cultivation of *Chlamydomonas reinhardtii*. *Journal of Applied Phycology* 31(3), 1545–1554.

7 Metabolites of Microalgae

Gao Chen
Shandong Academy of Agricultural Sciences

Abstract

Microalgae are primitive creatures and one of the oldest biological forms on the Earth. They existed in oceans at the beginning of the Earth's formation about 3 billion years ago, and can synthesize large amounts of lipids, proteins and carbohydrates by utilizing sunlight, carbon sources, water and inorganic salts under natural conditions. They can also grow heterotrophically in the absence of light. They have advantages of fast growth, wide variety, sustainable use of solar energy to produce biomass, high efficiency and low energy consumption, can adapt to extreme environment and simple cultivations, and so on. These characteristics determine that microalgae have important development value in the fields of medicine, food, aquaculture, chemical industry, energy, environmental protection, agriculture and aerospace. This chapter briefly introduces microalgae as chassis to produce natural metabolites and their applications and discusses the prospect of using metabolic engineering to transform microalgae to produce valuable secondary metabolites.

1. Introduction

Algae are primitive creatures and one of the oldest biological forms on Earth. They existed in the oceans at the beginning of Earth's formation about 3 billion years ago (Sathasivam et al., 2019). Algae can be divided into eight phyla, which are cyanophyta, grayphyla, Rhodophyta, green phyla, nudophyta, dinoflagellates, top complex phyla and cryptophyla. These organisms vary in size from a few microns to tens of meters, forming large algae usually visible to the naked eye. The smallest algae are single-celled, called microalgae, which can exist as single cells or in small colonies. Microalgae usually live in freshwater, salt water and land. They usually contain photosynthetic pigments to conduct photosynthesis, absorb carbon dioxide and produce oxygen.

Microalgae can not only use solar energy to produce organic matter, but also can grow heterotrophically in the absence of light (Mohan et al., 2020). Different kinds of microalgae can produce rich primary and secondary metabolites (Table 7.1). The primary metabolites of microalgae are biological macromolecules necessary for maintaining life activities, and also participate in other metabolic pathways such as energy generation and storage. Secondary metabolites appear in the specific metabolic pathways of a cell or organism (Demain and Fang, 2000). Secondary metabolic pathways do not exist independently. They are closely related to the synthetic pathways of primary metabolites, and are derived from the further extension of primary metabolism. Primary metabolism and secondary metabolism are generally controlled by genes. The production of secondary metabolites of microorganisms is also affected by environmental factors. The synthetic pathway of secondary metabolites forms a complex and huge metabolic network (Fig. 7.1). Microbes in a certain environment flow substances involved in metabolism according to a certain rule in the metabolic pathway, forming a metabolic material flow. The concentration change of metabolites or key enzymes is

Address for correspondence: Gxchen001@hotmail.com

Table 7.1. Types of active metabolites produced by microalgae.

Microalgae	Bioactive compounds	References
Spirulina	polysaccharide	(Amaro et al., 2013)
Spirulina platensis	phycocyanin, phenolic acid, tocopherol, neophytadiene, phytol, n-3 polyunsaturated fatty acids, oleic acid, linolenic acid, palmitoleic acid	(Plaza et al., 2009; Singh and Dhar, 2011; Ibanez and Cifuentes, 2013)
Spirulina	diacylglycerol	(Singh and Dhar, 2011; Ibanez and Cifuentes, 2013)
Haematococcus pluvialis	astaxanthin, xanthophyll, zeaxanthin, canthaxanthin, β-carotene, oleic acid	(Plaza et al., 2009; Singh and Dhar, 2011; Markou and Nerantzis, 2013)
Chlorella sp	carotenoid, sulfate of polysaccharide, sterol, n-3 polyunsaturated fatty acids	(Plaza et al., 2009; Priyadarshani and Rath, 2012; Amaro et al., 2013)
Chlorella vulgaris	canthaxanthin, astaxanthin, peptide, oleic acid, eicosapentaenoic acid	(Singh and Dhar, 2011)
Chlorella minutissima	zeaxanthin, xanthine	(Amaro et al., 2013)
Chlorella ellipsoidea	trans-betacarotene, cis-betacarotene, β-carotene, oleic acid, linolenic acid, palmitic acid	(Plaza et al., 2009; Palavra et al., 2011; Singh and Dhar, 2011)
Dunaliella	diacylglycerol	(Singh and Dhar, 2011)
Botryococcus braunii	linear alkadienes(C25,27,29,31), triene(C29)	(Palavra et al., 2011)
Chlorella zofingiensis	astaxanthin	(Markou and Nerantzis, 2013)
Chlorella protothecoides	xanthophyll, zeaxanthin, canthaxanthin	(De Jesus Raposo et al., 2013; Markou and Nerantzis, 2013)
Chlorella pyrenoidosa	xanthophyll, sulfate of polysaccharide	(Plaza et al., 2009)
Nostoc linckia and nostoc	borophycin	(Singh and Dhar, 2011)
Spongiaeforme	cryptophycin	(Singh and Dhar, 2011)

transmitted to the whole network through the distribution nodes of the metabolic network and has an impact on the metabolic network.

In view of the complexity of metabolic network, the basic research idea of natural product biology is to search for and screen relevant genes or verify functions through the phenotype of product groups; or to find new active metabolites and their regulatory mechanisms through the differences in gene expression under different conditions. At present, the commonly used method to analyze the natural metabolites of microalgae is mainly to extract and purify the secondary metabolites from the algae. With the development of DNA sequencing technology and genome technology, the existence of metabolites can also be preliminarily determined by detecting the genes related to the synthesis of secondary metabolites in algae. However, for the metabolites that are synthesized in microalgae only under certain stress conditions, this method has some limitations.

Compared with plants, microalgae overcome many limitations and can be used to produce algal biomass in large quantities. Through developing technologies, a large number of high-value bioactive compounds can be obtained (Hamed, 2016; Hashemian et al., 2019). Biomass of microalgae can not only be used to produce feed for fish and farm animals, biofuel (Schulze et al., 2014), but also to produce high-value bioactive substances required for medicine and cosmetics (Wichuk et al., 2014). In addition, microalgae can also be used in environmental engineering, especially in wastewater treatment and biological purification of CO_2 in flue gas of coal-fired power plants (Zhou et al., 2015), biological fixation (Santarelli et al., 2017), integration of anaerobic digestion and microalgae culture (Bona et al., 2020).

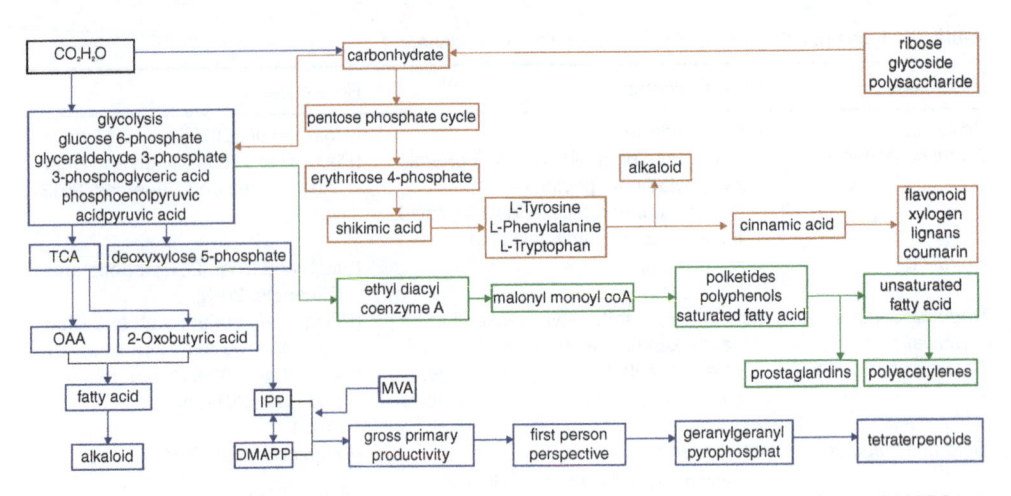

Fig. 7.1. TCA, tricarboxylic acid cycle; OAA, oxaloacetic acid; IPP, Isopentenyl pyrophosphate; DMAPP, dimethylallyl pyrophosphate; MVA, Mevalonic acid

This paper mainly introduces the natural metabolites of several microalgae.

2. Amino Acids, Peptides, Proteins

Chlorella, Spirulina, Palisada, Dunaliella, Chlamydomonas, Euglena and other organisms contain abundant proteins. Protein is necessary for the development and sustainability of life. They are molecular machines and building blocks in organisms, driving or playing most biological functions and endowing cells and tissues with structures and functions (Chou et al., 2012). Cyanoactivins are defined as small cyclic peptides (Martins and Vasconcelos, 2015) derived from ribosomes of bacterial symbiont *Prochlorophyta*, a cyanobacterial species, in marine sponge colonies. After heterocyclization, oxidation and/or isopentenylation, they are released from precursor proteins after modification.

There are many secondary metabolites in organisms, such as alkaloids, terpenoids and so on, which are generated by amino acids through different metabolic pathways. Many amino acids produced by algae can be used in production and life. Mycosporin-like amino acids (MAA) are a class of secondary metabolites, which are synthesized by many terrestrial and aquatic phototrophs (such as cyanobacteria, dinoflagellates and macroalgae) and have light protection functions (Matsumura and Ananthaswamy, 2004; Avenel-Audran et al., 2010; Moeller et al., 2021). MAA is a diversified group of colorless, uncharged and water-soluble nitrogen compounds with low molecular weight (400 Da). More than 30 types have been identified, and all MAAs have the same central structure, namely cyclohexene ketone or cyclohexene imine ring, which is responsible for absorbing ultraviolet light (Vega et al., 2021). Some studies have shown that MAA has two synthetic pathways, one is shikimic acid pathway (Fig. 7.2), which may be induced by ultraviolet radiation to synthesize MAA, and the other is pentose phosphate pathway, which is the main pathway to synthesize MAA. MAA has its own applications in various industries due to its ultraviolet absorption, antioxidant and therapeutic properties (Raj et al., 2021). MAA is considered as a free radical scavenger, which can absorb ROS (reactive oxygen species) to reduce cell damage during oxidative stress (Kageyama and Waditee-Sirisattha, 2019). It can be used as an anti-cancer agent by controlling cell proliferation or stimulating cancer cell apoptosis (leading to cell death). It is also known to have the ability to stimulate factors and signals involved in the wound healing pathway (Choi et al., 2015).

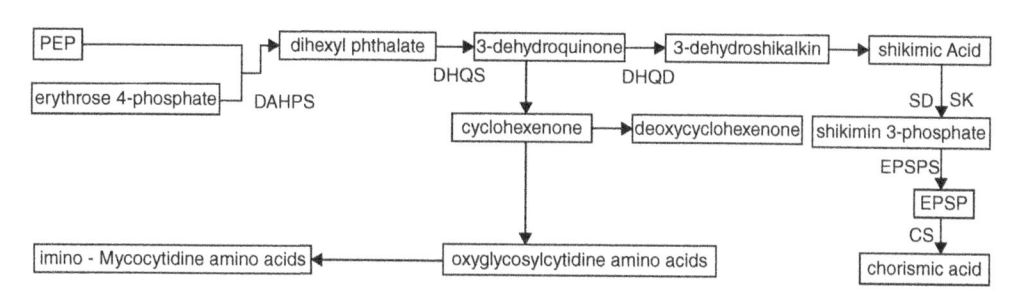

Fig. 7.2. PEP, Phosphoenolpyruvic acid; DAHPS, DAHPS synthesis enzyme; DHQS, 3-dehydroquinone synthetase; DHQD, 3-dehydroquinone dehydrogenase; SD, Shikimin hydroxylase; SK, Shikimin kinase; EPSP, 5-enolpyruvylshikimate-3-phosphate; EPSPS, 5-enolpyruvylshikimate-3-phosphate synthesis enzyme; CS, Shikimate synthetase

3. Carbohydrates

Carbohydrates are broad-category compounds that include sugars (monosaccharides) and their polymers (disaccharides, oligosaccharides and polysaccharides). Carbohydrates are the main components of microalgae cells, and their content exceeds 50% of the dry weight of algal cells. Carbohydrates in microalgae mainly include starch, glucose, cellulose, hemicelluloses and polysaccharides. After a series of decomposition reactions in the organism, sugar releases a lot of energy for the use of life activities. At the same time, some intermediates formed in the decomposition process can also be used as raw materials for the synthesis of lipids, proteins, nucleic acids and other biological macromolecules. The common hexose units in primary metabolism are glucose, galactose and mannose. Polysaccharides mostly exist in organisms as glycoconjugates, which are assembled and modified in the endometrial system (Nguema-Ona et al., 2014). Their synthesis involves three steps: first of all, the formation of activated nucleotide sugars in the cytosol, such as nucleoside diphosphate (NDP) sugars or nucleoside monophosphate (NMP) sugars, takes place (Bar-Peled and O'Neill, 2011). Nucleotide sugars are then actively transported to the endoplasmic reticulum (ER) and Golgi apparatus (GA), where they act as donor substrates for glycosyltransferases (GT), transferring specific sugars from their activated nucleotide forms to specific receptors, leading to the formation and extension of glycoconjugates.

Compared with the other three types of biomass compounds produced by microalgae, carbohydrates have the lowest energy content per unit mass, but they produce several important biofuels such as bioethanol, biobutanol and biohydrogen through biotechnology conversion technology. Bioethanol is one of the alternative energy sources to solve the current energy crisis. Compared with fossil fuel, bioethanol is biodegradable. It contains high oxygen, so it is flammable and can reduce carbon dioxide emissions by up to 18% (Dmpeipen and Dewa, 2015). The microalgae carbohydrates used for bioethanol production can be derived from microalgae, such as *Spirogyra* (Eshaq et al., 2011), with high natural carbohydrate content. It can also be derived from species known to have high growth rate or intracellular fermentation capacity, and one of the manipulation methods of biomass composition can be suitable for increasing carbohydrate content. Photo-drive carbon fixation of blue-green algae was first achieved in PCC7942 in 1999 by Deng et al. Over the next ten years, the research on the synthesis of ethanol by light-driven carbon fixation of cyanobacteria focused on another model freshwater algae strain, *Synechocystis* sp. PCC6803. In 2009, Dexter et al. and Hou Lv Xuefeng et al. (Deng and Coleman, 1999) demonstrated that by doubling the copy number of the Prbc pdcZM *slr1192* pathway in the genome of *Synechocystis* sp. PCC6803, the ethanol yield could be significantly increased, from 2g/L to 5.5g/L after 26 days of culture, with the production intensity reaching 212 mg/(L·d) (Gao et al., 2012).

Biobutanol is a four-carbon alcohol and its energy is 1.5 times higher than that of ethanol. In addition, butanol has several advantages over ethanol, including its low water solubility for which it can be used in conventional engines without modification (Grimaldi, 2011; Onay, 2020). Butanol has lower hygroscopicity, volatility and corrosivity than ethanol, so it is safer and more convenient to transport by pipeline, making it an ideal candidate for biofuel (Li et al., 2019). Another advantage of butanol is that it can be used as a drop additive for fuel and as an additive for gasoline (Xiao et al., 2020). The current research on the production strategy of biological butanol, on the one hand, is to apply advanced genome engineering technology to introduce heterologous genes or complete pathways into natural microalgae, to regulate the pathway of CO_2 fixation, and use its metabolites (ATP, NADPH) to produce butanol. On the other hand, phototrophic metabolism and fermentation metabolism are integrated to construct microalgae chimera, and CO_2 is directly converted into biological butanol by solar energy (Shanmugam et al., 2021).

4. Lipids

The most energy-rich compound in lipid algae biomass is lipid (37.6 kJg^{-1}), followed by proteins (16.7 kJg^{-1}) and carbohydrates (15.7 kJg^{-1}) (Wilhelm and Jakob, 2011). Under suitable environment, microalgae mainly synthesize not only glycolipids, phospholipids and sulfurlipids, but also are used to synthesize biofilms and maintain normal physiological and biochemical activities. Under the stress of adverse environmental factors, the synthesis of carbohydrates in algae is inhibited, whereas the synthesis of lipids is increased, and more energy is stored. The main component of stored lipids in microalgae is saturated fatty acid (SFA), which can be divided into saturated and monounsaturated fatty acids (Thompson, 1996). SFA can provide energy for the substances transferred here and it mainly exists in the form of triphthalglycerol (TAG). Fatty acids and TAG in algae are mainly synthesized through the following pathways (Fig. 7.3). The accumulation of microalgae oil is related to the key enzymes for oil production in its body, which are helpful for the synthesis and decomposition of TAG. Up till now, the genes of microalgae are modified by means of genetic engineering, greatly increasing the production of target products (Daniela et al., 2021). Knockout of the PtTL gene of *Phaeodactylum tricornutum* and the TpTL gene of seaweed affected the TAG lipase and increased the TAG accumulation (Trentacoste et al., 2013; Barka et al., 2016). Biodiesel is mainly synthesized by transesterification reaction in which TAG is more convenient.

Microalgae also produce a variety of polyunsaturated fatty acids, mainly ω-3 polyunsaturated fatty acids, which are essential fatty acids, such as β-linolenic acid (C18:3), EPA (C20:5) and DHA (C22:6); and ω-6 polyunsaturated fatty acids, such as ω-linoleic acid and linoleic acid (C18:2). In addition, microalgae also produce other less common polyunsaturated fatty acids, such as C16:2, C16:3 and C16:4 (Yao et al., 2015) or long-chain C24-C28 (Mansour et al., 2005). Many long-chain polyunsaturated fatty acids are associated with a variety of health benefits and need to be supplemented from natural sources from childhood to adulthood. Eicosapentaenoic acid (EPA) and docosahexaenoic acid (DHA) are widely believed to be beneficial to human health and mainly exist in marine microalgae species (Chauton et al., 2015). At present, a number of studies have been conducted to compare the oil content and oil yield of different microalgae species. The oil content and oil yield of different microalgae species are different. In general, the average oil yield of *Chlorella, Chlamydomonas reinhardtii* and *Microchloropsis* is high (Mata et al., 2010). Currently, the focus of microalgae research is to optimize the cultivation technology of microalgae biomass lipid content, while the obtained lipid is used for

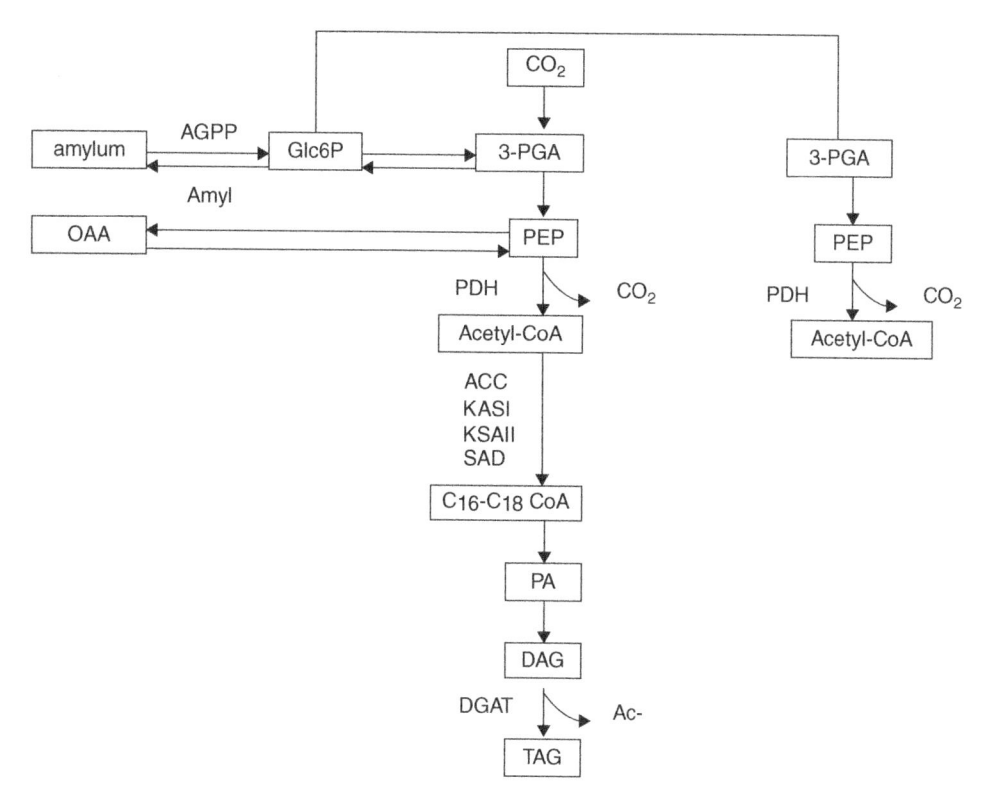

Fig. 7.3. AGPP, ADP-glucose pyrophosphorylase; Amyl, amylase; Glc6P, glucose-6-phosphate; 3-PGA, 3-phosphoglycerate; Acetyl-CoA, acetyl coenzyme A; C16-C18 CoA, C16-C18 coenzyme A; PA,Phosphatidic acid; DAG, diacylglycerol; DGAT, diacylglycerol acyltransferase; TAG, triglyceride; PDH, pyruvate dehydrogenase; ACC, acetyl-coa carboxylase; KASI, β-keto ester acyl ACP synthase I; KSAII, β-keto ester acyl ACP synthase II; SAD, stearoyl-ACP desaturase; OAA, oxaloacetic acid; TAG, triacylglyceride.

biodiesel production after esterification to fatty acid ethyl ester or methyl ester (McKendry, 2002). At the same time, algal biomass can also be used as raw materials to produce other biofuels through biotechnology and thermochemical conversion technology (Demirbas, 2010). The pretreated algal biomass can be used as raw material for acid production via fermentation, especially to produce biohydrogen together with volatile fatty acids (VFA) (Wieczorek et al., 2014; Venkata Mohan et al., 2016; Wang and Yin, 2018).

5. Pigments

Pigmented microalgae and cyanobacteria are photosynthetic organisms. Different microalgae produce different kinds of pigments to trap light for photosynthesis. These pigments are generally divided into three categories: chlorophyll, carotenoids and phycobiliproteins (Richmond and Hu, 2013). Chlorophyll and carotenoids are usually liposoluble molecules, while phycobilin is water-soluble. Chlorophyll is the most basic pigment, which can be divided into chlorophyll a, chlorophyll b and chlorophyll c. It is mainly responsible for oxygen, producing photosynthetic activity, and exists in all photosynthetic microalgae and cyanobacteria (Masojdek et al., 2007). Chlorophyll a is widely used as a colorant to replace artificial pigments because of its stability. This substance usually comes from higher plants, which also synthesize other kinds of chlorophyll. On the other hand, S. platensis only has chlorophyll a and is the largest source of chlorophyll. In Brazil, about 0.06 mg/g of chlorophyll in spinach is used as a natural green colorant (Gross, 1991), while 1.15 mg/g of this pigment is contained in *Spirulina* biomass (Henrikson, 1989). In addition to being used as food and drug colorants, chlorophyll also has wound-healing and anti-inflammatory properties (Ferruzi and Blakeslee, 2007).

Carotenoids are crucial for the survival of microalgae. They can protect cells from the attack of reactive oxygen species generated during photosynthesis and high light intensity and dissipate excess light in the form of heat through photochemical quenching process. Lycopene is the intermediate metabolite of carotenoid synthesis. Different carotenoids are synthesized by different enzymes. The main carotenoids with commercial value are β-carotene, lutein and astaxanthin. Lutein is known for its protective effect on macular degeneration (MD) of the eyes. Dietary lutein is important because it cannot be synthesized by humans. Astaxanthin is a well-known antioxidant. It can effectively prevent the oxidation of low-density proteins and can be used to prevent arteriosclerosis, coronary heart disease and ischemic brain development (Miki et al., 1998) diseases. It has the characteristics of heart protection, neuroprotection, anti-cancer and anti-diabetes. β-Carotene is provitamin A and also an antioxidant with cardioprotective effect (Spolaore et al., 2006). *Haematococcus pluvialis, Dunaliella salina, Chlorella, Palisada, Spirulina platensis, Staphylococcus blarney* and *diatoma* are used to produce β-carotene, lutein, astaxanthin, keratin and fucoidin (Ambati et al., 2018). The biosynthesis of microalgae carotenoids takes place in a single cell (Fig. 7.4). IPP and DMAPP are the necessary intermediate metabolites for its biosynthesis, and different biosynthesis types and synthesis sites are different. IPP and DMAPP can be synthesized through two independent non-homologous metabolic pathways: one is mevalonate (MVA) pathway; the second is the 2-C-methyl-D-eithritol-4-phase pathway (MEP), also known as the 1-deoxy-D-xylulose 5-phase (DXP) pathway or the non-MVA pathway (Zinglé et al., 2010; Miziorko, 2011). Most bacteria and cyanobacteria have only MEP pathway, while most eukaryotes and archaea have only MVA pathway. In microalgae, IPP is only synthesized in chloroplasts through MEP pathway, while related enzymes of MEP pathway are encoded by nuclear genes, and these enzymes are synthesized in the form of polypeptide precursors containing transport peptide sequences. After synthesis, transport peptides can locate related enzymes in chloroplasts to play a role (León et al., 2007).

In addition, phycobiliproteins are special pigments that only exist in cyanobacteria and red algae. They are water-soluble molecules and represent the main light absorbers in these organisms. The most famous phycobiliproteins are phycocyanin and phycoerythrin (Pagels et al., 2020). Phycocyanin is widely used as food pigment to replace the current synthetic pigment. Phycocyanin from *S. platensis* is used as a colorant for food and cosmetics (Santiago-Santos et al., 2004). C-phycocyanin is used as a natural protein dye in the food industry (Sekar and Chandramohan, 2008). Phycoerythrin from *Pseudomonas aeruginosa* and *S. platensis* is also used in colored candies, gelatin desserts, fermented dairy products, ice cream, sweet cake decoration, milkshakes and cosmetics. In addition to its coloring properties, phycoerythrin also has a yellow fluorescent color, which is used to produce transparent lollipops derived from sugar solutions, dry sugar drops for cake decoration, and soft drinks and alcoholic beverages (Dufoss et al., 2005). Due to the absorption spectrum, phycoerythrin has been used as the second color of fluorescent labeled antibody (De Rosa et al., 2003). It is reported that phycoerythrin labeled with streptavidin can be used to detect DNA and protein probes. Chlorophyll, carotenoids and phycobiliprotein, the three kinds of pigments, maximize the light-absorption capacity of microalgae and cyanobacteria in the whole visible light spectrum (Pagels et al., 2021). The natural pigments in microalgae not only have coloring potential and can replace artificial pigments, but they also have health benefits (anti-oxidation, anti-cancer and anti-inflammatory) and are widely used in

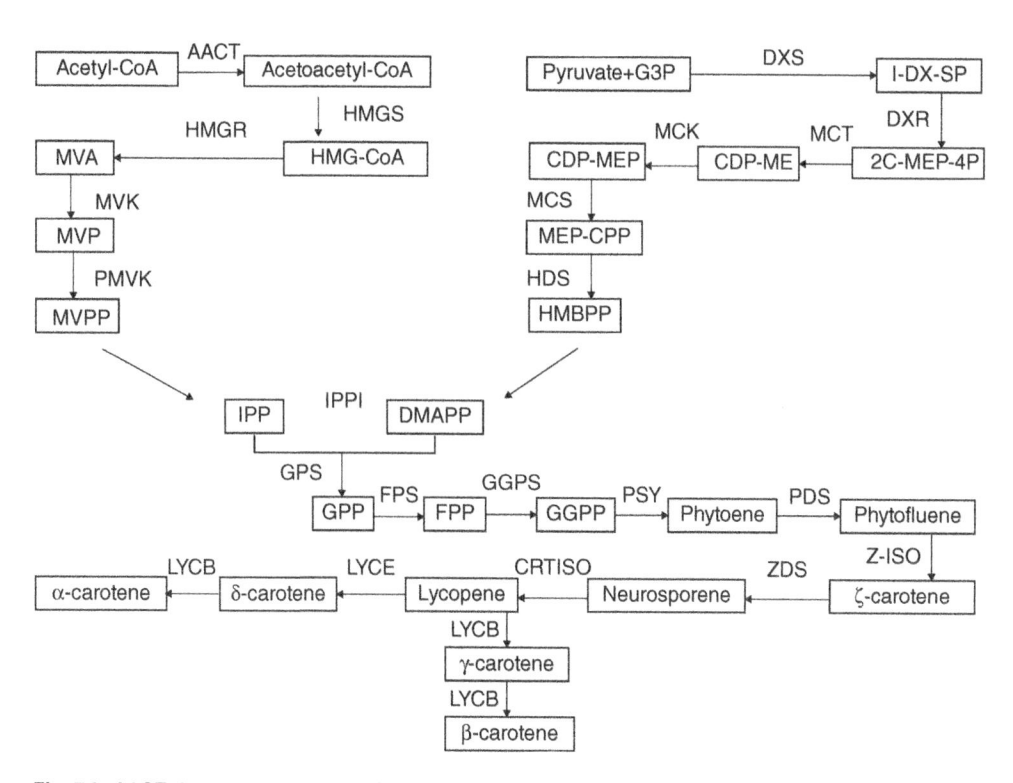

Fig. 7.4. AACT, Acetyl-coa acetyltransferase; HMGS, 3-hydroxy-3-methylglutaryl-CoA synthetase; HMGR, 3-hydroxy-3-methylglutaryl-CoA reductase; HMG-CoA, 3-hydroxy-3-methylglutaryl-CoA; MVA, mevalonic acid; MVK, mevalonate kinase; MVP, mevalonic acid-5-phosphate; PMVK, mevalonate phosphokinase; MVPP, Mevalonic acid-5-diphosphate; G3P, Glyceraldehyde 3-phosphate; DXS, 1-deoxyxylulose-5-phosphate synthase; DXR, 1-deoxyxylulose-5-phosphate reducing isomerase; MCT, 2-C-methyl-D-erythritol-4-cytidine phosphate transferase; MCK, 2-C-methyl-D-erythritol-4-cytidine diphosphate kinase; MCS, 2-C-methyl-D-erythritol-2, 4-cyclopyrophosphate synthase; HDS, 1-hydroxy-2-methyl-2-butenyl-4-phosphate synthase; I-DX-SP, 1-deoxyxylulose-5-phosphate; 2C-MEP-4P, 2-C-methyl-D-erythritol-4-phosphate; CDP-ME, Cytidine diphosphate-2-C-methyl-D-erythritose; CDP-MEP, 4-cytidine diphosphate-2-C-methyl-D-erythritol-2-phosphate; MEP-CPP, 2-C-methyl-D-erythritol-2,4-cyclic diphosphate; HMBPP, E-4-hydroxy-3-methylbutyl-2-enyldiphosphate; IPP, Isoprene pyrophosphate; IPPI, Isoprene pyrophosphate isomerase; DMAPP, dimethylpropene pyrophosphate; GPS, Geranylgeranyl diphosphate synthase; GPP, geranyl pyrophosphate; FPS, farnesyl pyrophosphate synthetase; FPP, farnesyl pyrophosphate; GGPP, geranylgeranyl pyrophosphate; PSY, phytoene synthase; PDS, phytoene desaturase; Z-ISO, zeta-carotene isomerase; ZDS, Zeta-carotene dehydrogenase; CRTISO, carotenoid isomerase; LYCE, Lycopene ε-cyclase; LYCB, Lycopene β-cyclase

food, pharmaceutical and cosmetic industries (Rodrigues et al., 2015; Begum et al., 2016). For example, lutein can be used as food additive and feed additive to deepen the color of egg yolk and brighten poultry feathers (Lin et al., 2015).

6. Summary and Prospects

The microalgae group has extremely high-level physiological and metabolic diversity, hence it produces a wide range of metabolites. In recent years, microalgae have been regarded as a new microbial photosynthetic platform with great potential to directly convert solar energy and CO_2 into various biological products. This production mode is called optical drive carbon fixation synthesis technology. The development of optical drive carbon fixation technology mainly has the following three trends.

Firstly, through extensive collection and screening of microalgae germplasm resources, analysis and identification of their intracellular metabolites, and excavation of the development model of microalgae optical drive carbon fixation synthesis route of compounds with high added value, it has the characteristics of relatively low technical threshold and high system maturity. So far, it is still the main research direction in the field of microalgae bioengineering. At present, many websites and platforms have included the metabolites of microorganisms (Van Santen et al., 2021). CyanoMetDB is a website that includes the secondary metabolites of cyanobacteria, including the natural metabolites of cyanobacteria that have been identified to date (Jones et al., 2021). Jones believes that the discovery of secondary metabolic networks of cyanobacteria is still on the rise, and the metabolites found are basically derivatives of natural metabolites that have been found. Some scholars believe that the combination of genomics and metabolomics will provide an effective method for further prediction and analysis of microbial metabolites (Avalon et al., 2022).

Secondly, for the development of natural microalgae germplasm resources, in addition to large-scale new germplasm screening to tap algae strains and products, we can 'activate' the silent metabolic pathways and functional modules of microalgae through special cultural conditions and strategies, and explore the changes of intracellular metabolites in combination with omics analysis technology, so as to achieve the accumulation mode of environmentally responsive high additional compounds, which will effectively expand the understanding and utilization of the microalgae metabolic network. At present, the widely used strategies include imposition of environmental stress, chemical regulator addition and co-culture. Studies have shown that high salinity can inhibit biomass, but it can promote algal cells to accumulate large amounts of oil. Miyasaka and Ikeda (1997) have reported that *Chlamydomonas reinhardtii* HS-5 accumulated a large amount of oil under high salt concentration. Similarly, Ben-Amotz and Avron (1973) showed that *Dunaliella salina* cells under salt stress synthesized a large amount of glycerol, and the intracellular glycerol content was positively correlated with the external salt concentration. Some studies have also suggested that any magnetic field (MF) applied to microalgae could change the conditions of microalgae culture and could be conducive to the production of biomolecules, because it interacts with microorganisms and affects their growth, resulting in changes in the concentration and composition of biomass and biomolecules (Santos et al., 2022).

Thirdly, with the development and application of synthetic biology tools, the development mode of microalgae optical drive carbon fixation synthesis technology that redirects photosynthetic carbon flow to natural or non-natural metabolites through the artificial design and reconstruction of metabolic pathways and the modification of the natural metabolic network of microalgae has attracted more and more attention. This model is mainly based on the analysis of the genetic background and metabolic network of microalgae, and then according to the requirements of the target product synthesis pathway, the introduction of foreign genes and the modification and transformation of the endogenous pathway. It is obvious that this model has the characteristics of strong targeting and expansion. At present, the combination of analytical chemistry and genome technology is a new stage in the development of natural product chemistry. The comparative integration of evolutionary studies and reverse biosynthesis predictions, the correlation between genome sequence and structure are helpful to understand the details of biosynthesis mechanisms and have a positive effect on the preliminary judgment of microbial synthesis of secondary metabolites with potential applications (Avalon et al., 2022).

Acknowledgments

The author sincerely thanks Miss Xue Cao and Mr Jiaye Zhang for their critical comments and suggestions.

References

Amaro, H.M., Barros, R., Guedes, A.C., Sousa-Pinto, I. and Malcata, F.X. (2013) Microalgal compounds modulate carcinogenesis in the gastrointestinal tract. *Trends in Biotechnology* 31, 92–98.

Ambati, R.R., Gogisetty, D., Aswathanarayana, R.G., Ravi, S., Bikkina, P.N. et al. (2018) Industrial potential of carotenoid pigments from microalgae: current trends and future prospects. *Critical Reviews in Food Science and Nutrition* 59, 1880–1902.

Avalon, N.E., Murray, A.E. and Baker, B.J. (2022) Integrated metabolomic-genomic workflows accelerate microbial natural product discovery. *Analytical Chemistry* 94, 35.

Avenel-Audran, M., Dutartre, H., Goossens, A., Jeanmougin, M., Comte, C. et al. (2010) Octocrylene, an emerging photoallergen. *Archives of Dermatology* 146, 753–757.

Barka, F., Angstenberger, M., Ahrendt, T., Lorenzen, W., Bode, H.B. et al. (2016) Identification of a triacylglycerol lipase in the diatom Phaeodactylum tricornutum. *Biochimica et Biophysica Acta (BBA)-Molecular and Cell Biology of Lipids* 1861, 239–248.

Bar-Peled, M. and O'Neill, M.A. (2011) Plant nucleotide sugar formation, interconversion, and salvage by sugar recycling. *Annual Review of Plant Biology* 62, 127–155.

Begum, H., Yusoff, F.M., Banerjee, S., Khatoon, H. and Shariff, M. (2016) Availability and utilization of pigments from microalgae. *Critical Reviews in Food Science and Nutrition* 56, 13.

Ben-Amotz, A. and Avron, M. (1973) The role of glycerol in the osmotic regulation of the halophilic alga Dunaliella parva. *Plant Physiology* 51(5), 875-878.

Bona, D., Papurello, D., Flaim, G., Cerasino, L., Biasioli, F. et al. (2020) Management of digestate and exhausts from solid oxide fuel cells produced in the dry anaerobic digestion pilot plant: microalgae cultivation approach. *Waste Biomass Valorization* 11, 6499–6514.

Chauton, M.S., Reitan, K.I., Norsker, N.H., Tveterås, R. and Kleivdal, H.T. (2015) A techno-economic analysis of industrial production of marine microalgae as a source of EPA and DHA-rich raw material for aquafeed: research challenges and possibilities. *Aquaculture* 436, 95–103.

Choi, Y.-H., Yang, D.J., Kulkarni, A., Moh, S.H. and Kim, K.W. (2015) Mycosporine-like amino acids promote wound healing through focal adhesion kinase (FAK) and mitogen-activated protein kinases (MAP kinases) signaling pathway in Keratinocytes. *Marine Drugs* 13, 7055–7066.

Chou, C.J., Affolter, M. and Kussmann, M. (2012) A nutrigenomics view of protein intake. *Progress in Molecular Biology and Translational Science* 108.

Daniela, C., Simone, L. and Sergio, E. (2021) Advanced applications for protein and compounds from microalgae. *Plants* 10(8).

De Jesus Raposo, M.F., De Morais, R.M.S.C. and De Morais, A.M.M.B. (2013) Health applications of bioactive compounds from marine microalgae. *Life Sciences* 93(15), 479–486.

De Rosa, S.C., Brenchley, J.M. and Roederer, M. (2003) Beyond six colours: a new era in flow cytometry. *Nature Medicine* 9, 112–117.

Demain, A.L. and Fang, A. (2000) The natural functions of secondary metabolites. *Advances in Biochemical Engineering/Biotechnology* 69.

Demirbas, A. (2010) Use of algae as biofuel sources. *Energy Conversion and Management* 51(12), 2738–2749.

Deng, M. and Coleman, J.R. (1999) Ethanol synthesis by genetic engineering in cyanobacteria. *Applied and Environmental Microbiology* 65(2), 523–528.

Dmpeipen, E.J. and Dewa, R.P. (2015) *Majalah BIAM* 11(2), 63–75.

Dufoss, L., Galaup, P., Yarnon, A., Arad, S.M., Blanc, P. et al. (2005) Microorganisms and microalgae as source of pigments for use: a scientific oddity or an industrial reality? *Trends in Food Science and Technology* 16, 389–406.

Eshaq, F.S., Ali, M.N. and Mohd, M.K. (2011) Production of bioethanol from next generation feed-stock alga Spirogyra species. *International Journal of Engineering, Science and Technology* 3(2), 1749–1755.

Ferruzi, M.G. and Blakeslee, J. (2007) Digestion, absorption, and cancer preventive activity of dietary chlorophyll derivatives. *Nutrition Research* 27, 1–12.

Gao, Z., Zhao, H., Li, Z., Tan, X. and Lu, X. (2012) Photosynthetic production of ethanol from carbon dioxide in genetically engineered cyanobacteria. *Energy and Environmental Science* 5(12), 9857–9865.

Grimaldi, J. (2011) *The Search for Environmentally Friendly Energy: Ethanol vs. Butanol.* Available at: https://www.aiche.org/chenected/2011/11/search-environmentally-friendly-energy-ethanol-vs-butanol#:~:text= Butanol%20has%20about%201.5%20times%20more%20volumetric%20energy%20content%20 than%20ethanol.&text=Another%20advantage%20over%20ethanol%20is,and%20a%20standalone%20 fuel%20source (accessed 20 June 2023).

Gross, J. (1991) Chlorophylls. In: *Pigments in Vegetables: Chlorophylls and Carotenoids*. AVI, New York, pp. 3–74.

Hamed, I. (2016) The evolution and versatility of microalgal biotechnology: a review. *Comprehensive Reviews in Food Science and Food Safety* 15, 1104–1123.

Hashemian, M., Ahmadzadeh, H., Hosseini, M., Lyon, S. and Pourianfar, H. (2019) Production of microalgae-derived high-protein biomass to enhance food for animal feedstock and human consumption. *Advanced Bioprocessing for Alternative Fuels, Biobased Chemicals, and Bioproducts*, 393–405.

Henrikson, R. (1989) *Earth Food Spirulina*. Ronore Enterprises Inc., Richmond, California.

Ibanez, E. and Cifuentes, A. (2013) Benefits of using algae as natural sources of functional ingredients. *Journal of the Science of Food and Agriculture* 93(4), 703–709.

Jones, M.R., Pinto, E., Torres, M.A., Dörr, F., Mazur-Marzec, H. *et al.* (2021) CyanoMetDB, a comprehensive public database of secondary metabolites from cyanobacteria. *Water Research* 196.

Kageyama, H. and Waditee-Sirisattha, R. (2019) Antioxidative, anti-inflammatory, and anti-aging properties of mycosporine-like amino acids: molecular and cellular mechanisms in the protection of skin-aging. *Marine Drugs* 17, 222.

León, R., Couso, I. and Ferná Ndez, E. (2007) Metabolic engineering of ketocarotenoids biosynthesis in the unicelullar microalga Chlamydomonas reinhardtii. *Journal of Biotechnology* 130(2), 143–152.

Li, Y., Tang, W., Chen, Y., Liu, J. and Chia-fon, F.L. (2019) Potential of acetone-butanol-ethanol (ABE) as a biofuel. *Fuel* 242, 673–686.

Lin, J.H., Lee, D.J. and Chang, J.S. (2015) Lutein production from biomass: marigold flowers versus microalgae. *Bioresource Technology* 184, 421–428.

Mansour, M.P., Frampton, D.M.F., Nichols, P.D., Volkman, J.K. and Blackburn, S.I. (2005) Lipid and fatty acid yield of nine stationary-phase microalgae: applications and unusual C24–C28 polyunsaturated fatty acids. *Journal of Applied Phycology* 17, 287–300.

Markou, G. and Nerantzis, E. (2013) Microalgae for high-value compounds and biofuels production: a review with focus on cultivation under stress conditions. *Biotechnology Advances* 31(8), 1532–1542.

Martins, J. and Vasconcelos, V. (2015) Cyanobactins from cyanobacteria: current genetic and chemical state of knowledge. *Marine Drugs* 13(11).

Masojdek, J., Koblek, M. and Torzillo, G. (2007) Photosynthesis in microalgae. In: Richmond, A. and Hu, Q. (eds) *Handbook of Microalgal Culture: Applied Phycology and Biotechnology*, 2nd edn. Wiley-Blackwell, Hoboken, New Jersey, pp. 21–36.

Mata, T.M., Martins, A.A. and Caetano, N.S. (2010) Microalgae for biodiesel production and other applications: a review. *Renewable and Sustainable Energy Reviews* 14, 217–242.

Matsumura, Y. and Ananthaswamy, H.N. (2004) Toxic effects of ultraviolet radiation on the skin. *Toxicology and Applied Pharmacology* 195, 298–308.

McKendry, P. (2002) Energy production from biomass (part 2): conversion technologies. *Bioresource Technology* 83(1), 47–54.

Miki, W.W., Hosada, K., Kqndo, K. and Itakura, H. (1998) Astaxanthin-containing drink. *Japanese Patent* #10155459.

Miyasaka, H. and Ikeda, K. (1997) Osmoregulating mechanism of the halotolerant green alga Chlamydomonas, strain HS-5. *Plant Science* 27(1), 91–96.

Miziorko, H.M. (2011) Enzymes of the mevalonate pathway of isoprenoid biosynthesis. *Archives of Biochemistry and Biophysics* 505(2), 131–143. DOI: 10.1016/j.abb.2010.09.028

Moeller, M., Pawlowski, S., Petersen-Thiery, M., Miller, I.B., Nietzer, S. *et al.* (2021) Challenges in current coral reef protection: possible impacts of UV filters used in sunscreens: a critical review. *Frontiers in Marine Science* 8, 383.

Mohan, S.V., Hemalatha, M., Chakraborty, D., Chatterjee, S. and Ranadheer, P. (2020) Algal biorefinery models with self-sustainable closed loop approach: trends and prospective for blue-bioeconomy. *Bioresource Technology* 295, 122128.

Nguema-Ona, E., Vicré-Gibouin, M., Gotté, M., Plancot, B., Lerouge, P. *et al.* (2014) Cell wall O-glycoproteins and N-glycoproteins: aspects of biosynthesis and function. *Frontiers on Plant Science* 5. DOI: 10.3389/fpls.2014.00499

Onay, M. (2020) The effects of indole-3-acetic acid and hydrogen peroxide on Chlorella zofingiensis CCALA 944 for bio-butanol production. *Fuel* 273, 117795.

Pagels, F., Salvaterra, D., Amaro, H.M. and Guedes, A.C. (2020) Pigments from microalgae. In: Jacob-Lopes, E., Queiroz, M.I., Maroneze, M.M. and Zepka, L.Q. (eds) *Handbook of Microalgae-Based Processes and Products*. Academic Press, Cambridge, Massachusetts, pp. 465–492.

Pagels, F., Pereira, R.N., Vicente, A.A. and Guedes, A.C. (2021) Extraction of pigments from microalgae and cyanobacteria: a review on current methodologies. *Applied Sciences* 11(11).

Palavra, A.M.F., Coelho, J.P., Barroso, J.G., Rauter, A.P., Fareleira, J.M.N.A. *et al.* (2011) Supercritical carbon dioxide extraction of bioactive compounds from microalgae and volatile oils from aromatic plants. *Journal of Supercritical Fluids* 60, 21–27.

Plaza, M., Herrero, M., Alejandro Cifuentes, A. and Ibánez, E. (2009) Innovative natural functional ingredients from microalgae. *Journal of Agricultural and Food Chemistry* 57(16), 7159–7170.

Priyadarshani, I. and Rath, B. (2012) Commercial and industrial applications of micro algae: a review. *Journal of Algal Biomass Utilization* 3(4), 89–100.

Raj, S., Kuniyil, A.M., Sreenikethanam, A., Gugulothu, P., Jeyakumar, R.B. *et al.* (2021) Microalgae as a source of mycosporine-like amino acids (MAAs): advances and future prospects. *International Journal of Environmental Research and Public Health* 18(23).

Richmond, A. and Hu, Q. (2013) *Handbook of Microalgal Culture: Applied Phycology and Biotechnology*, 2nd edn. Wiley-Blackwell, Hoboken, New Jersey.

Rodrigues, D.B., Menezes, C.R., Mercadante, A.Z., Jacob-Lopes, E. and Zepka, L.Q. (2015) Bioactive pigments from microalgae Phormidium autumnale. *Food Research International* 77, 273–279.

Santarelli, M., Briesemeister, L., Gandiglio, M., Herrmann, S., Kuczynski, P. *et al.* (2017) Carbon recovery and re-utilization (CRR) from the exhaust of a solid oxide fuel cell (SOFC): analysis through a proof-of-concept. *Journal of CO2 Utilization* 18, 206–221.

Santiago-Santos, M.C., Ponce-Noyola, T., Olvera-Ramírez, R., Ortega-López, J. and Cañizares-Villanueva, R.O. (2004) Extraction and purification of phycocyanin from Calothrix sp. *Process Biochemistry* 39, 2047–2052.

Santos, L.O., Silva, P.G.P., Machado, B.R., Sala, L. and Deamici, K.M. (2022) Update on the application of magnetic fields to microalgal cultures. *World Journal of Microbiology and Biotechnology* 38, 11.

Sathasivam, R., Radhakrishnan, R., Hashem, A. and Abd Allah, E.F. (2019) Microalgae metabolites: A rich source for food and medicine. *Saudi Journal of Biological Sciences* 26, 709–722.

Schulze, P.S., Barreira, L.A., Pereira, H.G., Perales, J.A. and Varela, J.C. (2014) Light emitting diodes (LEDs) applied to microalgal production. *Trends in Biotechnology* 32, 422–430.

Sekar, S. and Chandramohan, M. (2008) Phycobiliprotein as a commodity: trends in applied research, patents and commercialization. *Journal of Applied Phycology* 20, 113–136.

Shanmugam, S., Hari, A., Kumar, D., Rajendran, K. and Mathimani, T. (2021) Recent developments and strategies in genome engineering and integrated fermentation approaches for biobutanol production from microalgae. *Fuel* 285, 119052.

Singh, N.K. and Dhar, D.W. (2011) Microalgae as second generation biofuel: a review. *Agronomy for Sustainable Development* 31(4), 605–629.

Spolaore, P., Cassan, C.J., Duran, E. and Isambert, A. (2006) Commercial application of microalgae. *Journal of Bioscience and Bioengineering* 101, 87–96.

Thompson, G.A. (1996) Lipids and membrane function in green algae. *Biochimica et Biochimica et Biophysica Acta* 1302(1), 17–45.

Trentacoste, E.M., Shrestha, R.P., Smith, S.R., Glé, C., Hartmann, A. *et al.* (2013) Metabolic engineering of lipid catabolism increases microalgal lipid accumulation without compromising growth. *Proceedings of the National Academy of Sciences of the USA* 110, 19748–19753.

Van Santen, J.A., Kautsar, S.A., Medema, M.H. and Linington, R. (2021) Microbial natural product databases: moving forward in the multiomics era. *National Product Reports* 38(1), 264–278.

Vega, J., Schneider, G., Moreira, B.R., Herrera, C., Bonomi-Barufi, J. *et al.* (2021) Mycosporine-like amino acids from red macroalgae: UV-photoprotectors with potential cosmeceutical applications. *Applied Sciences* 11(11).

Venkata Mohan, S., Nikhil, G.N., Chiranjeevi, P., Reddy, C.N., Rohit, M.V. *et al.* (2016) Waste biorefinery models towards sustainable circular bioeconomy: critical review and future perspectives. *Bioresource Technology* 215, 2–12.

Wang, J. and Yin, Y. (2018) Fermentative hydrogen production using pretreated microalgal biomass as feedstock. *Microbial Cell Factories* 17(1), 22.

Wichuk, K., Brynjolfsson, S. and Fu, W. (2014) Biotechnological production of value-added carotenoids from microalgae. *Bioengineered* 5, 204–208.

Wieczorek, N., Kucuker, M.A. and Kuchta, K. (2014) Fermentative hydrogen and methane production from microalgal biomass (Chlorella vulgaris) in a two-stage combined process. *Applied Energy* 132, 108–117.

Wilhelm, C. and Jakob, T. (2011) From photons to biomass and biofuels: evaluation of different strategies for the improvement of algal biotechnology based on comparative energy balances. *Applied Microbiology and Biotechnology* 92(5), 909–919.

Xiao, H., Guo, F., Wang, R., Yang, X., Li, S. *et al.* (2020) Combustion performance and emission characteristics of diesel engine fueled with iso-butanol/biodiesel blends. *Fuel* 268, 117387.

Yao, L., Gerde, J.A., Lee, S.-L., Wang, T. and Harrata, K.A. (2015) Microalgae lipid characterization. *Journal of Agricultural and Food Chemistry* 63, 1773–1787.

Zhou, X., Yuan, S., Chen, R. and Ochieng, R.M. (2015) Sustainable production of energy from microalgae: review of culturing systems, economics, and modelling. *Journal of Renewable and Sustainable Energy* 7, 012701.

Zinglé, C., Kuntz, L., Tritsch, D. *et al.* (2010) Isoprenoid biosynthesis via the methylerythritol phosphate pathway: structural variations around phosphonate anchor and spacer of fosmidomycin, a potent inhibitor of deoxyxylulose phosphate reductoisomerase. *The Journal of Organic Chemistry* 75(10), 3203–3207.

8 Green Extraction of Bioactive Compounds from Microalgae by Ionic Liquids

Xiangxiang Zhang[1], Yali Zhu[1] and Quanyu Zhao[1]*

[1]*School of Pharmaceutical Science, Nanjing Tech University, Nanjing, China*

Abstract

Microalgae are an available source for the production of bioactive compounds such as foods, pharmaceuticals and so on. It is possible to use the microalgae powder directly, or fresh microalgae. It has been identified that some crude extracts of microalgae have anti-oxidant, anti-tumor and anti-viral activities. The bioactive compounds in microalgae include proteins, lipids, carbohydrates, pigments and some secondary metabolites. It is necessary to extract the target bioactive compounds in a green approach. Ionic liquids are green solvents for extraction of bioactive compounds in microalgae. Ionic liquid could be used as a cell wall disruption agent and integrated with organic solvent to isolate the bioactive compounds with high efficiency. Much more important, aqueous biphasic systems using ionic liquids have been applied for the extraction of lipids, proteins and pigments in microalgae. It is important how to select an ionic liquid assisting cell wall disruption or to construct an aqueous biphasic system. The current state of extraction of bioactive compounds in microalgae assisted by ionic liquids is reviewed in this chapter. In the future, the selectivity and efficiency should be improved for the extraction of bioactive compounds by ionic liquids. The environmental ecological effect of ionic liquids should be evaluated. Although ionic liquids are considered as green solvent, the possible inhibition effect on microalgae and other organisms should be investigated. The scale-up of bioactive compounds extraction will be improved by ionic liquids.

Keywords: microalgae; ionic liquids; cell wall deconstructing; bioactive compounds; aqueous biphasic systems

1. Introduction

Microalgae are photosynthetic autotrophic microorganisms, which can use light energy and inorganic substances (CO_2, nitrogen and phosphorus etc.) to synthesize a large number of valuable bioactive compounds, such as oils, proteins, carbohydrates, pigments and secondary metabolites. These active substances have anti-oxidant, anti-inflammatory, anti-tumor and immunomodulatory activities shown in Table 8.1. They can not only be applied as food to meet the needs of nutrients, but they can also be used in the pharmaceutical industry. They have the effects of reducing the risk of cardiovascular disease and fighting cancer cells. *Spirulina* has high protein content, which can be directly edible in the form of algae powder (Deng and Chow, 2010), filtrates (Smieszek et al., 2017) or water extract (Czerwonka et al., 2018). It was also confirmed that aqueous extract and methanol extract of *Chlorella stigmatophora* and *Phaeodactylum tricornutum* had anti-inflammatory activity (Guzman et al., 2001). In a further study, it was confirmed that the

Address for correspondence: zhaoqy@njtech.edu.cn

Table 8.1. Bioactive compounds from microalgae and their activities.

Microalgae	Bioactive compounds	Category	Activity	References
Spirulina platensis	C-phycocyanin	Protein	Immunomodulating activity	Chen *et al.* (2014)
Chlamydomonas hedleyi	Mycosporine-Like Amino Acids (MAAs)	Protein	Anti-inflammation activity	Suh *et al.* (2014)
Isochrysis galbana [not available]	Fatty acids	Lipid	Anti-inflammatory activity	Bonfanti *et al.* (2018)
	Fucoxanthin	Pigment	Anti-inflammatory activity and anti-oxidant activity	Rodriguez-Luna *et al.* (2018)
Cronbergia siamensis and *Sphaerospermopsis aphanizomenoides*	Flavonoids and carotenoids	Secondary metabolite and pigment	Anti-inflammatory and anti-oxidant activity	Badr *et al.* (2019)
Chlorella vulgaris	Extracellular polysaccharide	Carbohydrate	Anti-asthmatic activity	Barborikova *et al.* (2019)
Amphidinium carterae	Amphidinol 22	Secondary metabolite	Anti-fungal	Martinez *et al.* (2019)
Phaeodactylum tricornutum	Fucoxanthin	Pigment	Anti-proliferative and antioxidant activities	Neumann *et al.* (2019)
Porphyridium cruentum	Sulfated polysaccharides	Polysaccharides	Immunomodulatory and anti-oxidant activity	Casas-Arrojo *et al.* (2021)
P. cruentum (purpureum)	Exopolysaccharide	Carbohydrate	Immunostimulant	Risjani *et al.* (2021)
Chlorococcum sp. SABC 012504	Bioactive lipids	Lipids	Anti-inflammatory and anti-thrombotic activities	Shiels *et al.* (2021)

major bioactive component was crude polysaccharides (Guzman et al., 2003). In order to make more accurate use of a natural product in microalgae, it is necessary to extract these active substances by green and efficient approach, and then develop it into food or pharmaceuticals. Utilization of microalgae-based bioactive compounds is an important content of biorefinery of microalgae besides biodiesel and other biofuels development. In addition, it proposes a potential pathway for reducing CO_2 emission.

2. Bioactive Compounds in Microalgae

2.1 Lipids

Microalgae can accumulate high oil content under nitrogen-limited and other stress conditions. These lipids can be used as not only biodiesel but also many unsaturated fatty acids. The essential unsaturated fatty acids needed by the human body cannot be produced by the human body itself, and must be supplemented through dietary intake. They play a vital role in health. Microalgae, especially marine microalgae, are good sources of unsaturated fatty acids, such as linolenic acid, docosahexaenoic acid (DHA) and eicosapentaenoic acid (EPA) (Balakrishnan et al., 2019). These essential fatty acids are not only important nutrients in the human body but can also help prevent and treat a series of human diseases. EPA is an essential unsaturated fatty acid and plays a great role in reducing the blood viscosity and cholesterol level of the human body, reducing the incidence of heart disease and depression, preventing cardiovascular diseases such as cerebral thrombosis and hypertension, increasing membrane fluidity, and promoting the development of the nervous system. Therefore, it is suggested that adults need to supplement EPA every day. At present, the main sources of EPA are fish oil and seafood. However, the production of fatty acids in these organisms is affected by weather, environment and other factors. Therefore, in order to meet the gap and obstacles between the growing human demand for EPA and the current supply, it is necessary to find an appropriate and sustainable alternative source to produce EPA. It was shown that under suitable conditions, the contents of EPA were up to 39% and 57% of total fatty acids for *Nannochloropsis* sp. and *P. tricornutum*, respectively (Martins et al., 2013). The EPA contents per dry mass (g/kg) of both microalgae were higher than those in salmon, herring, tuna, codfish and other fishes (Schade et al., 2020). It is indicated that there are lots of bioactive lipid compounds in microalgae besides EPA and DHA. The anti-oxidant and anti-inflammatory activities were identified in the polar lipids in *Gloeothece* sp. (da Costa et al., 2020).

2.2 Protein

Microalgae contain a large amount of protein, and the proteins are similar to traditional protein sources (such as soybean, milk and animal meat). Microalgae protein is also recognized for its high quality, so that using microalgae to produce a series of protein products is a good choice. Although microalgae are rich in pigments that are widely used commercially, phycobiliprotein is the most widely used (Porav et al., 2020). Phycobiliprotein is a dark complex protein compound with water solubility. It has a natural coloring effect and has been widely used in food, cosmetics and pharmaceuticals. But its more important function is to capture the light energy of 495–550 nm wavelength and transfer it to chlorophyll a in the photosynthetic reaction center. In addition, other characteristics of phycobiliproteins are also very suitable for commercial applications. For example, they have a large number of chromophores which can be used as specific fluorescent probes with high sensitivity for fluorescent labeling; they have good water solubility, and can form very stable conjugates with many materials, such as antibodies, biotin, etc. This ability makes the potential development of phycobiliprotein more prominent. C-phycocyanin (Chen et al., 2014) and Mycosporine-Like Amino

Acids (MAAs) (Suh et al., 2014) were also isolated from *Spirulina platensis* and *Chlamydomonas hedleyi*, respectively. *Chlamydomonas reinhardtii* is the model microalga. Its genome has been sequenced and some genetic engineering tools have been developed. It is possible to make a genetic engineering modification to produce bioactive peptides with anti-hypertensive, opioid, anti-microbial, and hypocholesterolemic activities from *C. reinhardtii* (Campos-Quevedo et al., 2013).

2.3 Carbohydrate

Carbohydrate is also one of the main biological components of microalgae. Starch in microalgae is fixed by photosynthesis. Carbohydrates in microalgae could be substrate for ethanol fermentation. The bioactive carbohydrates in microalgae are mainly extracellular polysaccharides. Exopolysaccharide (EPS) extracted from marine microalgae, *Thraustochytriidae* sp. (KCTC 12770BP), inhibited cell growth of BG-1 ovarian, MCF-7 breast and SW-620 colon cancer cell lines *in vitro* although the dilution was 10^{-11} (Park et al., 2017). About 0.0625 mg/mL EPS isolated from marine microalgae, *Porphyridium marinum*, inhibited the growth of murine mammary carcinoma cells 4T1 (Gargouch et al., 2021). Crude polysaccharides of *Tribonema minus* (Huo et al., 2022) and *Rhodosorus* sp. SCSIO-45730 (Wang et al., 2022) had anti-oxidant activity. These carbohydrates are candidates for pharmaceuticals.

2.4 Pigments

Carotenoids are important nutrients for human beings. Their main role is the precursor of vitamin A in the human body and participation in human metabolism. For example, carotenoids are often used in food and health products as food colorants and additives. However, only very few are used commercially (β-carotene, astaxanthin and lutein).

Lutein, as an edible natural pigment, plays an important role in medicine, food, animal feed, cosmetics and others, because of its rich nutritional value and wide range of applications (Patel et al., 2022). According to recent studies, lutein can not only be used for macular degeneration (Steiner et al., 2018), but also be added to cosmetics and skincare products as an anti-oxidant to delay aging (Mitri et al., 2011). Lutein is closely related to light-capture complexes in photosynthesis of green plants. Although the commercial source of lutein is still extracted from marigold (Lin et al., 2015), it has gradually replaced marigold as a high-quality natural source of lutein due to the higher content of lutein in microalgae, short growth cycle, and less occupation of cultivated land. The microalgae that can be used for commercial production of lutein include *Chlorella* (Khoo et al., 2021a) and *Scenedesmus* (Minhas et al., 2016).

β-carotene is the most common pigment in nature and the most stable natural pigment. Two-thirds of the vitamin A in the human body is made of β-carotene and is synthesized, and it can also assist photosynthesis in green plants. The content of β-carotene in *Dunaliella* was up to 12.9% (Sui et al., 2021). Astaxanthin is another carotenoid that can be extracted from microalgae. Astaxanthin is not only widely used in drugs, cosmetics and health products, but is also an effective anti-oxidant. It plays an important role in regulating human hormone levels, resisting inflammation and enhancing human immunity. At present, the algal strain for commercial production of astaxanthin is *Haematococcus pluvialis*. Fucoxanthin is also an important pigment as an anti-oxidant (Neumann et al., 2019). It is always isolated from marine macroalgae and microalgae (Peng et al., 2011). The potential applications in cancer therapy and others were investigated (Neumann et al., 2019; Meresse et al., 2020).

2.5 Secondary metabolites

Besides major biological components of microalgae, there are amounts of secondary metabolites in microalgae. Some of them are bioactive. A stigmasterol ($C_{29}H_{48}O$) was isolated from marine microalgae, *Navicula in-*

certa (Kim *et al.*, 2014). About 20 µM of this stigmasterol led to 54% toxicity to HePG2 cell. It has potential to be used as an anti-cancer pharmaceutical against live cancer. Much more effort should be made for the isolation of secondary metabolites and evaluation of their bioactivities. Amphidinol 22 was extracted from *Amphidinium carterae*, which was a newly identified natural product (Martinez *et al.*, 2019). The concentrations of these secondary metabolites in microalgae may be very low. It presents new direction for drug development. These secondary metabolites could be synthesized through organic chemistry pathway. Alternatively, they could be extracted from microalgae after a large-scale cultivation.

3. Ionic Liquids

Ionic liquids consist of anions and cations as liquid at room temperature. In addition, they have many properties of conventional organic solvents, such as excellent solvation performance, variable viscosity property, wide temperature range, non-volatile and excellent chemical and thermal stability (Wang *et al.*, 2010). Ionic liquids are ideal substitutes for traditional volatile solvents and environmentally friendly liquids. They are a new generation of green design solvents in chemical research. When salt or ionic liquid is added to the solution of hydrophilic ionic liquid, two phases are formed. It is widely used in the separation and purification of natural compounds, and has attracted more and more attention.

Ionic liquids are composed of organic cations, and inorganic or organic anions. Some ionic liquids used for extraction of bioactive compounds in microalgae are shown in Fig. 8.1. Common cations include tetramethylguanidinium (Chiappe *et al.*, 2016), 1-ethyl-3-methylimidazolium (Desai *et al.*, 2016; Orr *et al.*, 2016; Wahidin *et al.*, 2016; Choi *et al.*, 2019), cholinium (Chua *et al.*, 2019), 1-butyl-3-methylimidazolium (Liu *et al.*, 2019; Zhu *et al.*, 2021a,b), tetrakis (hydroxymethyl) phosphonium (Olkiewicz *et al.*, 2015), 1-octyl-3-methylimidazolium (Chang *et al.*, 2018; Santos *et al.*, 2018), choline (Ruiz *et al.*, 2018a, b, 2020a), and tributyl-1-tetradecylphosphonium (Suarez Ruiz *et al.*, 2020), shown in Tables 8.2 and 8.3. Anions include methyl carbonate/bicarbonate (Chiappe *et al.*, 2016), ethyl sulfate (Orr *et al.*, 2016), hydrogen sulfate (Choi *et al.*, 2019), arginate (Chua *et al.*, 2019), dibutylphosphate (Desai *et al.*, 2016), dihydrogenphosphate (Ruiz *et al.*, 2018a, b, 2020a), chloride (Olkiewicz *et al.*, 2015; Liu *et al.*, 2019), and bromide (Chang *et al.*, 2018; Zhu *et al.*, 2021a,b). According to the different solubility of ionic liquids in water, ionic liquids can also be divided into hydrophilic ionic liquids and hydrophobic ionic liquids. Lots of new kinds of ionic liquids have been developed. The possible extraction capabilities for bioactive compounds in microalgae should be explored by experiments.

The combination of various cations and anions makes the constituent ions asymmetric and large in structure, which is the key for ionic liquids to distinguish salts in the traditional sense. For example, in the case of salt, the relatively strong ion interaction between closely packed symmetrical ions is the reason why the melting point is higher than that of ionic liquids. Ionic liquids are often called 'designable solvents' and 'future solvents' because they can meet the needs of industry and scientific research by changing the types of cations and anions, the length of alkane chains in the structure, and the types of substituent functional groups. At the same time, due to the mild reaction conditions and green process, ionic liquids have higher efficiency and less risk than traditional organic solvents, which have attracted more and more attention. The ionic liquids used for extractions of bioactive compounds in microalgae are shown in Table 8.2 and 8.3.

4. Ionic Liquid-assisted Cell Wall Crush before Extraction

At present, many methods have been applied for extracting the active substances of microalgae.

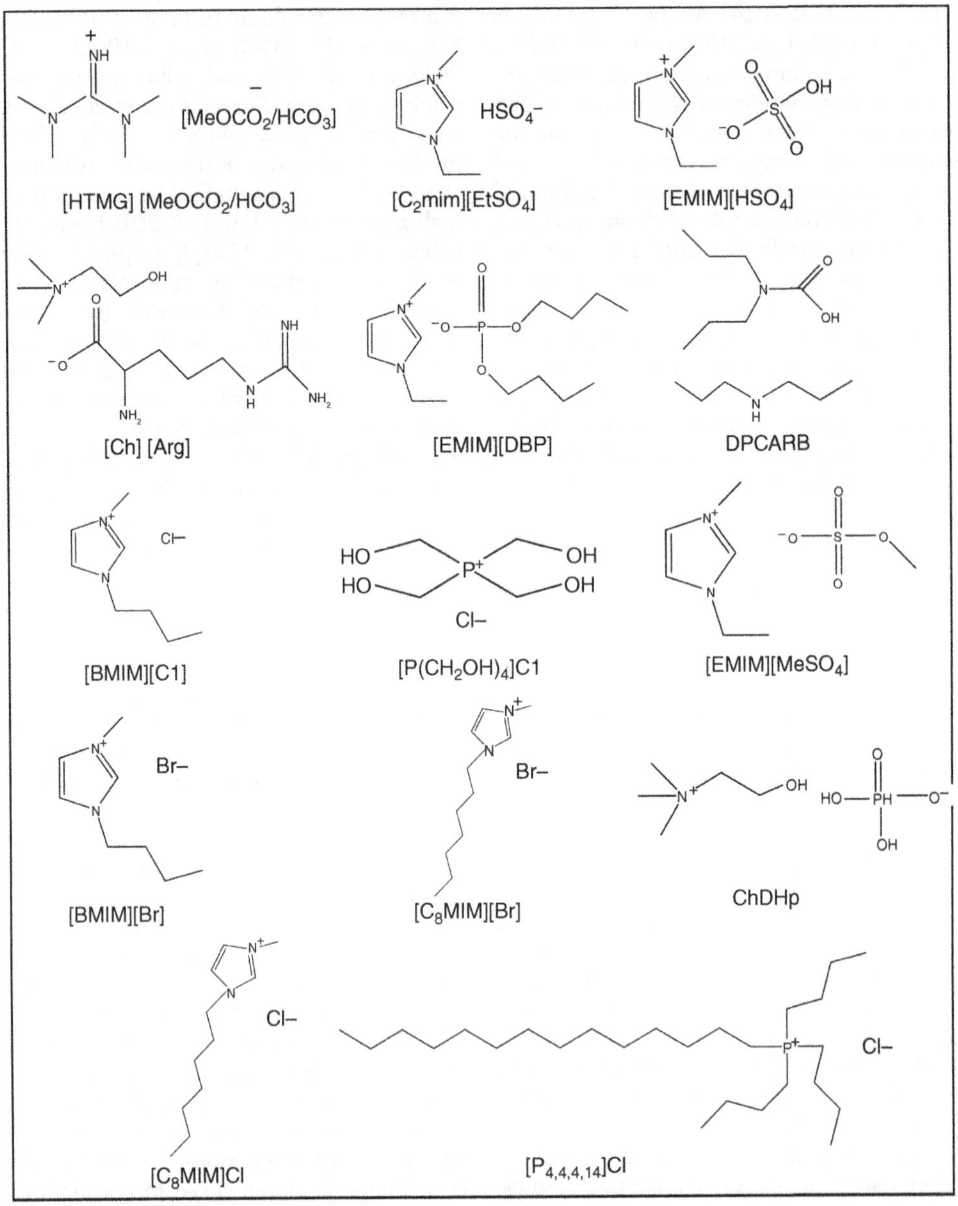

Fig. 8.1. Lists of some ionic liquids using for bioactive compounds extraction.
[HTMG][MeOCO$_2$/HCO$_3$], tetramethylguanidinium methyl carbonate/bicarbonate; [C$_2$mim][EtSO$_4$], 1-ethyl-3-methylimidazolium ethyl sulfate; [EMIM][HSO$_4$], 1-ethyl-3-methylimidazolium hydrogen sulfate; [Ch][Arg], Cholinium arginate; [EMIM][DBP], 1-Ethyl-3-methylimidazolium dibutylphosphate; DPCARB, dipropylammonium dipropylcarbamate; [BMIM]Cl, 1-Butyl-3-methylimidazolium Chloride; [P(CH$_2$OH)$_4$]Cl, tetrakis (hydroxymethyl)phosphonium chloride; [EMIM][MeSO$_4$], 1-butyl-3-methylimidazolium methyl sulfate; [BMIM]Br, 1-Butyl-3-methylimidazolium Bromide; [C$_8$MIM]Br, 1-octyl-3-methylimidazolium bromide; ChDHp, choline dihydrogenphosphate; [C$_8$MIM]Cl, 1-methyl-3-octylimidazolium chloride; [P$_{4,4,4,14}$]Cl, tributyl-1-tetradecylphosphonium chloride.

Table 8.2. Ionic liquid-assisted cell wall disruption for bioactive compound extraction.

Microalgae	Algal status	Ionic liquids	Target chemical	Efficiency (%)	References
Scenedesmus obliquus	Wet	[HTMG][MeOCO$_2$/HCO$_3$]	Lipid	88	Chiappe et al. (2016)
Chlorella vulgaris	Wet	[C$_2$mim][EtSO$_4$]**	Lipid	32	Orr et al. (2016)
H. pluvialis	Dried	[EMIM][HSO$_4$]*	Astaxanthin	> 99	Choi et al. (2019)
			Lipid	82	
Nannochloropsis sp.	Dried	[Ch][Arg]	Lipid	94.1	Chua et al. (2019)
H. pluvialis	Dried	[EMIM][DBP]	Astaxanthin	> 70	Desai et al. (2016)
Chlorella sorokiniana	Dried	DPCARB**	Lutein	97	Khoo et al. (2021a)
Chaetoceros calcitrans	Dried	DPCARB*	Fucoxanthin	17.51 mg/g	Khoo et al. (2021b)
H. pluvialis	Dried	[BMIM]Cl*	Astaxanthin	> 80	Liu et al. (2019)
C. vulgaris	Dried	[P(CH$_2$OH)$_4$]Cl	Lipid	8.1	Olkiewicz et al. (2015)
Nannochloropsis oculata	Dried	[P(CH$_2$OH)$_4$]Cl	Lipid	12.8	Olkiewicz et al. (2015)
N. oculata	Wet	[P(CH$_2$OH)$_4$]Cl	Lipid	14.6	Olkiewicz et al. (2015)
Nannochloropsis sp.	Dried	[EMIM][MeSO$_4$]	Biodiesel	36.79	Wahidin et al. (2016)
Chlorella sp.	Dried	[BMIM]Br	Lutein	98.06	Zhu et al. (2021a)
Chlorella sp.	Wet	[BMIM]Br	Lutein	97.73	Zhu et al. (2021a)

*, mixture of ionic liquid and water; **, mixture of ionic liquid + methanol

Table 8.3. Bioactive compound extraction by aqueous two-phase systems with ionic liquids.

Microalgae	Ionic liquids	Polymer or salt	Target chemical	Efficiency (%)	References
Spirulina platensis	C8MIM-Br	K_3PO_4	C-phycocyanin	99	Chang et al. (2018)
Neochloris oleoabundans	ChDHp	PEG 400	Free glucose	99	Ruiz et al. (2020a)
			Starch	70	
Neochloris oleoabundans	ChDHp	PEG 400	Protein	61	Ruiz et al. (2018a)
			Lutein	97.3	
			Chlorophyll	51.6	
-	ChDHp	PEG 400	Rubisco	79.6	Ruiz et al. (2018b)
Isochrysis galbana	$[C_8min]Cl$	K_3PO_4	Proteins,	100	Santos et al. (2018)
			Carbohydrate	71.21	
Neochloris oleoabundans	$[P_{4,4,4,14}]Cl$	PEG 8000 + NaPA 8000	Proteins	100	Suarez Ruiz et al. (2020)
			Carbohydrates	80	
			Lutein	98	

For example, lipids in microalgae are generally extracted with low-boiling-point volatile organic solvents. Although this method is simple and effective, the toxicity of organic reagents is also a hidden danger of this method. Most organic solvents have high toxicity. If there are some organic solvent residues in subsequent purification and separation procedures, the relevant products will also have a certain risk coefficient. Among microalgae, which can be used as high-quality biofuels, the most commonly used extraction method is solvent extraction. Although extraction can get the target products, it also has the defects of being time-consuming, laborious and low yield. For the extraction of pigments from microalgae, although the traditional extraction methods are still used at present, the defects of complex extraction methods and long cycle will lead to a large amount of energy loss and pigment degradation in microalgae cells. Although many green and safe extraction methods have appeared in recent years, such as microwave-assisted extraction (MAE), ultrasonic-assisted extraction (UAE) and supercritical fluid extraction (SFE), their extraction effects need to be further investigated. Recently, some new technologies, such as surfactants and convertible solvents, have been applied to extract microalgae active substances, but they often target a single product, reducing or ignoring the value of other molecules. Most of the research on microalgae is focused on a single compound, and almost all other compounds are discarded, resulting in a great waste of resources. For example, organic solvents are often used to separate hydrophobic compounds such as lipids and pigments from broken microalgae, while water-soluble compounds such as proteins and carbohydrates will be discarded. Moreover, if only a single component is extracted, the biomass production and downstream process of microalgae are too expensive. The extraction of multi-component compounds is particularly important. Several microalgae components can be extracted at the same time to supply different markets. Therefore, it is particularly important to develop a technology that is mild, economical, low-energy consumption, high efficiency, high safety and can separate valuable active molecules from microalgae.

Because the cell wall of microalgae is very solid (de Carvalho et al., 2020), and the active compounds accumulated in microalgae are in the cells, in order to extract the active compounds it is necessary to break the cells and release the contents for extraction. The methods of cell disruption include ball-milling, ultrasonic method, ionic liquid (Singh et al., 2013; Shankar et al., 2017) and enzyme (Ye et al., 2020). Different disruption methods also have different advantages and disadvantages according to the substances extracted. For example, the ball-milling method is a very common cell disruption method, which can completely crush the cell wall of microalgae, with good effect, and is widely used. However, when using homogenizer for grinding, it requires a higher rotation speed. For each working cycle, it is necessary to ice-bath the algae liquid for a while before the next cycle, so as to reduce the loss of microalgae active compounds, which is time-consuming and laborious. Ultrasonic disruption is also a widely used cell disruption method with simple operation and good repeatability. However, because ultrasonic method is an energy-consuming method, there may be the loss of some active compounds in the ultrasonic process, which will affect the efficiency of subsequent extraction. The experimental conditions of enzyme disruption are relatively mild, and the degree of cell wall disruption can be controlled, but the price of enzyme disruption is high, so it is not easy to be used on a large scale. If the users want to recover the enzyme, it is necessary to increase the steps of purifying the enzyme, and it is not suitable for protein extraction, which brings great difficulties to further purification. In recent years, some studies have focused on the disruption of microalgae cell walls by ionic liquids. This method is simple and efficient, but the exact disruption mechanism is not fully understood.

As shown in Table 8.2, ionic liquids could be available for bioactive compounds from both wet and dried algal biomass. Freeze-drying is also one method for cell wall disruption so that the performances of cell disruption by ionic liquid for wet and dried microalgae are different. Although the extraction efficiency of dried microalgae was a little higher than that of wet microalgae (Zhu et al., 2021a), the lutein content of dried microalgae was lower than that of wet microalgae. It is indicated that there is loss of lutein in the freeze-drying process. In addition, freeze-drying needs energy consumption. Therefore, using wet microalgae is a better choice for bioactive compounds extraction. Freeze-drying is beneficial for storage of algal biomass. If the cultivation of microalgae and extraction of bioactive compounds are not conducted in the same place, freeze-drying is necessary. It is difficult to store wet microalgae, and the bioactive compounds in wet microalgae may not be stable.

Ionic liquid disrupts the hydrogen bond network in algal cell wall (Chua et al., 2019; Ye et al., 2020). Water (Choi et al., 2019; Zhu et al., 2021a; Khoo et al., 2021b) or methanol (Orr et al., 2016; Khoo et al., 2021a) was added to help break down cell wall. It was considered that water was not free in aqueous solution of ionic liquid. There are multi-steps in the cell wall disruption and extraction process. Ionic liquid-assisted cell wall disruption is an important step. The choice of ionic liquid is not always rational. The related mechanism is unclear. It is recognized that the tensioactive properties are essential for the cell disruption of microalgae and extraction of bioactive compounds (Suarez Ruiz et al., 2020).

5. Ionic Liquid–assisted Aqueous Two-phase Extraction

Aqueous two-phase system (ATPS) is made of polymers, salts and ionic liquids, which can separate a variety of products at the same time. It is an efficient, mild and scalable separation technology with high biocompatibility. In some cases, it is also denoted as aqueous biphasic systems (ABS). It is used for the extraction and purification of biological molecules (Fig. 8.2). ATPS consists of polymers and salts (or ionic liquids) separated into two immiscible phases in aqueous media. Because most of the components in ATPS are water and those with low interfacial tension, it can not only provide them with a mild environment when extracting biological molecules, but also the method of extracting substances using aqueous two-phase is not complex and is easy to operate, and it also has significant efficiency in the purification of different biological materials such as cells, viruses, nucleic acids, enzymes carbohydrates, lipids and proteins. Purification methods may destroy the structure or activity of proteins. Suitable ATPS is beneficial for extraction of proteins and preserves the bioactivities of proteins. The extraction efficiencies were almost 100% (Chang et al., 2018; Santos et al., 2018). Besides proteins, ATPS was also utilized to isolate carbohy-

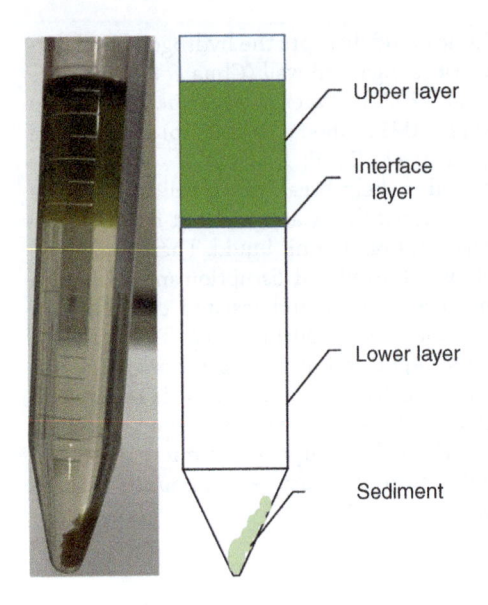

Fig. 8.2. Typical aqueous two-phase system for extraction of bioactive compounds in microalgae.

drates (Santos et al., 2018; Ruiz et al., 2020a) and pigments (Ruiz et al., 2018a). It is proved that ATPS is effective to extract proteins, carbohydrates and pigments in a highly efficient way.

In recent years, there have also been some relevant reports on the application of aqueous two-phase technology in extracting active substances from microalgae. At the same time, aqueous two-phase technology is different from traditional methods in extracting target substances. Firstly, ATPS should be constructed. In general, polymer, ionic liquid and water are three components in it. The concentrations of these three components are critical. In the ternary-phase diagrams, there are single-phase region and ATPS region. With the development of each aqueous two-phase system, the data of two-phase equilibrium and the corresponding thermodynamic characterization are particularly important (Martins et al., 2016; Jimenez et al., 2020; Hamzehzadeh and Fardshirayeh, 2021). Ternary-phase diagram is helpful to construct the ATPS. If the ATPS is constructed, the partition coefficients of target compounds in two phases should be evaluated (Suarez Ruiz et al., 2020). High partition coefficient leads to high extraction efficiency of the target compounds. Aqueous two-phase technology can achieve the extraction of multiple target products.

Extraction of bioactive compounds by ATPS should be integrated with cell wall disruption methods including bead milling, sonication (Di Caprio et al., 2021) and ionic liquid (Suarez Ruiz et al., 2022). It was shown that bead milling was better than sonication for the cell disruption of *Tetradesmus obliquus* (Di Caprio et al., 2021). In a multiple steps process, $[P_{4,4,4,14}]$Cl aqueous solution was first used for enhancing cell disruption of *Neochloris oleoabundans*. Secondly, PEG 8000 and NaPA8000 were added to construct ATPS. Finally, the proteins, carbohydrates, lutein and chlorophylls were purified by ultrafiltration and organic solvent extractions. The ionic liquid and polymers could be recycled. Based on the biorefinary concept, it is better to utilize much more biological molecules in microalgae (Suarez Ruiz et al., 2022). Therefore, researchers tried to extract multiple bioactive compounds in microalgae (Ruiz et al., 2018a; Suarez Ruiz et al., 2020; Ruiz et al., 2020b). In the opinion of unit operation, the whole process including cell disruption, fractionation and purification will be optimized. The core unit is fractionation by ATPS. The biological molecules will be in two phases or interface. The selectivity of multiple bioactive compounds in one separator is different (Suarez Ruiz et al., 2020). Then, one main compound was isolated in it. The rest will be isolated by other steps (Ruiz et al., 2020b).

6. Discussion

Nowadays, a large number of organic solvents are widely used in the chemical and food industries. Some organic solvents with potential environmental risks accumulate in water and soils. These environmental pollutants pose a serious threat to human health and social development. The choice of environmentally friendly and green solvents has attracted a great deal of attention. Recently, some ionic liquids have been found to be toxic to microorganisms, plants etc. If these ionic liquids are discharged into the environment, due to their high stability and low biodegradability, they may cause water pollution and present negative effects on the ecological environment. Whether ionic liquids are green solvents for the extraction of lipids and bioactive compounds is an important topic. New prepared ionic liquids emerge. To choose ionic liquids as cell disruption agents, in addition to the high extraction efficiency of bioactive compounds, it is necessary to investigate whether they are environmentally friendly.

Microalgae are important primary producers widely distributed in water. Microalgae are also an important part of the food chain in the aquatic ecosystem. Because they are sensitive to chemical substances, microalgae can be used as the object of environmental monitoring (Quraishi et al., 2017). The toxicological effects of pollutants on microalgae can not only reveal the threat to algal communities, but they also clarify the potential harm to higher organisms. *Dunaliella salina, Chlorella* and *Anabaena* are model microalgae for the detection of ecotoxicity of ionic liquids and other compounds. Therefore, three kinds of algae were selected, namely freshwater algae *Anabaena, Chlorella pyrenoidosa* and green algae *D. salina* in seawater (Zhu et al., 2021b). The acute toxicity of the four ionic liquids, namely [EMIM] EtOSO$_3$, [BMIM] Cl, [BMIM] Br and [EMIM] Cl, was evaluated. The growth inhibition rate and chlorophyll content of these three microalgae were detected within 96 hours, oxidative stress response and anti-oxidant enzyme activity. It was indicated that the toxicity of imidazolium salt with anion bromide to a phytoplankton *Selenastrum capricornutum* was related to alkyl-chain length of ionic liquids (Cho et al., 2007). In addition, long alkyl-chains led to higher toxcity than short ones. It is necessary to establish structure-ecotoxicity relationship of ionic liquids (Santos et al., 2015; Vanderveen et al., 2015) besides experimental investigations. It is helpful to evaulate the ecotoxicity.

7. Conclusions

Microalgae are a source for bioactive compounds including proteins, lipids, carbohydrates and secondary metabolites. Ionic liquids are green solvents for the extraction of bioactive compounds in microalgae. They are effective for cell wall disruption. In the aqueous two-phase system, ionic liquid is also a solvent. Multiple products extraction is essential in the field of biorefinery. It should improve the selectivity and efficiency during the extraction process. In addition, the choice of ionic liquid is the first step. New ionic liquids will be investigated in the future.

Acknowledgments

The authors acknowledge the support received from the National Natural Science Foundation of China (22038007).

References

Badr, O.A.M., El-Shawaf, I.I.S., El-Garhy, H.A.S., Moustafa, M.M.A. and Ahmed-Farid, O.A. (2019) The potent therapeutic effect of novel cyanobacterial isolates against oxidative stress damage in redox rats. *Journal of Applied Microbiology* 126(4), 1278–1289.

Balakrishnan, J., Dhavamani, S., Sadasivam, S.G., Arumugam, M., Vellaikumar, S. et al. (2019) Omega-3-rich *Isochrysis* sp. biomass enhances brain docosahexaenoic acid levels and improves serum lipid profile and antioxidant status in Wistar rats. *Journal of the Science of Food and Agriculture* 99(13), 6066–6075.

Barborikova, J., Sutovska, M., Kazimierova, I., Joskova, M., Franova, S. et al. (2019) Extracellular polysaccharide produced by *Chlorella vulgaris* – chemical characterization and anti-asthmatic profile. *International Journal of Biological Macromolecules* 135, 1–11.

Bonfanti, C., Cardoso, C., Afonso, C., Matos, J., Garcia, T. et al. (2018) Potential of microalga *Isochrysis galbana:* bioactivity and bioaccessibility. *Algal Research-Biomass Biofuels and Bioproducts* 29, 242–248.

Campos-Quevedo, N., Rosales-Mendoza, S., Teresita Paz-Maldonado, L.M., Martinez-Salgado, L., Carlos Guevara-Arauza, J. et al. (2013) Production of milk-derived bioactive peptides as precursor chimeric proteins in chloroplasts of *Chlamydomonas reinhardtii*. *Plant Cell Tissue and Organ Culture* 113(2), 217–225.

Casas-Arrojo, V., Decara, J., de los Angeles Arrojo-Agudo, M., Perez-Manriquez, C. and Abdala-Diaz, R.T. (2021) Immunomodulatory, antioxidant activity and cytotoxic effect of sulfated polysaccharides from *Porphyridium cruentum*. (S.F.Gray) Nageli. *Biomolecules* 11(4), 488.

Chang, Y.-K., Show, P.-L., Lan, J.C.-W., Tsai, J.-C. and Huang, C.-R. (2018) Isolation of C-phycocyanin from *Spirulina platensis* microalga using ionic liquid based aqueous two-phase system. *Bioresource Technology* 270, 320–327.

Chen, H.-W., Yang, T.-S., Chen, M.-J., Chang, Y.-C., Wang, E.I.C. et al. (2014) Purification and immunomodulating activity of C-phycocyanin from *Spirulina platensis* cultured using power plant flue gas. *Process Biochemistry* 49(8), 1337–1344.

Chiappe, C., Mezzetta, A., Pomelli, C.S., Iaquaniello, G., Gentile, A. et al. (2016) Development of cost-effective biodiesel from microalgae using protic ionic liquids. *Green Chemistry* 18(18), 4982–4989.

Cho, C.W., Pham, T.P., Jeon, Y.C., Vijayaraghavan, K., Choe, W.S. et al. (2007) Toxicity of imidazolium salt with anion bromide to a phytoplankton *Selenastrum capricornutum*: effect of alkyl-chain length. *Chemosphere* 69(6), 1003–1007.

Choi, S.-A., Oh, Y.-K., Lee, J., Sim, S.J., Hong, M.E. et al. (2019) High-efficiency cell disruption and astaxanthin recovery from *Haematococcus pluvialis* cyst cells using room-temperature imidazoliumbased ionic liquid/water mixtures. *Bioresource Technology* 274, 120–126.

Chua, E.T., Brunner, M., Atkin, R., Eltanahy, E., Thomas-Hall, S.R. et al. (2019) The ionic liquid cholinium arginate is an efficient solvent for extracting high-value *Nannochloropsis* sp. lipids. *ACS Sustainable Chemistry & Engineering* 7(2), 2538–2544.

Czerwonka, A., Kalawaj, K., Slawinska-Brych, A., Lemieszek, M.K., Bartnik, M. et al. (2018) Anticancer effect of the water extract of a commercial *Spirulina* (*Arthrospira platensis*) product on the human lung cancer A549 cell line. *Biomedicine & Pharmacotherapy* 106, 292–302.

da Costa, E., Amaro, H.M., Melo, T., Guedes, A.C. and Domingues, M.R. (2020) Screening for polar lipids, antioxidant, and anti-inflammatory activities of *Gloeothece* sp. lipid extracts pursuing new phytochemicals from cyanobacteria. *Journal of Applied Phycology* 32(5), 3015–3030.

de Carvalho, J.C., Magalhaes, A.I., Jr, de Melo Pereira, G.V., Medeiros, A.B.P., Sydney, E.B. et al. (2020) Microalgal biomass pretreatment for integrated processing into biofuels, food, and feed. *Bioresource Technology* 300 122719.

Deng, R. and Chow, T.-J. (2010) Hypolipidemic, antioxidant, and antiinflammatory activities of microalgae Spirulina. *Cardiovascular Therapeutics* 28(4), e33–e45.

Desai, R.K., Streefland, M., Wijffels, R.H. and Eppink, M.H.M. (2016) Novel astaxanthin extraction from *Haematococcus pluvialis* using cell permeabilising ionic liquids. *Green Chemistry* 18(5), 1261–1267.

Di Caprio, F., Chelucci, R., Francolini, I., Altimari, P. and Pagnanelli, F. (2021) Extraction of microalgal starch and pigments by using different cell disruption methods and aqueous two-phase system. *Journal of Chemical Technology & Biotechnology* 97(1), 67–78.

Gargouch, N., Elleuch, F., Karkouch, I., Tabbene, O., Pichon, C. et al. (2021) Potential of exopolysaccharide from *Porphyridium marinum* to contend with bacterial proliferation, biofilm formation, and breast cancer. *Marine Drugs* 19(2), 66.

Guzman, S., Gato, A. and Calleja, J.M. (2001) Antiinflammatory, analgesic and free radical scavenging activities of the marine microalgae *Chlorella stigmatophora* and *Phaeodactylum tricornutum*. *Phytotherapy Research* 15(3), 224–230.

Guzman, S., Gato, A., Lamela, M., Freire-Garabal, M. and Calleja, J.M. (2003) Anti-inflammatory and immunomodulatory activities of polysaccharide from *Chlorella stigmatophora* and *Phaeodactylum tricornutum*. *Phytotherapy Research* 17(6), 665–670.

Hamzehzadeh, S. and Fardshirayeh, M. (2021) Effect of ionic liquid [C_4C_1im]Br on the formation of aqueous biphasic systems composed of peg and biodegradable salts, and on the partition of l-tryptophan. *Fluid Phase Equilibria* 535, 112971.

Huo, S., Wang, H., Chen, J., Hu, X., Zan, X. et al. (2022) A preliminary study on polysaccharide extraction, purification, and antioxidant properties of sugar-rich filamentous microalgae *Tribonema minus*. *Journal of Applied Phycology*. DOI: 10.1007/s10811-021-02630-w

Jimenez, Y.P., Roman Freijeiro, C., Soto, A. and Rodriguez, O. (2020) Phase equilibrium for polymer/ionic liquid aqueous two-phase systems. *Fluid Phase Equilibria* 506, 112387.

Khoo, K.S., Chong, Y.M., Chang, W.S., Yap, J.M., Foo, S.C. et al. (2021a) Permeabilization of *Chlorella sorokiniana* and extraction of lutein by distillable CO_2-based alkyl carbamate ionic liquids. *Separation and Purification Technology* 256, 117471.

Khoo, K.S., Ooi, C.W., Chew, K.W., Foo, S.C. and Show, P.L. (2021b) Bioprocessing of *Chaetoceros calcitrans* for the recovery of fucoxanthin using CO_2-based alkyl carbamate ionic liquids. *Bioresource Technology* 322, 124520.

Kim, Y.-S., Li, X.-F., Kang, K.-H., Ryu, B. and Kim, S.K. (2014) Stigmasterol isolated from marine microalgae *Navicula incerta* induces apoptosis in human hepatoma HepG2 cells. *BMB Reports* 47(8), 433–438.

Lin, J.H., Lee, D.J. and Chang, J.S. (2015) Lutein production from biomass: marigold flowers versus microalgae. *Bioresource Technology* 184, 421–428.

Liu, Z.-W., Yue, Z., Zeng, X.-A., Cheng, J.-H. and Aadil, R.M. (2019) Ionic liquid as an effective solvent for cell wall deconstructing through astaxanthin extraction from *Haematococcus pluvialis*. *International Journal of Food Science and Technology* 54(2), 583–590.

Martinez, K.A., Lauritano, C., Druka, D., Romano, G., Grohmann, T. et al. (2019) Amphidinol 22, a new cytotoxic and antifungal amphidinol from the dinoflagellate *Amphidinium carterae*. *Marine Drugs* 17(7), 385.

Martins, D.A., Custodio, L., Barreira, L., Pereira, H., Ben-Hamadou, R. et al. (2013) Alternative sources of n-3 long-chain polyunsaturated fatty acids in marine microalgae. *Marine Drugs* 11(7), 2259–2281.

Martins, M.A.R., Neves, C.M.S.S., Kurnia, K.A., Carvalho, P.J., Rocha, M.A.A., Santos, L.M.N.B.F. et al. (2016) Densities, viscosities and derived thermophysical properties of water-saturated imidazolium-based ionic liquids. *Fluid Phase Equilibria* 407, 188–196.

Meresse, S., Fodil, M., Fleury, F. and Chenais, B. (2020) Fucoxanthin, a marine-derived carotenoid from brown seaweeds and microalgae: a promising bioactive compound for cancer therapy. *International Journal of Molecular Sciences* 21(23), 9273.

Minhas, A.K., Hodgson, P., Barrow, C.J., Sashidhar, B. and Adholeya, A. (2016) The isolation and identification of new microalgal strains producing oil and carotenoid simultaneously with biofuel potential. *Bioresource Technology* 211, 556–565.

Mitri, K., Shegokar, R., Gohla, S., Anselmi, C. and Mueller, R.H. (2011) Lipid nanocarriers for dermal delivery of lutein: preparation, characterization, stability and performance. *International Journal of Pharmaceutics* 414(1–2), 267–275.

Neumann, U., Derwenskus, F., Flister, V.F., Schmid-Staiger, U., Hirth, T. et al. (2019) Fucoxanthin, a carotenoid derived from phaeodactylum tricornutum exerts antiproliferative and antioxidant activities in vitro. *Antioxidants* 8(6), 183.

Olkiewicz, M., Caporgno, M.P., Font, J., Legrand, J., Lepine, O. et al. (2015) A novel recovery process for lipids from microalgae for biodiesel production using a hydrated phosphonium ionic liquid. *Green Chemistry* 17(5), 2813–2824.

Orr, V.C.A., Plechkova, N.V., Seddon, K.R. and Rehmann, L. (2016) Disruption and wet extraction of the microalgae *Chlorella vulgaris* using room-temperature ionic liquids. *ACS Sustainable Chemistry & Engineering* 4(2), 591–600.

Park, G.-T., Go, R.-E., Lee, H.-M., Lee, G.-A., Kim, C.-W. et al. (2017) Potential anti-proliferative and immunomodulatory effects of marine microalgal exopolysaccharide on various human cancer cells and lymphocytes in vitro. *Marine Biotechnology* 19(2), 136–146.

Patel, A.K., Albarico, F.P.J.B., Perumal, P.K., Vadrale, A.P., Nian, C.T. et al. (2022) Algae as an emerging source of bioactive pigments. *Bioresource Technology* 351, 126910.

Peng, J., Yuan, J.-P., Wu, C.-F. and Wang, J.-H. (2011) Fucoxanthin, a marine carotenoid present in brown seaweeds and diatoms: metabolism and bioactivities relevant to human health. *Marine Drugs*, 9(10), 1806–1828.

Porav, A.S., Bocaneala, M., Falamas, A., Bogdan, D.F., Barbu-Tudoran, L. et al. (2020) Sequential aqueous two-phase system for simultaneous purification of cyanobacterial phycobiliproteins. *Bioresource Technology* 315, 123794.

Quraishi, K.S., Bustam, M.A., Krishnan, S., Aminuddin, N.F., Azeezah, N. et al. (2017) Ionic liquids toxicity on fresh water microalgae, *Scenedesmus quadricauda*, *Chlorella vulgaris* and *Botryococcus braunii*: selection criterion for use in a two-phase partitioning bioreactor (TPPBR). *Chemosphere* 184, 642–651.

Risjani, Y., Mutmainnah, N., Manurung, P., Wulan, S.N., Yunianta. (2021) Exopolysaccharide from *Porphyridium cruentum* (purpureum) is not toxic and stimulates immune response against vibriosis: the assessment using zebrafish and white shrimp *Litopenaeus vannamei*. *Marine Drugs* 19(3), 133.

Rodriguez-Luna, A., Avila-Roman, J., Luisa Gonzalez-Rodriguez, M., Cozar, M.J., Rabasco, A.M. et al. (2018) Fucoxanthin-containing cream prevents epidermal hyperplasia and uvb-induced skin erythema in mice. *Marine Drugs* 16(10), 378.

Ruiz, C.A.S., Emmery, D.P., Wijffels, R.H., Eppink, M.H.M. and van den Berg, C. (2018a) Selective and mild fractionation of microalgal proteins and pigments using aqueous two-phase systems. *Journal of Chemical Technology and Biotechnology* 93(9), 2774–2783.

Ruiz, C.A.S., van den Berg, C., Wijffels, R.H. and Eppink, M.H.M. (2018b) Rubisco separation using biocompatible aqueous two-phase systems. *Separation and Purification Technology* 196, 254–261

Ruiz, C.A.S., Baca, S.Z., van den Broek, L.A.M., van den Berg, C., Wijffels, R.H. et al. (2020a) Selective fractionation of free glucose and starch from microalgae using aqueous two-phase systems. *Algal Research-Biomass Biofuels and Bioproducts* 46, 101810.

Ruiz, C.A.S., Kwaijtaal, J., Peinado, O.C., van den Berg, C., Wijffels, R.H. et al. (2020b) Multistep fractionation of microalgal biomolecules using selective aqueous two-phase systems. *ACS Sustainable Chemistry & Engineering* 8(6), 2441–2452.

Santos, J.H.P.M., Trigo, J.P., Maricato, E., Nunes, C., Coimbra, M.A. et al. (2018) Fractionation of *Isochrysis galbana* proteins, arabinans, and glucans using ionic-liquid-based aqueous biphasic systems. *ACS Sustainable Chemistry and Engineering* 6(11), 14042–14053.

Santos, J.I., Goncalves, A.M.M., Pereira, J.L., Figueiredo, B.F.H.T., e Silva, F.A. et al. (2015) Environmental safety of cholinium-based ionic liquids: assessing structure-ecotoxicity relationships. *Green Chemistry* 17(9), 4657–4668.

Schade, S., Stangl, G.I. and Meier, T. (2020) Distinct microalgae species for food-part 2: comparative life cycle assessment of microalgae and fish for eicosapentaenoic acid (EPA), docosahexaenoic acid (DHA), and protein. *Journal of Applied Phycology* 32(5), 2997–3013.

Shankar, M., Chhotaray, P.K., Agrawal, A., Gardas, R.L., Tamilarasan, K. et al. (2017) Protic ionic liquid-assisted cell disruption and lipid extraction from fresh water *Chlorella* and *Chlorococcum* microalgae. *Algal Research-Biomass Biofuels and Bioproducts* 25, 228–236.

Shiels, K., Tsoupras, A., Lordan, R., Nasopoulou, C., Zabetakis, I. et al. (2021) Bioactive lipids of marine microalga *Chlorococcum* sp. SABC 012504 with anti-inflammatory and anti-thrombotic activities. *Marine Drugs* 19(1), 28.

Singh, D., Puri, M., Wilkens, S., Mathur, A.S., Tuli, D.K. et al. (2013) Characterization of a new zeaxanthin producing strain of *Chlorella saccharophila* isolated from New Zealand marine waters. *Bioresource Technology* 143, 308–314.

Smieszek, A., Giezek, E., Chrapiec, M., Murat, M., Mucha, A. et al. (2017) The influence of *Spirulina platensis* filtrates on Caco-2 proliferative activity and expression of apoptosis-related microRNAs and mRNA. *Marine Drugs* 15(3), 65.

Steiner, B.M., McClements, D.J. and Davidov-Pardo, G. (2018) Encapsulation systems for lutein: a review. *Trends in Food Science & Technology* 82, 71–81.

Suarez Ruiz, C.A., Martins, M., Coutinho, J.A.P., Wijffels, R.H., Eppink, M.H.M. et al. (2020) *Neochloris oleoabundans* biorefinery: integration of cell disruption and purification steps using aqueous biphasic systems-based in surface-active ionic liquids. *Chemical Engineering Journal* 399, 125683.

Suarez Ruiz, C.A., Cabau-Peinado, O., van den Berg, C., Wijffels, R.H. and Eppink, M.H.M. (2022) Efficient fractionation of lipids in a multiproduct microalgal biorefinery by polymers and ionic liquid-based aqueous two-phase systems. *ACS Sustainable Chemistry & Engineering* 10(2), 789–799.

Suh, S.-S., Hwang, J., Park, M., Seo, H.H., Kim, H.-S. et al. (2014) Anti-inflammation activities of mycosporine-like amino acids (MAAs) in response to UV radiation suggest potential anti-skin aging activity. *Marine Drugs* 12(10), 5174–5187.

Sui, Y., Mazzucchi, L., Acharya, P., Xu, Y., Morgan, G. et al. (2021) A comparison of β-carotene, phytoene and amino acids production in *Dunaliella salina* DF 15 (CCAP 19/41) and *Dunaliella salina* CCAP 19/30 using different light wavelengths. *Foods* 10(11), 2824.

Vanderveen, J.R., Patiny, L., Chalifoux, C.B., Jessop, M.J. and Jessop, P.G. (2015) A virtual screening approach to identifying the greenest compound for a task: application to switchable-hydrophilicity solvents. *Green Chemistry* 17(12), 5182–5188.

Wahidin, S., Idris, A. and Shaleh, S.R.M. (2016) Ionic liquid as a promising biobased green solvent in combination with microwave irradiation for direct biodiesel production. *Bioresource Technology* 206, 150–154.

Wang, N., Dai, L., Chen, Z., Li, T., Wu, J. et al. (2022) Extraction optimization, physicochemical characterization, and antioxidant activity of polysaccharides from *Rhodosorus* sp. SCSIO-45730. *Journal of Applied Phycology* 34(1), 285–299.

Wang, Y., Xu, X., Yan, Y., Han, J. and Zhang, Z. (2010) Phase behavior for the [Bmim]BF$_4$ aqueous two-phase systems containing ammonium sulfate/sodium carbonate salts at different temperatures: experimental and correlation. *Thermochimica Acta* 501(1–2), 112–118.

Ye, Z., Tan, X.-H., Liu, Z.-W., Aadil, R.M., Tan, Y.-C. *et al.* (2020) Mechanisms of breakdown of *Haematococcus pluvialis* cell wall by ionic liquids, hydrochloric acid and multi-enzyme treatment. *International Journal of Food Science and Technology* 55(9), 3182–3189.

Zhu, Y., Li, X., Wang, Y., Ren, L. and Zhao, Q. (2021a) Lutein extraction by imidazolium-based ionic liquid-water mixture from dried and fresh *Chlorella* sp. *Algal Research* 60, 102528.

Zhu, Y., Zhong, X., Wang, Y., Zhao, Q. and Huang, H. (2021b) Growth performance and antioxidative response of *Chlorella pyrenoidesa, Dunaliella salina,* and *Anabaena cylindrica* to four kinds of ionic liquids. *Applied Biochemistry Biotechnology* 193(6), 1945–1966.

9 Metabolomics Analysis of Algae: Current Status and Future Perspectives

Yanhua Li[1,2] and Feng Ge[1,2]*

[1]State Key Laboratory of Freshwater Ecology and Biotechnology, Institute of Hydrobiology, Chinese Academy of Sciences, Wuhan, China; [2]Key Laboratory of Algal Biology, Institute of Hydrobiology, Chinese Academy of Sciences, Wuhan, China

Abstract

Algae are the most important primary producers in the ocean and play an important role in biogeochemical cycles. The use of metabolomics in algae has increased in recent years. It has helped to understand the interaction between algae and the environment at a metabolic level and to discover high-value metabolites produced by algae. We describe the use of metabolomics in the study of algae and the workflow of metabolomics from sample preparation, analytic methods, data analysis and metabolite annotation. The chapter aims to summarize the current status of metabolomics studies and recent publications regarding metabolomics in algae, and discusses the novel developments and applications for the analysis of metabolomics in algae. Challenges, opportunities and future perspectives in the metabolomics analysis in algae are also discussed.

Keywords: algae; metabolomics; liquid chromatography-mass spectrometry (LC-MS); gas chromatography-mass spectrometry (GC-MS); nuclear magnetic resonance spectroscopy (NMR)

1. Introduction

Algae are major contributors to global biodiversity, with estimates ranging from 30,000 to more than 1 million species. As a result of their diversity, algae play a vital role in the ecosystem. Algal species are interconnected with other organisms in biogeochemical cycles, food chains and symbiotic associations. However, environmental changes such as global warming and pollution caused by human activities are likely to affect the growth environment of algae and perturb natural patterns of algal species.

Metabolites are the end-products of biological information transmission and regulation processes, changes in their types and quantities are regarded as the final response of biological systems to genetic or environmental changes, and are also the comprehensive results of the combined effects of genotype and environment (Fiehn, 2002). The exposure of algae to environmental conditions will lead to changes of metabolic process since metabolism is the first to respond. Metabolomics examines changes in metabolite levels from the perspectives of time, perturbation or stimulation, and studies metabolic pathways (Nicholson et al., 2002). Therefore, metabolomics can effectively describe the dynamic response laws of the body's metabolic process, and help us

*Address for correspondence: gefeng@ihb.ac.cn

better reveal the complex interactions and essential causes in the organism.

This paper reviews the workflow of metabolomics, from sample preparation and extraction, to analytic methods, data analysis and metabolite annotation. It also provides the researches of metabolomics in algae, and discusses the prospect and challenges of metabolomics in algae.

2. Sample Preparation and Extraction

The importance of sample preparation must be underscored as it is the gateway to metabolomics analysis. Once samples are collected, they generally need to be quenched of the metabolism immediately to prevent unwanted metabolic changes and maintain sample integrity. Protocols designed for plant or animal tissue usually use liquid nitrogen (De Vos *et al.*, 2007; Lu and Xu, 2008) to flash freeze samples, or acetone/dry ice bath can be used instead when liquid nitrogen is not available. For some samples, that may require washing or centrifugation prior to flash freezing, cell metabolism can be quickly quenched with cold methanol (Sellick *et al.*, 2009) or acetonitrile to minimize metabolic changes and kept at low temperature (e.g. 4°C) for processing as quickly as possible. After flash-freezing, samples should be stored at −80°C until metabolomics analysis and avoid freeze-thaw cycles that can be achieved by preparing multiple aliquots at the sample collection stage.

2.1 Sample extraction

The type of sample determines the choice of extraction method. Algae metabolites extraction method usually follows the protocols used in plant metabolomics. As there is no single solvent that can dissolve all compounds, it probably needs different solvents to do extraction. Methods for polar metabolites extraction have been reported in some metabolomic studies (Fan *et al.*, 1993, 2006; Bolten *et al.*, 2007). These methods usually adopt cold acids, methanol and water mixture to quench and reduce the extraction of lipids, proteins, nucleic acids and carbohydrates. For non-polar and weakly polar metabolites (such as lipids) extraction, methanol is suitable for the extraction of polar lipids such as phospholipids (PLs), whereas chloroform/methanol (2:1, v/v) mixtures are more suitable for extraction of neutral lipids such as triglyceride (Rujoi *et al.*, 2003; Yappert *et al.*, 2003). In addition, the two-phase solvent (chloroform/methanol/water (2:1:1, v/v/v)) is very useful for extraction of both non-polar and polar metabolites (Kim and Verpoorte, 2010).

2.2 Preparation sample before analysis

NMR-based metabolomics studies usually analyze dissolved samples; therefore the samples extracted by a proper method can be directed into the NMR. The only additional step is filtering samples to remove insoluble materials and paramagnetic ions. In addition, pH is a critical factor because spectral artefacts of some metabolites (such as organic acid, amino acid and phosphorylated compounds) could form. Therefore, adjusting pH of all samples similar to the standard is critical prior to NMR analysis.

In GC-MS analysis, extracts usually need to be derivatized, except the volatile metabolites, which can be analyzed directly. After derivatization, the metabolites such as fatty acids, organic acids and sugars can be detected by GC-MS. As the water can damage GC-MS, the extracts should be dried by lyophilization or nitrogen to remove water before derivatization.

The presence of salt is detrimental to electrospray ionization (ESI) source in LC-MS metabolomics, and the lipid attached to the column can affect chromatographic separation. Removal of salts and lipids before injection is important for metabolomic analysis. Solid-phase extraction with Diaion HP-20 resins (Houssen and Jaspars, 2012) or C18 is a commonly used method for desalting and removing lipids.

3. Analytic Methods

Nuclear magnetic resonance (NMR) and mass spectrometry (MS) are the two major analytical techniques employed for metabolomics studies (Lindon and Nicholson, 2008). Both of these two techniques have high throughput, high sensitivity and high precision, which can deliver high-quality spectroscopic or structural information of metabolite. MS is usually used by coupling to liquid chromatography (LC) or gas chromatography (GC), with chromatography as the separation unit and mass spectrometry for metabolite identification. Additionally, other minor analytical techniques, such as supercritical fluid chromatography coupled to mass spectrometry (SFC-MS), capillary electrophoresis coupled to mass spectrometry (CE-MS) and Fourier transform infrared spectroscopy (FT-IR) have been used in some studies. Nevertheless, these minor techniques, which have not been widely used, are limited as complementary tools to the major analytical techniques.

3.1 Nuclear magnetic resonance spectroscopy

NMR spectroscopy is a widely used analytical technique that can provide information on the molecular structure of pure compounds or complex mixtures for structural elucidation. There are several advantages of NMR-based metabolomics which include high reproducibility, high throughput, less sample preparation and non-destructive. As a tool for structural analysis, its unbiased analysis allows it to be equally sensitive to all compounds. Although the detection sensitivity of NMR is slightly inferior compared with mass spectrometry, and it is not suitable for trace analysis, experiments and dynamic detection can be performed close to physiological conditions at a selected temperature and buffer range. In addition, NMR is suitable for the detection of compounds that are difficult to ionize or demand derivatization for MS (Markley et al., 2017). 1D ^{1}H-NMR is the most common analytic technique used in metabolomics study due to the high sensitivity of ^{1}H, while there is very low sensitivity of ^{13}C or ^{15}N because of the very low natural abundance.

3.2 Liquid chromatography (LC)-based methods with mass spectrometry (MS)

LC-MS is the most widely used analytic platform in metabolomics. It has the advantages of high throughput, high sensitivity and wide application range. Compared with GC-MS, sample pretreatment of LC-MS is simple and does not require derivatization of low-volatility compounds. Therefore, LC-MS is more suitable for compounds with high boiling point, strong polarity, poor thermal stability and large molecular weight.

There are several ion sources such as Electrospray Ionization (ESI), Atmospheric Pressure Chemical Ionization (APCI), Atmospheric Pressure Photoionization (APPI), Matrix Assisted Laser Desorption/Ionization (MALDI) and Fast Atom Bombardment (FAB). Electrospray ionization (ESI) is the most commonly used ionization technique of LC-MS in metabolic studies (Gaskell, 1997); both positive and negative ion mode should be performed to obtain a comprehensive profile. It is a soft ionization technique that can utilize a high electric field to produce a large number of complete molecular ions by charge exchange in solution (Lu et al., 2008).

LC can couple with many types of mass analyzers, such as single-quadrupole, triple quadrupole (QQQ), time-of-flight (TOF), ion traps, orbitrap and Fourier transform ion cyclotron resonance (FT-ICR). In addition, there are some hybrid systems that combine two basic types of mass spectrometers, such as quadrupole-time-of-flight mass spectrometry (Q-TOF) (Morris et al., 1996), quadrupole-linear ion trap (Q-Trap) (Hopfgartner et al., 2003), or ion trap FT mass spectrometers (Syka et al., 2004). Q-TOF or linear ion trap FT-MS are suitable for metabolic profiling studies due to their capability for sensitive full scan mode and exact masses, while for targeted metabolites studies, QQQ and Q-Trap are frequently used under multiple reaction monitoring (MRM) mode. These high-resolution

(3–5 ppm mass error) mass spectrometers and the resulting accurate mass can significantly reduce the number of possible structures of putative metabolites and improve the identification certainty of metabolites.

3.3 Gas chromatography-mass spectrometry (GC-MS)

GC-MS-based metabolomic analysis techniques are mainly used for the detection of volatile and thermally stable metabolites and those that can be volatile by derivatization. This means that the number of metabolites that can be detected by GC-MS is far less than LC-MS, but the main advantage of GC-MS is the high sensitivity and reproducible analytical method with established databases for metabolite identification. The mass spectrometry fragmentations obtained by the electron ionization (EI) source with good repeatability and the relatively complete standard spectral database, such as NIST, make the analysis process simple and fast. As it can detect a large number of trace small molecular compounds having mass/ratio (m/z) less than 600 au with less matrix effects, it is suitable for metabolomic studies of primary metabolites such as amino acid, lipids, sterol and carbohydrates in the samples.

4. Data Analysis

4.1 Preprocessing

Once the metabolomics data are acquired, the first step is to transform the raw files into a suitable format that can be used for multivariate data analysis.

For MS data, the software msConver (Adusumilli and Mallick, 2017) converts from most proprietary formats to mzML (Hamacher, 2011) and mzXML (Pedrioli et al., 2004). The initial stages of data preprocessing are similar for LC–MS and GC–MS metabolomics. Typically, the pipeline consists of baseline correction, noise filtering, peak picking, deconvolution, peak matching, peak alignment and annotation (Want and Masson, 2011).

For LC-MS preprocessing, there are some open access software tools for whole workflow to simplify data preprocessing, such as XCMS (Mahieu et al., 2016), MZmine2 (Pluskal et al., 2010), mzMatch (Scheltema et al., 2011), OpenMS (Reinert and Kohlbacher, 2010), MS-DIAL (Tsugawa et al., 2015), MetaboAnalyst (Chong et al., 2019). In addition, many instrument manufacturers have produced their own software, such as Progenesis QI (Waters), MassProfiler Professional (Agilent) and MarkerView (SCIEX).

AMDIS (Meyer et al., 2010) and XCMS (Benton et al., 2008) were the two most common software that were used for GC-MS metabolomics data preprocessing. Some other GC-MS-specific preprocessing software are MetaboliteDetector (Hiller et al., 2009), MET-IDEA (Broeckling et al., 2006), MeltDB (Kessler et al., 2013), metaMS (Wehrens et al., 2014) and MSeasy (Nicolè et al., 2012).

The preprocessing of NMR data is different from MS data at the first stage, which consists of zerofilling, apodization, Fourier transformation and phase correction (Smolinska et al., 2012). The subsequent steps of baseline correction, deconvolution, binning, peak alignment, scaling and normalization (Brennan, 2014) are the same as for MS, the only difference may be the algorithm used. TopSpin software, which is produced by Bruker instruments, is the most widely used software for NMR metabolomics data preprocessing (Weber et al., 2017). There are a number of open-source software available for NMR metabolomics analysis, most of them written by using Matlab, R, Jave, or Python code (Li, 2020), such as Automics (Wang et al., 2009), KIMBLE (Verhoeven et al., 2018), Metaboanalyst (Chong et al., 2019), NMRProcFlow (Jacob et al., 2017), Workflow4Metabolomics (W4M) (Giacomoni et al., 2015), ASCIS (Lefort et al., 2019).

4.2 Statistical analysis

After data preprocessing, the next step is multivariate statistical analysis. It can be classified into two categories: supervised and unsupervised data analysis (Brennan, 2014). For unsupervised data analysis, the goal is to reduce the dimensionality of the date and overview the data. Principal Component Analysis (PCA) and Hierarchical Clustering Analysis (HCA) are two well-known unsupervised multivariate methods (Cho et al., 2008). On the other hand, Partial Least Squares Discriminant Analysis (PLS-DA) and Orthogonal Partial Least Squares Discriminant Analysis (OPLS-DA) are the most widely used supervised techniques that can be used to identify discriminatory metabolites or features that are perturbed in the biological system.

4.3 Annotation

Metabolite annotation is a time-consuming and complex task in metabolomics analysis (Weber et al., 2017). Schymanski et al. (2014) divided MS metabolite identification into five confidence levels:

Level 1. Confirmed structure by reference standard. It needs to be compared with the standard under the same experimental conditions (Salek et al., 2013). Due to the lack of standards of many metabolites, they cannot be identified to Level 1.

Level 2. Probable structure by library spectrum match (2a) or diagnostic evidence (2b). That is the highest level of metabolite identification. There are some online databases, such as mzCloud, METLIN (Smith et al., 2005) and MassBank (Horai et al., 2010), which contain authenticated MS/MS spectra that can be used for library spectrum matches. Whilst much metabolomics software, such as CFM-ID (Allen et al., 2014), FingerID (Heinonen et al., 2012), MAGMa (Ridder et al., 2013), MetFrag (Ruttkies et al., 2016) and MyCompoundID (Li et al., 2013), can perform automatic database matching.

Level 3. Tentative candidate (s). Several possible metabolite structures can be inferred from some databases (such as HMDB (Wishart et al., 2022), METLIN (Smith et al., 2005), KEGG (Kanehisa, 2004)) or software (MI-PACK (Weber and Viant, 2010) and PUTMEDID-LCMS (Brown et al., 2011), ProbMetab (Silva et al., 2014) and MetAssign-mzMatch (Daly et al., 2014)), but it cannot be explicit to a single structure.

Level 4. Unequivocal molecular formula. Only the molecular formula can be obtained, there is not enough information to infer the structure. Software like CAMERA (Kuhl et al., 2012), MZedDB (Draper et al., 2009), Rdisop and SIRIUS (Kim et al., 2016) can be used to annotate ionization products.

Level 5. Exact mass (m/z). Only the m/z of interest can be obtained.

5. Metabolomics Application in Algae

Metabolomics has been widely used in drug research, disease diagnosis, ecotoxicology, plants and microorganisms etc., but its application in algae is not universal. In recent years, with the increase in the popularity of algae research and the rapid development of metabolomics technology, a number of studies use a metabolomics approach to discover secondary metabolites produced with medicinal or economic value, study ecotoxicology and stress resistance of algae.

5.1 Application in stress-resistance research

The metabolic response of algae to environmental stress is the basis for the study of high biomass and economic products, so the metabolomic studies of algae under environmental induction are gradually increasing. Renberg et al. (2010) have studied the major metabolite changes during acclimation to limiting CO_2 in *Chlamydomonas reinhardtii* by GC-TOF-MS; 128 metabolites were detected with significant differences. Under the condition of transferring *C. reinhardtii* from high concentration to low concentration of CO_2, the contents of 5 metabolites involved in photorespiration, 11 amino acids, and 1 lipid increased, while 6 amino acids and 21 lipids decreased. A model has been revealed that photorespiration is increasing during early induction of the carbon-concentrating. Another example is the metabolite changes of *Ettlia oleoabundans* under different nutrient conditions (Matich *et al.*, 2016). The authors combined GC-MS/MS targeted and LC-Q-TOF-MS untargeted metabolomics. It showed that 202 metabolites were changed under nutrient stress. Under nutrient-deficiency conditions, triacylglycerols and sulfoquinovosyl diacylglycerols were increased, chlorophylls were depleted, and diacylglycerols were differentially regulated at the molecular level.

5.2 Application in the research of economical high-value-added products

Algae can synthesize a large amount of lipids, proteins and carbohydrates by utilizing sunlight, carbon source, water and inorganic salts. The application of metabolomics to the study of secondary metabolites produced by algae with high economic value, such as fatty acids, carotenoids, steroids, polysaccharides and steroids, can provide theoretical support for explaining the metabolic mechanism and directional substance accumulation. Cheng *et al.* (2012) combined metabolomics with OPLS-DA to reveal the metabolite profile of *Synechocystis* sp. PCC6803, *Anabaena* sp. PCC7120 and *Scenedesmus obliquus* by GC-TOF-MS. Seventy-four metabolites were detected and the study reveals that adding 2 mmol L^{-1} ethanolamine can significantly increase the lipid content. On the other hand, metabolomics has also been applied to the discovering of new activity compounds in algae. For example, the polysaccharides with the anticoagulant action were identified from the green alga *Ulothrix flacca* using the metabolomics approach. In this typical target metabolome analysis, the authors used the analytic platform (such as LC, GC, FTIR and 1D/2D NMR) to isolate and identify the polysaccharides structure, that is an ulvan, a sulfated polysaccharide composed mainly by l-rhamnose and d-glucuronic acid. In addition, the researchers also verified the ulvan's anti-coagulant activity that is less than heparin (Li *et al.*, 2018).

5.3 Application in ecotoxicology research

In the past few years, a number of studies use the metabolomics approaches to the ecotoxicology of algae. Felline *et al.* (2019) studied the metabolic response of the algae *Fucus virsoides* to glyphosate. Glyphosate is a broad-spectrum herbicide which has been widely used on land but frequently detected in freshwater. Even when exposed to the lowest glyphosate concentration (0.5 mg/ml), the metabolome of *F. virsoide* is changed. This is mainly related to the deregulation of shikimate pathway, resulting synthesis of essential aromatic amino acids, which ultimately affects the synthesis of proteins and secondary metabolites. In addition, Kluender *et al.* (2009) analyzed the metabolite profiles of *Scenedesmus vacuolatus* exposed to promethazine, a photosystem II-inhibitor, and revealed the damage of promethazine on intracellular energy metabolism with an activation of catabolism processes and the inhibition of carbohydrates biosynthesis.

6. Perspectives

The studies of metabolomics in algae have been increased over the past decades. It has deepened the understanding of the metabolic process and the function of metabolites. Moreover, it provides a new approach for discovery of high-value compounds from algae. Although metabolomic analysis platforms have been advancing and new analytical tools have emerged, such as CE-MS, SFC-MS and IR, it is still impossible to detect all metabolites in a single analysis. We need to combine multiple analytic platforms to get as much information on metabolites as possible. The metabolite annotation remains a challenge for metabolomics research. High-quality NMR and MS spectra data are critical for metabolite identification. At the same time, the continuous development and expansion of the database has also increased the coverage of metabolite matching. With the development of mass analyzers with faster acquisition speed, higher resolution and accuracy, and the continuous improvement of the database, we believe that the application of metabolomics in algae will be more widespread.

Conflicts of interest

The authors declare that they have no conflicts of interest.

Acknowledgments

This work was supported by the Strategic Priority Research Program of Chinese Academy of Sciences (Grant No. XDB31040103), the National Natural Science Foundation of China (Grant No. 31870756) and the Chinese Academy of Sciences (Grant No. QYZDY-SSW- SMC004).

References

Adusumilli, R. and Mallick, P. (2017) Data conversion with ProteoWizard msConvert. In: Comai, L., Katz, J.E. and Mallick, P. (eds) *Proteomics: Methods in Molecular Biology*. Springer, New York, pp. 339–368.

Allen, F., Pon, A., Wilson, M., Greiner, R. and Wishart, D. (2014) CFM-ID: a web server for annotation, spectrum prediction and metabolite identification from tandem mass spectra. *Nucleic Acids Research* 12, 94–99.

Benton, H.P., Wong, D.M., Trauger, S.A. and Siuzdak, G. (2008) XCMS^2 : processing tandem mass spectrometry data for metabolite identification and structural characterization. *Analytical Chemistry* 80, 6382–6389.

Bolten, C.J., Kiefer, P., Letisse, F., Portais, J.-C. and Wittmann, C. (2007) Sampling for metabolome analysis of microorganisms. *Analytical Chemistry* 79, 3843–3849.

Brennan, L. (2014) NMR-based metabolomics: from sample preparation to applications in nutrition research. *Progress in Nuclear Magnetic Resonance Spectroscopy* 83, 42–49.

Broeckling, C.D., Reddy, I.R., Duran, A.L., Zhao, X. and Sumner, L.W. (2006) MET-IDEA: data extraction tool for mass spectrometry-based metabolomics. *Analytical Chemistry* 78, 4334–4341.

Brown, M., Wedge, D.C., Goodacre, R., Kell, D.B., Baker, P.N. et al. (2011) Automated workflows for accurate mass-based putative metabolite identification in LC/MS-derived metabolomic datasets. *Bioinformatics* 27, 1108–1112.

Cheng, J.-S., Niu, Y.-H., Lu, S.-H. and Yuan, Y.-J. (2012) Metabolome analysis reveals ethanolamine as potential marker for improving lipid accumulation of model photosynthetic organisms. *Journal of Chemical Technology and Biotechnology* 87, 1409–1418.

Cho, H.-W., Kim, S.B., Jeong, M.K., Park, Y., Gletsu, N. et al. (2008) Discovery of metabolite features for the modelling and analysis of high-resolution NMR spectra. *International Journal of Data Mining and Bioinformatics* 2, 176–192.

Chong, J., Wishart, D.S. and Xia, J. (2019) Using MetaboAnalyst 4.0 for comprehensive and integrative metabolomics data analysis. *Current Protocols in Bioinformatics* 68, 1–128.

Daly, R., Rogers, S., Wandy, J., Jankevics, A., Burgess, K.E.V. et al. (2014) MetAssign: probabilistic annotation of metabolites from LC–MS data using a Bayesian clustering approach. *Bioinformatics* 30, 2764–2771.

De Vos, R.C., Moco, S., Lommen, A., Keurentjes, J.J., Bino, R.J. et al. (2007) Untargeted large-scale plant metabolomics using liquid chromatography coupled to mass spectrometry. *Nature Protocols* 2, 778–791.

Draper, J., Enot, D.P., Parker, D., Beckmann, M., Snowdon, S. et al. (2009) Metabolite signal identification in accurate mass metabolomics data with MZedDB, an interactive m/z annotation tool utilising predicted ionisation behaviour 'rules'. *BMC Bioinformatics* 10, 227.

Fan, T.W.M., Colmer, T.D., Lane, A.N. and Higashi, R.M. (1993) Determination of metabolites by 1H NMR and GC: analysis for organic osmolytes in crude tissue extracts. *Analytical Biochemistry* 214, 260–271.

Fan, T.W.M., Bandura, L.L., Higashi, R.M. and Lane, A.N. (2006) Metabolomics-edited transcriptomics analysis of Se anticancer action in human lung cancer cells. *Metabolomics* 1, 325–339.

Felline, S., Del Coco, L., Kaleb, S., Guarnieri, G., Fraschetti, S. et al. (2019) The response of the algae *Fucus virsoides* (Fucales, Ochrophyta) to Roundup® solution exposure: a metabolomics approach. *Environmental Pollution* 254, 112977.

Fiehn, O. (2002) Metabolomics: the link between genotypes and phenotypes. In: Town, C. (ed.) *Functional Genomics*. Springer, Dordrecht, The Netherlands, pp. 155–171.

Gaskell, S.J. (1997) Electrospray: principles and practice. *Journal of Mass Spectrometry* 32, 677–688.

Giacomoni, F., Le Corguille, G., Monsoor, M., Landi, M., Pericard, P. *et al.* (2015) Workflow4Metabolomics: a collaborative research infrastructure for computational metabolomics. *Bioinformatics* 31, 1493–1495.

Hamacher, M. (ed.) (2011) *Data Mining in Proteomics: From Standards to Applications*. Humana Press, New York and Heidelberg.

Heinonen, M., Shen, H., Zamboni, N. and Rousu, J. (2012) Metabolite identification and molecular fingerprint prediction through machine learning. *Bioinformatics* 28, 2333–2341.

Hiller, K., Hangebrauk, J., Jäger, C., Spura, J., Schreiber, K. *et al.* (2009) MetaboliteDetector: comprehensive analysis tool for targeted and nontargeted GC/MS based metabolome analysis. *Analytical Chemistry* 81, 3429–3439.

Hopfgartner, G., Husser, C. and Zell, M. (2003) Rapid screening and characterization of drug metabolites using a new quadrupole-linear ion trap mass spectrometer. *Journal of Mass Spectrometry* 38, 138–150.

Horai, H., Arita, M., Kanaya, S., Nihei, Y., Ikeda, T. *et al.* (2010) MassBank: a public repository for sharing mass spectral data for life sciences. *Journal of Mass. Spectrometry* 45(7), 703–714.

Houssen, W.E. and Jaspars, M. (2012) Isolation of marine natural products. In: Sarker, S.D. and Nahar, L. (eds) *Natural Products Isolation: Methods in Molecular Biology*. Humana Press, Totowa, New Jersey, pp. 367–392.

Jacob, D., Deborde, C., Lefebvre, M., Maucourt, M. and Moing, A. (2017) NMRProcFlow: a graphical and interactive tool dedicated to 1D spectra processing for NMR-based metabolomics. *Metabolomics* 13, 36.

Kanehisa, M. (2004) The KEGG resource for deciphering the genome. *Nucleic Acids Research* 32, 277D–280.

Kessler, N., Neuweger, H., Bonte, A., Langenkamper, G., Niehaus, K. *et al.* (2013) MeltDB 2.0-advances of the metabolomics software system. *Bioinformatics* 29, 2452–2459.

Kim, H.K. and Verpoorte, R. (2010) Sample preparation for plant metabolomics. *Phytochemical Analysis* 21, 4–13.

Kim, S., Thiessen, P.A., Bolton, E.E., Chen, J., Fu, G. *et al.* (2016) PubChem substance and compound databases. *Nucleic Acids Research* 44, D1202–D1213.

Kluender, C., Sans-Piché, F., Riedl, J., Altenburger, R., Härtig, C. *et al.* (2009) A metabolomics approach to assessing phytotoxic effects on the green alga *Scenedesmus vacuolatus*. *Metabolomics* 5, 59–71.

Kuhl, C., Tautenhahn, R., Böttcher, C., Larson, T.R. and Neumann, S. (2012) CAMERA: an integrated strategy for compound spectra extraction and annotation of liquid chromatography/mass spectrometry data sets. *Analytical Chemistry* 84, 283–289.

Lefort, G., Liaubet, L., Canlet, C., Tardivel, P., Père, M.-C. *et al.* (2019) ASICS: an R package for a whole analysis workflow of 1D 1H NMR spectra. Kelso, J. (ed.) *Bioinformatics* 35, 4356–4363.

Li, L., Li, R., Zhou, J., Zuniga, A., Stanislaus, A.E. *et al.* (2013) MyCompoundID: using an evidence-based metabolome library for metabolite identification. *Analytical Chemistry* 85, 3401–3408.

Li, P., Wen, S., Sun, K., Zhao, Y. and Chen, Y. (2018) Structure and bioactivity screening of a low molecular weight ulvan from the green alga *Ulothrix flacca*. *Marine Drugs* 16, 281.

Li, S. (ed.) (2020) *Computational Methods and Data Analysis for Metabolomics*. Springer US, New York.

Lindon, J.C. and Nicholson, J.K. (2008) Spectroscopic and statistical techniques for information recovery in metabonomics and metabolomics. *Annual Review of Analytical Chemistry* 1, 45–69.

Lu, X. and Xu, G. (2008) LC-MS metabonomics methodology in biomarker discovery. In: Wang, F. (ed.) *Biomarker Methods in Drug Discovery and Development: Methods in Pharmacology and Toxicology™*. Humana Press, Totowa, New Jersey, pp. 291–315.

Lu, X., Zhao, X., Bai, C., Zhao, C., Lu, G. *et al.* (2008) LC–MS-based metabonomics analysis. *Journal of Chromatography B* 866, 64–76.

Mahieu, N.G., Genenbacher, J.L. and Patti, G.J. (2016) A roadmap for the XCMS family of software solutions in metabolomics. *Current Opinion in Chemical Biology* 30, 87–93.

Markley, J.L., Brüschweiler, R., Edison, A.S., Eghbalnia, H.R., Powers, R. *et al.* (2017) The future of NMR-based metabolomics. *Current Opinion in Biotechnology* 43, 34–40.

Matich, E.K., Butryn, D.M., Ghafari, M., del Solar, V., Camgoz, E. *et al.* (2016) Mass spectrometry-based metabolomics of value-added biochemicals from *Ettlia oleoabundans*. *Algal Research* 19, 146–154.

Meyer, M.R., Peters, F.T. and Maurer, H.H. (2010) Automated mass spectral deconvolution and identification system for GC-MS screening for drugs, poisons, and metabolites in urine. *Clinical Chemistry* 56, 575–584.

Morris, H.R., Paxton, T., Dell, A., Langhorne, J., Berg, M. *et al.* (1996) High sensitivity collisionally-activated decomposition tandem mass spectrometry on a novel quadrupole/orthogonal-acceleration time-of-flight mass spectrometer. *Rapid Communications in Mass Spectrometry* 10, 889–896.

Nicholson, J.K., Connelly, J., Lindon, J.C. and Holmes, E. (2002) Metabonomics: a platform for studying drug toxicity and gene function. *Nature Reviews Drug Discovery* 1, 153–161.

Nicolè, F., Guitton, Y., Courtois, E.A., Moja, S., Legendre, L. et al. (2012) MSeasy: unsupervised and untargeted GC-MS data processing. *Bioinformatics* 28, 2278–2280.

Pedrioli, P.G.A., Eng, J.K., Hubley, R., Vogelzang, M., Deutsch, E.W. et al. (2004) A common open representation of mass spectrometry data and its application to proteomics research. *Nature Biotechnology* 22, 8.

Pluskal, T., Castillo, S., Villar-Briones, A. and Orešič, M. (2010) MZmine 2: modular framework for processing, visualizing, and analyzing mass spectrometry-based molecular profile data. *BMC Bioinformatics* 11, 395.

Reinert, K. and Kohlbacher, O. (2010) OpenMS and TOPP: open source software for LC-MS data analysis. In: Hubbard, S.J. and Jones, A.R. (eds) *Proteome Bioinformatics: Methods in Molecular Biology*. Humana Press, Totowa, New Jersey, pp. 201–211.

Renberg, L., Johansson, A.I., Shutova, T., Stenlund, H., Aksmann, A. et al. (2010) A metabolomic approach to study major metabolite changes during acclimation to limiting CO_2 in *Chlamydomonas reinhardtii*. *Plant Physiology* 154, 187–196.

Ridder, L., van der Hooft, J.J.J., Verhoeven, S., de Vos, R.C.H., Bino, R.J. et al. (2013) Automatic chemical structure annotation of an LC–MS n based metabolic profile from green tea. *Analytical Chemistry* 85, 6033–6040.

Rujoi, M., Jin, J., Borchman, D., Tang, D. and Yappert, M.C. (2003) Isolation and lipid characterization of cholesterol-enriched fractions in cortical and nuclear human lens fibers. *Investigative Opthalmology and Visual Science* 44, 1634.

Ruttkies, C., Schymanski, E.L., Wolf, S., Hollender, J. and Neumann, S. (2016) MetFrag relaunched: incorporating strategies beyond in silico fragmentation. *Journal of Cheminformatics* 8, 16.

Salek, R.M., Steinbeck, C., Viant, M.R., Goodacre, R. and Dunn, W.B. (2013) The role of reporting standards for metabolite annotation and identification in metabolomic studies. *GigaScience* 2, 13.

Scheltema, R.A., Jankevics, A., Jansen, R.C., Swertz, M.A. and Breitling, R. (2011) PeakML/mzMatch: a file format, Java library, R library, and tool-chain for mass spectrometry data analysis. *Analytical Chemistry* 83, 2786–2793.

Schymanski, E.L., Jeon, J., Gulde, R., Fenner, K., Ruff, M. et al. (2014) Identifying small molecules via high resolution mass spectrometry: communicating confidence. *Environmental Science and Technology* 48, 2097–2098.

Sellick, C.A., Hansen, R., Maqsood, A.R., Dunn, W.B., Stephens, G.M. et al. (2009) Effective quenching processes for physiologically valid metabolite profiling of suspension cultured mammalian cells. *Analytical Chemistry* 81, 174–183.

Silva, R.R., Jourdan, F., Salvanha, D.M., Letisse, F., Jamin, E.L. et al. (2014) ProbMetab: an R package for Bayesian probabilistic annotation of LC–MS-based metabolomics. *Bioinformatics* 30, 1336–1337.

Smith, C.A., Maille, G.O., Want, E.J., Qin, C., Trauger, S.A. et al. (2005) METLIN: a metabolite mass spectral database. *Therapeutic Drug Monitoring* 27, 747–751.

Smolinska, A., Blanchet, L., Buydens, L.M.C. and Wijmenga, S.S. (2012) NMR and pattern recognition methods in metabolomics: from data acquisition to biomarker discovery: a review. *Analytica Chimica Acta* 750, 82–97.

Syka, J.E.P., Marto, J.A., Bai, D.L., Horning, S., Senko, M.W. et al. (2004) Novel linear quadrupole ion Trap/FT mass spectrometer: performance characterization and use in the comparative analysis of histone H3 post-translational modifications. *Journal of Proteome Research* 3, 621–626.

Tsugawa, H., Cajka, T., Kind, T., Ma, Y., Higgins, B. et al. (2015) MS-DIAL: data-independent MS/MS deconvolution for comprehensive metabolome analysis. *Nature Methods* 12, 523–526.

Verhoeven, A., Giera, M. and Mayboroda, O.A. (2018) KIMBLE: a versatile visual NMR metabolomics workbench in KNIME. *Analytica Chimica Acta* 1044, 66–76.

Wang, T., Shao, K., Chu, Q., Ren, Y., Mu, Y. et al. (2009) Automics: an integrated platform for NMR-based metabonomics spectral processing and data analysis. *BMC Bioinformatics* 10, 83.

Want, E. and Masson, P. (2011) Processing and analysis of GC/LC-MS-based metabolomics data. *Metabolic Profiling*. Humana Press, New York, pp. 277–298.

Weber, R.J.M. and Viant, M.R. (2010) MI-Pack: increased confidence of metabolite identification in mass spectra by integrating accurate masses and metabolic pathways. *Chemometrics and Intelligent Laboratory Systems* 104, 75–82.

Weber, R.J.M., Lawson, T.N., Salek, R.M., Ebbels, T.M.D., Glen, R.C. *et al.* (2017) Computational tools and workflows in metabolomics: an international survey highlights the opportunity for harmonisation through Galaxy. *Metabolomics* 13, 12.

Wehrens, R., Weingart, G. and Mattivi, F. (2014) metaMS: an open-source pipeline for GC–MS-based untargeted metabolomics. *Journal of Chromatography B* 966, 109–116.

Wishart, D.S., Guo, A., Oler, E., Wang, F., Anjum, A. *et al.* (2022) HMDB 5.0: the Human Metabolome Database for 2022. *Nucleic Acids Research* 50: D622–D631.

Yappert, M.C., Rujoi, M., Borchman, D., Vorobyov, I. and Estrada, R. (2003) Glycero-versus sphingo-phospholipids: correlations with human and non-human mammalian lens growth. *Experimental Eye Research* 76, 725–734.

10 The Need for Taxonomic Revision of *Phormidium treleasei* Gomont and Its Potential Use in Biotechnology[1]

Kaitlin Simmons and Qingfang He*

Department of Biology, University of Arkansas at Little Rock, Little Rock, Arkansas, USA

Abstract

The taxonomic nomenclature for cyanobacteria is constantly under revision as the new technology emerges. Scientists, at first, based classification solely upon appearance. Later, it was found, through DNA-DNA hybridization or DNA sequencing, that some of these were genetically not the same species. A study conducted by the authors of this chapter aims to identify one such species that has been revised through the years: *Phormidium treleasei* Gomont. This species is currently accepted as *Leptolyngbya treleasii* Gomont, as of 1988, by Anagnostidis and Komárek. This reclassification was solely based on descriptions of collections found in Poland (Starmach, 1966) and Austria (Claus, 1961), and therefore is not reliable, since the original location, where it was found (Hot Springs, Arkansas, in 1897) was not resampled or redescribed. This location was sampled again starting in 2021 by the authors of this chapter to determine what genus it belongs to through a polyphasic analysis, and to document the associated cyanobacteria and other thermal microbes. Since these organisms rely on a community structure to survive, it is important to study all organisms' presence to determine their respective roles. Biofilm is a structure that some bacteria can form. It is a thin and strong net of filaments, often made up of multiple species of cyanobacteria and bacteria. The production of the biofilm has been explored by scientists in producing more natural polymers and biofuels. The *Phormidium* and *Leptolyngbya* genera are two that can produce such a film. Efforts to create a polymer from the biofilm-making species collected are being explored. The community structure of biofilms and mats often change due to abiotic and biotic factors. Studying the seasonal growth of the microbes, as well as measuring pH, temperature and dissolved oxygen content in Display Springs is important for a better understanding of the effects of these factors on microbes in thermal alkaline mineral springs. Included in this chapter is a review of past and present taxonomic views for cyanobacteria, the history of *Phormidium treleasei*, factors influencing growth of cyanobacteria in Display Springs, and possible biotechnological applications for this type species.

1. How Taxonomic Classifications of Cyanobacteria were First Determined

The Botanical Code and the Bacteriological Code are two independent systems that scientists have tried to implement when naming cyanobacteria. The first classifications of any genus of cyanobacteria were made using the Botanical Code and were based solely on morphology. This approach, however, did not give an

---

*Address for correspondence: qfhe@ualr.edu
[1]Research approved by: Hot Springs National Park Services

accurate representation of the genetics of these genera. Once bacteriologists had performed biochemical and cytological studies on axenic strains, Cyanophyceae was changed to Cyanobacteria (Palinska and Surosz, 2014). Cyanobacteria are prokaryotes, however, they still have some characteristics that photosynthetic eukaryotes also have, like chlorophyll *a* and thylakoids. Microbiologists use the Bacteriological Code to describe axenic cultures of cyanobacteria. These names are constantly under revision to ensure that they are the most accurate and approved nomenclature. With scientists using two different systems over the last 100 years, this has caused the misidentification and misclassification of many cultures found in nature and in the laboratory.

In 1985, Anagnostidis, K. and Komárek, J. started an updated and revised list for the taxonomic criteria of cyanobacteria. This system combined the Bacteriological and Botanical approaches. The overall nomenclature is dependent on the Botanical approach while molecular and bacteriological information is also utilized. Nomenclatural suggestions by Geitler (1932) are predominantly followed. This work is still in progress through Jiří Komárek, with his most recent guide to cyanobacterial taxonomy being published in 2014. Komárek's research contributions include ultrastructural properties that can help distinguish species morphologically and help provide a basis for genus criteria. Such properties include cell proportions, division patterns, how trichomes disintegrate, presence of aerotopes, and motility.

1.1 Early history of *Leptolyngbya* and *Phormidium* genera classifications

Many different taxonomic schemes have been proposed throughout the years by various researchers, including Gomont (1892), Geitler (1925), Frémy (1929), Rippka *et al.* (1979), and Anagnostidis and Komárek (1985). Rippka's system classifies cyanobacteria based on field populations. This system continued by following Geitler's (1932) taxonomic nomenclature. All cultures studied through this system, however, were axenic, therefore they lacked representation of the natural morphological and ecological diversity once taken from nature. Using Rippka's system, the order Oscillatoriaceae was not able to be distinguished from one another enough, to be broken down into individual genera. Instead, complex groups LLP (*Lyngbya-Phormidium-Pleconema*) A and B were created (Marquardt and Palinska, 2006). This system was adopted (with a few modifications) to include descriptions from natural populations described in the first two editions of *Bergey's Manual of Systematic Bacteriology* (Castenholz and Waterbury, 1989; Castenholz *et al.*, 2001). In the 2001 edition, Castenholz added a new genera, *Leptolyngbya* Komárek & Anagnostidis 1988). Many new genera were added, some of them previously were included in the LPP-group B. Anagnostidis and Komárek's new classification system had revised the list to include classifications based on the ultrastructural properties the cyanobacteria possessed. This categorization caused the split of the genera into 18 newly defined genera.

The *Phormidium* genus was classified based on radially oriented thylakoids that were in transversal sections of the cells (Marquardt and Palinska, 2006). *Geiterinema* was placed as a subgenus of *Phormidium*, based on cells that were similar in size, with rapid gliding motility by rotation. These cells also had peripheral thylakoids. Cells of *Phormidium*, *Plectonema* and *Lyngbya* genera that were narrow, with no motility, were moved to the new genus *Leptolyngbya* Komárek & Anagnostidis (1988). *Leptolyngbya* and *Phormidium*, currently, were both in the order Oscillatoriaceae. It was not until 2014 that *Leptolyngbya* was moved to the order Synechococcales. This was because the criteria for characterizing cyanobacterial cells had changed, so genera had to be rearranged to fit that system. The change was based on the characteristic that *Leptolyngbya* cells have parietal thylakoid patterns (Komárek *et al.*, 2014).

1.2 The history of *Phormidium treleasei*

Throughout the years, the name for *Phormidium treleasei* has been changed multiple times and one researcher even argued for it to stay in the genus *Phormidium*. Table 10.1 summarizes these various nomenclatural changes. *P. treleasei* was first found in 1897, by botanist William Treleasei. He found this species within Hale Springs (Display Springs now) in Hot Springs National Park, Arkansas. He then sent his collection to Maurice Gomont, a renowned French phycologist. Gomont, in 1899, first described this species within the *Bulletin de la société botanique* (Vol. 46). In this volume, *P. treleasei's* growth was described to have an 'olive-green color thallus that was expanded, lamellar, and had several superimposed layers like papyrus'. The filaments he described were 'sticky, mucosal, parallel, very thin, straight, and rigid'. The sheath was very thin and colorless. Trichomes were described as a 'pale bluish-green, straight at the apex, somewhat attenuated, and not constricted at the [cross walls]'. The trichome lengths were '0.6–0.8 µm thick, very long and articulated... [Cells] were up to eleven times longer in diameter, [and] apical compartment was rounded with no calyptra'. *P. treleasei* was the most tenuous species he had observed within that genus (Gomont, 1899, pp. 37–38). It should be noted that this text was only available in its original language (Latin/French); some words, when translated, did not represent terms used commonly in phycology. These clarifications are indicated by brackets for the reader's benefit.

Phycologist Dr Francis Drouet, however, placed *P. treleasei* under the new name of *Schizothrix calcicola*, based on a sample collected from Hot Springs NP, Arkansas (Drouet, F., 1968). His revisions were based on morphological features that were not affected by the environment. Later, it was found through DNA-DNA hybridizations that some species he put in the same taxa were not the same genotypically (Wilmotte, 1994; Palinska and Surosz, 2014).

Compère (1974) listed the cyanobacteria *Lyngbya treleasei* (Gomont), without description, as being one of the specimens collected from Lake Chad in Nigeria. In 1988, *P. treleasei* was renamed to *Leptolyngbya treleasii* (Gomont) Anag. et Kom by Anagnostidis and Komárek in their book *Principles of a Modern Classification*. The book provided a description and drawing of *Leptolyngbya treleasii*. This name change was based on information after Claus (1961) from Starmach (1966). The description Anagnostidis and Komárek give for this species is like Gomont's, with a few exceptions. One being the thallus was described as an olive green to blue-green. Secondly, they describe the trichomes as not attenuated at the ends, while Gomont's description claims they were slightly attenuated. The reason for the differences could be that Anagnostidis and Komárek based their revision on Claus (1961) who described a new form of *P. treleasei* in Austria (*Phormidium treleasei* Gom. fa. *breviarticulata* fa. *nova*). This species was different from the type due to shorter cells.

Many believed that Drouet's broad categorization should not be subjugated to some species (Wilmotte, 1994; Meyer, 1996;

Table 10.1. History of nomenclatures for *Phormidium treleasei* Gomont.

Name	Contributor	Date/Was description included?	Location of collection
Phormidium treleasei	Gomont	1899/Yes	Hot Springs, Arkansas
Schizothrix calcicola	Drouet	1968/No	Hot Springs National Park, Arkansas
Lyngbya treleasei (Gomont)	Compère	1974/No	Lake Chad, Nigeria
Leptolyngba treleasii (Gomont)*	Anagnostidis & Komárek	1988/Yes	Europe

*Based after Claus, G. (1961) who describes a new form of this species from Austria with shorter cells. Information from Starmach, K. (1966), describing it in Poland, was also used as a basis for reclassification.

Palinska and Surosz, 2014). After studying *P. treleasei* from Display Springs in Hot Springs, Arkansas, during 1981, Meyer, R.L., (1996) rejected Drouet's classification of *P. treleasei* within the *Schizothrix calcicola* taxon because it did not take into consideration some important taxonomic characteristics. Meyer also contended that the reclassification of it into the *Leptolyngbya* taxon should include a synonym to *P. treleasei*. He, however, implored that it should be retained as a species of *Phormidium*, his reason being that a few things do separate these genera from one another, such as reproduction, cell size and colony morphology (Meyer, 1996).

Leptolyngbya forms as flaky clusters while *Phormidium* forms as a thallus (non-differentiated plant body) on substrates. In *P. treleasei*, the thylakoids are distributed throughout cytoplasm. This is different from *Leptolyngbya*, which only have thylakoids located in the peripheral lamella. Cyanophycean granules and pigmentation were nearly evenly distributed throughout the cytoplasm of *P. treleasei*. In addition, *P. treleasei* reproduces in ways that are like both *Phormidium* and *Leptolyngbya* genera. To reproduce, some unicellular filamentous cyanobacteria will start to form necridia. These are dead cells that will cause the main filament of cyanobacteria to branch off. These branches can break off and grow into longer filaments which are known as hormogonia (Bryant and Tandeau de Marsac, 1994). *P. treleasei's* vegetative reproduction involves disintegration of the trichomes, which then produce hormogonia (characteristic of *Leptolyngbya*). However, necridia cells were still present, as seen in *Phormidium*. The trichomes of *P. treleasei* were noted to be unconstricted or slightly constricted, as is characteristic of *Phormidium* (Komárek and Johansen, 2015). Since the species gathered from Hot Springs, Arkansas, has so many similarities to the *Phormidium* and *Leptolyngbya* genera, a reevaluation of this type species is being done by the authors through a polyphasic analysis.

P. treleasei was first thought to be found only in Arkansas, Canada* (Tilden, 1910), and a few other locations around the world. In the years following the initial discovery of this species, it has been found in Wyoming*, California*, Death Valley (Inman, 1940) and Alaska (Setchell, 1903), the only non-thermal spring. Globally, it has been found in Spain (Seoane, 1991), China*, New Zealand*, Iceland (Geitler, 1932), Venezuela (Steyermark, 1957), Austria (Claus, 1961) and West Africa (Geitler, 1932). All locations with asterisks were listed on the Smithsonian Institute Collections Search Center website. The last time this species was recorded in Hot Springs, Arkansas, was in 2006, by Smith *et al*. It had been found within a mat located in the Display pool and was listed under the name *Leptolyngbya treleasii* (Gomont) Anagnostidis & Komárek. When Meyer (1996) and Smith *et al*. (2013) sampled the Display Springs, they made identifications based on microscopy only. Even though they followed identifications accepted at the time, the morphological and genetic blueprint of these organisms needed to be analyzed.

2. The Need for Revisions in the Taxonomy System Using a Polyphasic Approach

The classical system is so outdated and hinders progress in the phylogenetics of cyanobacteria. Reword: Therefore, it is beneficial to have a narrowly defined and allegedly monophyletic genus, which should contain few species, rather than a broad and poorly defined polyphyletic genera that may not be related (Komárek *et al*., 2014).

Komárek *et al*. (2014) proposed a system that encompasses both the morphological and genetic approaches. In this system, each genus name given is a representation of the environment they are found in and any specialized structures they may have. It is important to name cyanobacterial species according to their natural environments for future identification of different cyanobacterial species, to be accurate. For example, gas vesicles are an adaptation for cyanobacteria living a planktic lifestyle. There can be differences in vesicles' spacing, appearance and formation depending on the growth environments. Gas vesicles for some species cover part of the trichome, and some gas vesicles only arise in specific situations (Komárek, 2018). All coincidences or interactions should be documented as they are useful in determining if what was witnessed was situational morphology or if it was a true representation of its natural morphology.

Morphology alone, however, cannot be used to determine common ancestry between cyanobacteria because many species of cyanobacteria are polyphyletic. Even though two species may look the same, genetically they could be completely different and arise from two different distinct lineages. This is especially true for morphologically simple cyanobacteria that have a large variability, like *Leptolyngbya* and *Phormidium*.

To avoid misidentification, often indicated by identical taxonomic strains being found in multiple places in a phylogenetic scheme (Komárek, 2018), species previously described need to be revised using new molecular methodologies. Current polyphasic approaches, however, do have problems that affect the identification of species. Many of the databases have identifications based on the old morphological system, this causes misidentification when new sequences are run for comparison against the old ones. If findings through both morphology and genetics give different results, the terms cryptospecies, morphospecies or ecospecies are useful when observing the future diversity of that cyanobacteria. An example of a case where these terms would be useful is when attempting to name species within the genus *Leptolyngbya* or *Phormidium*. After DNA sequencing was performed on some species of *Leptolyngbya*, it was found that more than 15 taxa were of separate lineages (Komárek, 2018). These all, however, morphologically, had very narrow and simple filaments about 0.5–3.5 μm wide and all developed sheaths and had a parietal pattern of thylakoids. In a study done by Keshari et al. (2022), the genus *Leptolyngbya* was determined to be a cryptospecies. This indicates that *Leptolyngbya* can have species that look like one another; however, it is not genetically the same species. There have been over 140 described cyanobacteria within the genus *Leptolyngbya* (Komárek, J. and Johansen, J.R., 2015). *Phormidium* is another such genus that has been found to be polymorphic. It is very common and grows in soil, on rocks, as well as in standing or running water. The distribution of this genus is worldwide, with over 200 species described (Komárek and Kaštovský, 2003). Komárek (2018) divided *Phormidium* into eight different groups based on their morphology that loosely matched the molecular results. The original concept of the genus was moved to group VIII. The other groups were named based on different morphological markers like sheath, trichomes and cell size.

The modern system of naming cyanobacteria proposed by Komárek et al. (2014) follows botanical nomenclature rules as well as molecular results. In the new system, some species were split up to create new genera. Genera in this system are separated based on thylakoid patterns first (irregular or parietal), then they are further separated by cellular division patterns, and finally by specialized cells. For example, if a cyanobacteria of interest had an irregular pattern of thylakoids as well as irregular cell division, you would check the order Pleurocapsales (V). The family could either be *Hydrococcaceae* or *Xenococcaceae* depending on other morphological features like baeocytes or pseudofilaments. Thylakoid patterns have been deemed the most important feature inside a cell because they allow for a stable and reliable taxonomic classification of orders and families (Palkinska and Surosz, 2014).

Since laboratory conditions often change the appearance of cyanobacterial cells, the most reliable way of identifying and categorizing cyanobacterial samples is through 16s rRNA sequencing. This is because the 16s rRNA gene is highly conserved across the population of cyanobacteria and has not changed much over that time (Palkinska and Surosz, 2014; Komárek, 2018; Tang et al., 2018). Targeting this gene allows for a list of OTUs (operational taxonomic units) to be compiled from a collected sample. These units help establish a way of identifying differences between and among samples. Using this method, situational morphology does not impede the results, allowing for a stable identification system to be established. The cpcBA (phycocyanin intergenic spacer and flanking regions) and ITS sequences (internal transcribed spacer region between the 16s-23s rRNA) have been used to distinguish the genus *Phormidium* from others. The lengths and sequences of these regions are similar among genera and therefore can be used as markers (Teneva et al., 2005).

Once the sequences have been uploaded to GenBank, the BLAST program can be used to compare sequences found to ones already sequenced over the years. This helps determine the percentage identity of each strain to one another. Sequence clusters that differ by more than 95%, which also can be distinguished by definite morphological and ecological markers, are reasonable criteria to base separation of genera from one another (Komárek, 2016). Using genomic analysis programs with packages like Biotools, a dendrogram can be created to visualize the different phylogenetic lineages. Other quantitative genomic analysis programs can provide heat maps for all strains identified. This is helpful in determining if they are more prone to genetic changes. It also helps identify how much genetic diversity is in each population (Tang et al., 2018).

Globally, there is not a list of accepted species of cyanobacteria. Databases that provide taxonomic information do not all correspond to one another. This is partly because some species have been reclassified and often under different principles. There needs to be only one method for classifying cyanobacteria to help future phycologists and bacteriologists (Komárek et al., 2014).

3. Ecology of Display Springs in Hot Springs National Park, Arkansas

Hot Springs, Arkansas, was made into a National Park in 1921. There were over 70 open springs flowing freely on the side of Hot Springs Mountain. The Hot Springs water is half meteoric in origin and possibly new water (Bedinger et al., 1974). Rainwater travels down the faults to around 6000–8,000 feet and is slowly heated by the Earth's crust. This is a slow process that takes about 4400 years. The average temperature of all the springs is about 62°C (U.S. Department of the Interior, 2022). Hot Springs National Park's alkaline thermal springs were found to differ from other known systems within the US by their low conductivity, low concentration of dissolved ions and specific chemical composition (Meyer, 1981).

One of the few springs that is open to the public was known as the Hale Spring. Many researchers have been attracted to this area because of the algae and bacterial communities capable of surviving high temperatures. Hale Spring is where *Phormidium treleasei* Gomont 1899, was first collected, therefore it is the 'type' species. Now known as Display Springs, it is made up of the two springs: Noble and Lamar, both of which emerge at the base of Hot Springs Mountain. The springs are high in magnesium and calcium bicarbonates. Limestone can be actively seen growing in pockets of the rocks where the spring outfall for Lamar comes out. The combination of magnesium and calcium bicarbonates helps keep the pH of the springs between 7.5 and 7.9 (Haywood and Weed, 1902). The Hot Springs water also contains quite a bit of silica. In springs where they are found in high amounts, silicates are known to increase microbial populations.

3.1 Limiting factors for cyanobacteria within Display Springs

The communities of cyanobacteria in hot springs can be greatly affected by temperature, pH, mineral composition, climate and/or other microbiota of the spring. In turn, the formation of biofilms and mats are also influenced by these factors. Tang et al. (2018) found that one of the main driving forces for bacterial community structures was temperature. Temperatures higher than 65°C are not habitable for cyanobacteria within the springs they studied. Cyanobacteria are adaptive organisms that can tolerate high environmental stress situations. Speculations for them being absent after a specific temperature include competitive exclusion of species and/or an imbalance in metabolites due to pH and temperature. In previous studies, oxygenic photosynthesis was found to only work up until around 73–75°C; this, in turn, inhibits cyanobacterial growth at high temperatures (Pedersen, D. and Miller, S., 2017).

Driving forces such as pH have been deemed an important factor when studying community diversity elsewhere. For instance, pH values at either extreme limit the diversity of cyanobacterial populations because they cause the dominance of only species that could adapt to inhabit that niche. The mineral content of the spring is also important since high pH can cause leaching of minerals from rocks. This can cause an imbalance in minerals that cyanobacteria need for photosynthesis, such as bicarbonates (Tang et al., 2018). By studying these ranges of growth, the thermotolerance for each species can be determined and is useful in explaining the disappearance or sudden appearance of certain species.

Many researchers have contributed knowledge of the microorganisms and limnological features of Display Springs. Meyer (1981) concluded that the factor driving microbial distribution was most likely temperature. Another possible factor mentioned was the influence of stable shade and sun on the growth of microbial mats. Seasonal changes in light intensity, light duration and shade could cause mat compositions and mat size to change. Another study was done in 2006 by Smith et al. on the microbiota of Hot Springs National Park. Display Springs was one of the sites sampled. Smith et al. (2013) did a regression analysis on the total number of species at each site in comparison to temperature. The analysis only supported a portion of the temperature theory (45.3%). It did not explain 54.7% of why the populations were found in certain areas. Their data showed that there must be other factors that influence microbial distribution within thermal springs other than temperature. The Display pool has had the consistent lowest temperature of 50–51°C since 1969, no matter the season (Tankersley III, J., 1970; Meyer, R.L., 1981; Smith, T., et al., 2013; author's field data from 2021). This could have caused *Phormidium treleasei* to migrate to cooler spring sites that were consistent, since the thermal tolerance range for *Phormidium treleasei* was determined to be 38–52°C (Meyer, 1981). This could be why Smith et al. (2013) only found it in the Display pool.

Within the years 2021/22, Display Springs has had constant olive green biofilms within it; however, mat formations fluctuated. Lime green bubbly biofilms could be seen from time to time, year round, but were not constant (Fig. 10.1). The only site that has a consistent mat, year round, was the tufa rock formation of Lamar outfall. Upon microscopic examination of sites in Display Springs (Fig. 10.2) it resembles very closely to how Meyer, R.L., (1996) described *Phormidium treleasei* Gomont.

4. Biofilms/Microbial Mats of Hot Springs

Biofilms are the beginning stages of microbial mats. They exist in hot springs, lakes, the ocean, and even in the desert (Castenholz, 2009). In the beginning stages, a thin film of cohesive microbes will form on the surface of rocks, sticks, leaves and anything else that surrounds them. This film is very cohesive and is often hard to break apart depending on the microorganisms within. Some biofilms will become mats if adequate nutrients are available. Biofilms and microbial mats are a conglomerate of cyanobacteria and other microorganisms. In hot springs, the populations making them up are often, but not limited to, cyanobacteria, algae, diatoms and proteobacteria. The composition of the mat layers depends on the temperature and pH of the hot springs they inhabit. In alkaline hot springs, the top layer is exposed to the most light, and is composed of cyanobacteria. Anaerobic cyanobacteria that are reddish-to-orange mainly comprise the lower dark zones of mats in these types of springs. These anaerobes rely on the organic compounds and electromagnetic radiation that are produced, or unused, by the top layer of aerobic cyanobacteria.

Fig. 10.1. Gas bubbles trapped in a biofilm from Nobel Spring flowing into Display pool (17 July 2022).

Algal mats in hot springs are less common because algae cannot survive in temperatures above 45°C and the cooler parts where they can, are prone to predation by flies and small crustaceans (Castenholz, 2009).

Microbes living in hot springs experience a stable supply of resources. This allows them to inhabit these waters for a long time. In turn, researchers can study the evolutionary mechanisms cyanobacteria have acquired throughout that time (Keshari *et al.*, 2022). Within hot springs that are alkaline, calcium carbonate springs, microbial mats can become embedded in the precipitation. This is known as travertine or 'tufa'. Tufa can form without the aid of cyanobacteria. However, when present, they do shape the morphology to some degree because of their natural metabolic processes (Fig. 10.3). With CO_2 being released, combined with carbonate springs, precipitation forms that causes tufa to grow with and/or on cyanobacteria, algae and mosses. In most other conditions, mats are gelatinous or leathery (Castenholz, 2009).

These micro-ecosystems are a prime example of how biochemical cascades work and how nutrient recycling is possible within a niche. Cyanobacteria can grow from atmospheric CO_2 and N_2 and produce oxygen. This organism's ability to also fix nitrogen is very important, as it provides ammonia, nitrite and nitrates for plants that normally are not accessible. Plants can then use the nutrients to synthesize biological compounds. Just like plants, cyanobacteria can create their own food using water and light. This allows cyanobacteria to produce a stable cycle of nutrients and therefore regenerate cells constantly.

Biofilms are made of exopolysaccharides (EPSs), a conglomerate of mostly heteropolysaccharides, with small amounts of peptides, DNA and fatty acids. The biofilms that are thicker and more compacted are created from capsular polysaccharides (CPS) produced by cells that have sheaths or capsules around them. The EPSs provide structural integrity to the film and serve as a means of protection from abiotic and biotic stresses. Biofilms and mats are resistant to drought conditions because of the gelatinous envelope around the cells, capable of

Fig. 10.2. Cyanobacteria found in Display fountain. Species looks like what Meyer, R.L., (1996) described as *Phormidium treleasei* Gomont.

water-absorption amounts greater than their dry weight. When placed under stressful conditions, more EPSs will be produced to adjust accordingly (Rossi and De Philippis, 2015). For example, temperatures between 50°C and 75°C in hot springs cause biofilms/mats to produce more EPSs, which protects the structure and allows a chance at prolonged habitation of that biofilm or mat. These types of biofilms are desirable since they are capable of withstanding high-stress environments.

There are some molecules that are uniquely used by cyanobacterial EPSs. The presence of sulfate groups and uronic acids accounts for the anionic and sticky properties of the macromolecules. Sulfate groups are found in eukaryotic and archaic cells, but not among other prokaryotic cells. The negatively charged surface of EPSs causes them to have a high affinity for cations. Other prokaryotes create EPSs but they do not contain pentoses like cyanobacteria cells do. Rhamnose and fucose peptides, along with ester-linked acetyl groups, create a hydrophobic property to the film that contributes to its ability to stick to solid surfaces. Different strains of cyanobacteria will create different monosaccharides. Some of the most common ones found in cyanobacteria are rhamnose, fucose, glucose, xylose, mannose, arabinose, galactose, xylose, glucuronic acid and glacturonic acid. Glucose is a major component of polymers, as well as galactose, and, therefore, normally is present in larger amounts than others listed (Rossi and De Philippis, 2015).

EPSs contribute to the Exocellular Polymeric Matrix (EPM) of microbial biolayers. The EPM provides adhesive properties that allow attachment to substrates as well as nutrient and metal ion sequestering. The matrix is constantly under reconstruction and is dependent on the microbial activity residing in it, to replace old products with new synthesized materials. This creates a self-sustaining niche that has many biotechnology applications.

4.1 *Phormidium* and *Leptolyngbya* biotechnology applications

Cyanobacteria have been an interest of researchers for their metabolic properties that could eventually be used in biofuel production and creation of natural polymers. In the lab, conditions can be controlled to maximize yields because cyanobacterial biofilms are easy to manipulate and require low amounts of nutrients. However, the things limiting the use of cyanobacteria for industrial biofuel production are low biomass, low product titers, low reaction rates, and the fluctuating synthesis of products (Bozan et al., 2022). Figure 10.4 shows a biofilm from the Noble Spring channel grown in the laboratory with medium D after 53 days, at temperatures between 45°C and 50°C, and light at 5 K Lux for 11 hours and darkness for 13 hours.

With the discrepancy of which genus *Phormidium treleasei* Gomont belongs in, it is

Fig. 10.3. Tufa formations with bacterial growth on them in Lamar Spring outfall.

Fig. 10.4. Biofilm of a culture collected from the Noble Spring channel after 53 days of growth.

hard to say which biotechnology application this species would be ideal for. Studies on the biofilm-making properties of this type species are being explored currently by the authors of this chapter. Since the taxonomy for this species is still in question, applications that both genera can be used for have been included in this section. Undesirable features can also be deleted to make the species more ideal for the applications.

Bioactive compounds isolated from *Phormidium* species have shown a wide range of antimicrobial properties. In one study, it was shown to inhibit the growth of Gram-negative and Gram-positive bacteria (Bloor and England 1989). *Phormidium* also contains compounds that help aid in suppressing growth of fungi such as fisherellin A, carazostatin, tolytoxin and toyocamycin (Abed et al., 2009). Anti-viral extracts have been isolated from *Phormidium tenue* and have shown an affinity for HIV eradication (Rajeev and Xu, 2004). Another promising use of the *Phormidium* genus is its aid in suppressing algal growth. By nature, cyanobacteria possess compounds that allow them to outgrow algae, and therefore limit the amount of nutrients available for the algae to use.

Biofilms produced by *Phormidium* sp. can be used for bioremediation applications such as destruction of organic pollutants and heavy metal sequestration from wastewater. The process used to introduce cyanobacterial cells as naturally as possible, while still controlling growth, is the passive immobilization technique. The spongy loofa biomass and chitosan polysaccharide gels have been used as immobilization carriers for cells due to their biodegradability and ability to help provide an initial structure for biofilms to grow on, in areas affected by contaminants. Once nutrients like N and P are removed from the wastewater, the resulting N/P-rich biomass can be collected and used for fertilizer (Vasilieva et al., 2016). It was shown that within 24 hours, *Phormidium* sp. immobilized on a chitosan gel were able to use up to 95% of inorganic nitrogen and 87% of phosphate (de la Noüe and Proulx, 1988).

With the massive accumulation of plastic waste, it has become important to search for biodegradable alternatives. The naturally occurring polyhydroxyalkanoates (PHAs) hold great potential in this regard. Bacteria including cyanobacteria use these compounds to store energy or as carbon sources when they lack enough nutrients to reproduce, therefore, PHAs are naturally biodegradable. One specific monomer that has been studied in *Leptolyngbya* sp. is poly(3-hydroxybutyrate) (PHB), which has low oxygen permeability as well as insolubility in water, and yet is much like polypropylene plastics. Under aerobic conditions, PHB can be completely degraded to CO_2 and water. PHBs are produced by bacteria when there is a depletion of N, P or an accumulation of C sources. To test the capability of *Leptolyngbya* sp. NIVA-CYA 255 in production of large amounts of PHB, Kettner et al. (2022) investigated PHB formation in a three-stage cultivation process, containing a growing stage, a macronutrient-depleted phototrophic stage, and a subsequent mixotrophic stage. Cultivation in N and P deficiency supplemented with acetate resulted in an intracellular concentration of 32.3 wt% PHB. The authors also found that cells have the highest carbon-storage capacities (PHB and glycogen), under the nitrogen-limitation conditions. They proposed that N depletion is the choice of strategy for PHB production in *Leptolyngbya* sp. NIVA-CYA 255, since it can be conveniently achieved, and the process can be easily monitored using LipidGreen2-fluorescence. While this experiment demonstrated that *Leptolyngbya* is a promising genus for producing high levels of PHBs for large-scale production under controlled conditions, some limitations need to be overcome to use PHBs on an industrial scale. The cultivation cost is one of the major obstacles. To keep cultivation costs to a minimum, wastewater can be used to provide nutrients for the cyanobacteria. Once the recovery of the PHBs is optimized, it can then be used for synthesis of biodegradable plastics.

In another study, cyanobacterium BTA 287 was isolated from the north-eastern region of India and identified as *Leptolyngbya* sp. using 16S rDNA technique (Tiwari et al., 2019). When the strain was cultured in BG 11 medium supplemented with 25 mM of NaCl for 15 days at 28°C under 6 k Lux with

14:10 day:night ratio, the cultures produced unsaturated and saturated fatty acid in a ratio of 1:84. The saponification value, iodine value, cetane number, degree of unsaturation, long-chain saturation factor and cold filter plugging point and other parameters of the lipid extracted from BTA 287 were found to meet the globally accepted standard of biodiesel, indicating the suitability of *Leptolyngbya* genus as a promising feedstock for biodiesel.

The chapter by no means provides a comprehensive review of potential applications of cyanobacteria in hot springs; it is our hope that it will provide some food for thought and stimulate the interest in research on local cyanobacteria resources.

References

Abed, R.M., Dobretsov, S. and Sudesh, K. (2009) Applications of cyanobacteria in biotechnology. *Journal of Applied Microbiology* 106(1), 1–12.

Anagnostidis, K. and Komárek, J. (1985) Modern approach to the classification system of cyanophytes 1—Introduction. *Archiv für Hydrobiologie*, Supplement 291–302.

Anagnostidis, K. and Komárek, J. (1988) Modern approach to the classification system of cyanophytes. 3 – Oscillatoriales. *Algological Studies/Archiv für Hydrobiologie*, Supplement Volumes Nos 50-53, 327–472.

Bedinger, M.S., Pearson, F.J. Jr, Reed, J.E., Sniegocki, R.T. and Stone, C.G. (1979) The waters of Hot Springs National Park, Arkansas—their nature and origin. *Geological Survey Professional Paper* 1044-C:C1.

Bloor, S. and England, R.R. (1989) Elucidation and optimization of the medium constituents controlling antibiotic production by the cyanobacterium *Nostoc muscorum*. *Enzyme and Microbial Technology* 13, 76–81.

Bozan, M., Schmid, A. and Bühler, K. (2022) Evaluation of self-sustaining cyanobacterial biofilms for technical applications. *Science Direct, Biofilm* 4, 100073.

Bryant, D.A. and Tandeau de Marsac, N. (1995) Differentiation of hormogonia and relationships with other biological processes. In: Bryant, D.A. (ed.) *The Molecular Biology of Cyanobacteria*, Vol. 2. Kluwer Academic Publishers, The Netherlands, pp. 825–842.

Castenholz, R.W. (2009) Mats, microbial. *Encyclopedia of Microbiology* 3 edition. Elsevier/Academic Press, Amsterdam, pp. 278–292.

Castenholz, R.W. and Waterbury, J. (1989) Oxygenic photosynthetic bacteria. Group I. Cyanobacteria. *Bergey's Manual® of Systematic Bacteriology*. Springer, New York, 1710–1789.

Castenholz, R.W., Wilmotte, A., Herdman, M. et al. (2001) Phylum BX. Cyanobacteria. *Bergey's Manual® of Systematic Bacteriology*. Springer, New York, pp. 473–599.

Claus, G. (1961) Contributions to the knowledge of the blue-green algae of the Salzlackengebiet in Austria. In: *International Review of Hydrobiology* 46, 514–541.

Compère, P. (1974) Cyanophycées de La Région Du Lac Tchad, Taxons, Combinaisons et Noms Nouveaux. *Bulletin du Jardin Botanique National de Belgique / Bulletin van de National Plantentuin van België* 44, 17–21.

de la Noüe, J. and Proulx, D. (1988) Biological tertiary treatment of urban wastewaters with chitosan-immobilized *Phormidium*. *Applied Microbiology and Biotechnology* 29(2–3), 292–297.

Drouet, F. (1968) *Revision of the Classification of Oscillatoriaceae*. The Academy of Natural Sciences, Fulton Press, Lancaster, Pennsylvania.

Frémy, P. (1929) Les Nostocacées de la Normandie. *Not Mem Doc Soc Agric Archeol Hist nat Manche* 41, 197–228.

Geitler, L. (1925) Cyanophyceae. In: Pascher, A. (ed.) *Die Süsswasser-Flora, Deutschlands, Österreichs und der Schweiz* 12. G Fischer, Jena, pp. 1–450.

Geitler, L. (1932) *Cyanophyceae*. Leipzig: Akademische Verlagsgesellschaft m.b.h (Dr. L. Rabenhorst's Kryptogamen-flora von Deutschland, Österreich und der Schweiz. Zweite, vellständig neu bearb. Aufl. 14. Bd. Die Algen).

Gomont, M. (1899) Communication on some new Oscillaria. *Bulletin de la société botanique* 46.

Gomont, M. (1892) Monographie des Oscillariées (Nostocaceae homocystées). *Annales des Sciences Naturelles, Serie Botanique* 15, 265–368.

Haywood, J.K. and Weed, W.H. (1902) *The Hot Springs of Arkansas: Report of an Analysis of the Waters of the Hot Springs on the Hot Springs Reservation, Garland County, Ark.* 57th Congress, Document No. 282. Interior Department. Washington, DC.

Inman, O.L. (1940) Studies on the chlorophylls and photosyntheis of thermal algae from Yellowstone National Park, California, and Nevada. *Journal of General Physiology*, 23(6), 663–665.

Keshari, N., Zhao, Y., Das, S.K. et al. (2022) Cyanobacterial community structure and isolates from representative hot springs of Yunnan Province, China using an integrative approach. *Frontiers in Microbiology* 13, 1–17.

Kettner, A., Noll, M., and Griehl, C. (2022) *Leptolyngbya* sp. NIVA-CYA 255, a promising candidate for poly(3-hydroxybutyrate) production under mixotrophic deficiency conditions. *Biomolecules* 12(4), 504.

Komárek, J. (2016) A polyphasic approach for the taxonomy of cyanobacteria: principles and applications. *European Journal of Phycology* 51(3), 346–353.

Komárek, J. (2018) Several problems of the polyphasic approach in the modern cyanobacterial system. *Hydrobiologia* 811(1), 7–17.

Komárek, J. and Anagnostidis, K. (2005) *SiiBwasserflora von Mitteleuropa Cyanoprokaryota II (Oscillatoriales)* Band 19/2. Elsevier, Berlin.

Komárek, J. and Johansen, J.R. (2015) Filamentous Cyanobacteria. In: Wehr, J., Sheath, R. and Kociolek, P (eds) *Freshwater Algae of North America*, 2nd edn. Elsevier, pp. 135–235.

Komárek J. and Kaštovský J. (2003) Coincidences of structural and molecular characters in evolutionary lines of cyanobacteria. *Algolocical Studies*. 148, 305–325.

Komárek J., Kaštovský J., Mareš J. and Johansen J. R. (2014) Taxonomic classification of cyanoprokaryotes (cyanobacterial genera) 2014, using a polyphasic approach. *Preslia* 86, 295–335.

Marquardt, J. and Palinska, K.A. (2006) Genotypic and phenotypic diversity of cyanobacteria assigned to the genus *Phormidium* (Oscillatoriales) from different habitats and geographical sites. *Archives of Microbiology* 187(5), 397–413.

Meyer, R.L/ (1981) *A Qualitative Evaluation of the Algae Associated with the Springs and Formations in Hot Springs National Park*. Unpublished work. University of Arkansas, Fayetteville, Arkansas.

Meyer, R.L. (1996) The rediscovery of *Phormidium treleasei* Gomont. In: Prasad, J.A. (ed.) *Contributions in Phycology*. Vol. XV. Nova Hedwigia, Beiheft, Stuttgart, Germany, pp. 101–104.

Palinska, K.A. and Surosz, W. (2014) Taxonomy of cyanobacteria: a contribution to consensus approach. *Hydrobiologia* 740(1), 1–11.

Pedersen, D. and Miller, S. (2017) Photosynthetic temperature adaptation during niche diversification of the thermophilic cyanobacterium Synechococcus A/B clade. *ISME Journal* 11, 1053–1057.

Rajeev, K.J. and Xu, Z. (2004) Biomedical compounds from marine organisms. *Marine Drugs* 2, 123–146.

Rippka, E., Deruelles, J. and Waterbury, N.B. (1979) Generic assignments, strain histories and properties of pure cultures of cyanobacteria. *Journal of General Microbiology* 111, 1–61.

Rossi, F. and De Philippis, R. (2015) Role of cyanobacterial exopolysaccharides in phototrophic biofilms and in complex microbial mats. *Life (Basel, Switzerland)* 5(2), 1218–1238.

Seoane, A. (1991) Algas de fuentes termales del NW de España: Baños de Molgas y Caldas de Partovia. *Acta botánica malacitana* 1, 27–30.

Setchell, W. A. and Gardner, N. L. (1903) Algae of Northwestern America. *University of California Publications in Botany* 1. University of California Press, Berkeley, pp. 165–418, pls. 17-27.

Smith, T., Manoylov, K. and Packard, A. (2013) Algal extremophile community persistence from Hot Springs National Park (Arkansas, U.S.A.). *International Journal on Algae* 15, 65–76.

Smithsonian Institute Collections Search Center website. Available at: https://collections.si.edu/search/results.htm?q=phormidium+treleasei (accessed 10 September 2022).

Starmach, K. (1966) *Cyanophyta-sinice. glaucophyta-glaukofity*. Warszawa: Panstwowe wydawn, naukowe (Flora słodkowodna Polski, t. 2).

Steyermark, J. (1957) *Contributions to the flora of Venezuela*, Fieldiana. Botany series, Vol. 28, part 2, no. 2. Chicago Natural History Museum, Chicago, Illinois.

Tang, J., Liang, Y., Jiang, D. et al. (2018) Temperature-controlled thermophilic bacterial communities in hot springs of western Sichuan, China. *BMC Microbiology* 18, 134.

Tankersley, J. (1970) Thermophilic blue-green algae in Hot Springs National Park. MSc thesis, East Texas State University, Commerce, Texas.

Teneva, I., Dzhambazov, B., Mladenov, R. and Schirmer, K. (2005) Molecular and phylogenetic characterization of *Phormidium* species (Cyanoprokaryota) using the cpcB-IGS-cpcA locus. *Journal of Phycology* 41(1), 188–194.

Tilden, J.E. (1910) *Minnesota Algae Vol. 1: The Myxophyceae of North America and Adjacent Regions Including Central America, Greenland, Bermuda, the West Indies and Hawaii*. University of Minnesota, Minneapolis, Minnesota.

Tiwari, O.N., Bhunia, B., Bandyopadhyay, T.K. and Oinam, G. (2019) Strategies for improved induction of lipid in *Leptolyngbya* sp. BTA 287 for biodiesel production. *Fuel* 256, 115896.

U.S. Department of the Interior (2022) *Hot Springs Geology*. National Parks Service. Available at: https://www.nps.gov/hosp/learn/nature/hotsprings.htm (accessed 9 September 2022).

Vasilieva, S.G., Lobakova, E.S., Lukyanov, A.A. *et al.* (2016) Immobilized microalgae in biotechnology. *Moscow University Biological Sciences Bulletin* 71, 170–176.

Wilmotte, A. (1994) Molecular evolution and taxonomy of cyanobacteria. In: Bryant, D.A. (ed.) *The Molecular Biology of Cyanobacteria*. Kluwer Academic, Boston, MA, pp. 1–25.

11 Algae-based Aquaculture Wastewater Treatment and Resource Utilization

Pengfei Cheng[1,2], Chun Wang[1], Yahui Bo[1], Jiameng Guo[1], Xiaotong Song[1], Shengzhou Shan[1], Chengxu Zhou[1], Xiaojun Yan[3]* and Roger Ruan[2]*

[1]*College of Food and Pharmaceutical Sciences, Ningbo University, Ningbo, Zhejiang, China;* [2]*Center for Biorefining and Department of Bioproducts and Biosystems Engineering, University of Minnesota-Twin Cities, Saint Paul, Minnesota, USA;* [3]*Key Laboratory of Marine Biotechnology of Zhejiang Province, Ningbo University, Ningbo, Zhejiang, China*

1. Introduction

Aquaculture plays an important role in food security and human nutrition. However, the recent, rapid development of aquaculture also brings potential environmental challenges. Aquaculture wastewater is rich in nutrients and contaminants, such as nitrogen, phosphorus, heavy metals, antibiotics and organic matter. If discharged into surrounding natural bodies of water without effective treatment, these components could easily destroy the ecological balance and lead to eutrophication of the water bodies. Compared with other technologies for wastewater treatment, using microalgae to purify aquaculture wastewater can effectively remove these pollutants, especially nitrogen and phosphorus, and simultaneously fix CO_2 to achieve emission reductions. Moreover, the harvested microalgae can be used as aquatic bait and feed additives, as well as raw materials for the development of high-value-added products such as pharmaceuticals, food and biomass fuels. This chapter introduces the role that microalgae can play in treating aquaculture wastewater, and also focuses on the feasibility of mixotrophic microalgae to treat aquaculture wastewater to utilize the resources in this waste stream, and provide a technical reference for green, ecological treatment.

2. The Role of Microalgae in Aquaculture

2.1 Water quality control

The most direct methods of controlling water quality in aquaculture include reducing the stocking density and changing the water frequently. However, in aquaculture, the application of these two methods is limited due to low profit margins and high costs. Wastewater reuse treatment is a possible way to reduce the frequency of water replacement and control the operating costs of aquaculture systems (Altmann *et al.*, 2016). The commonly used technologies for aquaculture wastewater treatment include anaerobic treatment and aerobic treatment, which are used for wastewater removal and water purification. In a study by Mirzoyan *et al.* (2008), aquaculture sludge was anaerobic

*Addresses for correspondence: xiaojunyan@hotmail.com; ruanx001@umn.edu

treated to produce methane, and the sludge mass was reduced by 70% (Mirzoyan et al., 2008). However, these conventional treatment techniques are not highly recommended from a nutrient recovery point of view (Lu et al., 2019b). For example, organic carbon in aquaculture wastewater is converted to CO_2 and CH_4 by aerobic and anaerobic treatment, and the resources in the wastewater cannot be fully utilized. Constructed wetlands are considered an environmentally friendly way to recover nutrients and improve water quality from aquaculture wastewater, but their commercialization is hindered by their large footprint and high management costs. Longo et al. (2016) reported that maintaining high dissolved oxygen (DO) levels through aeration is a technically feasible way to increase stocking densities, but the high electrical power consumption of aeration increases aquaculture costs (Longo et al., 2016). Therefore, to date, ecofriendly, cost-effective water quality control technologies need to be widely used in aquaculture.

2.2 Aquaculture wastewater treatment

Traditional aquaculture provides food for human beings, but produces large amounts of wastewater, threatening global sustainable development. Misuse of antibiotics and water substitution or treatment can lead to safety concerns and increase aquaculture costs. To overcome environmental and economic problems in the aquaculture industry, a great deal of effort has been devoted to the application of microalgae in wastewater remediation, biomass production and water quality control. In this review, the technologies required for microalgae-assisted aquaculture and their recent advances are systematically described. The problems caused by aquaculture wastewater discharge are reviewed in depth, and the principles of microalgae-assisted aquaculture are introduced. Contents include nutrient assimilation mechanisms, algal culture systems (ditch ponds and degrading algal biofilms), wastewater pretreatment, algal-bacterial cooperation, harvesting techniques (fungi-assisted harvesting and flotation), algal species selection, and development of high-value-added microalgae as aquaculture feed. Given the limitations of recent studies, to further reduce the negative impact of aquaculture wastewater on global sustainability, future directions for industrial applications of microalgae-assisted aquaculture are proposed.

Microalgae can effectively absorb nutrients in eutrophic water bodies and have been shown to be a good method for wastewater remediation (Leng et al., 2018; Wang et al., 2015). The enormous performance of microalgae in nutrient assimilation has been widely observed in the remediation of food-industry wastewater, agricultural waste streams, municipal sewage and many other types of wastewaters (Lu et al., 2015; Wang et al., 2015). In recent years, an increasing number of studies have confirmed the beneficial role of microalgae in aquaculture wastewater treatment (Gao et al., 2016; Ansari et al., 2017). In addition to treating wastewater, microalgae can synthesize value-added components, including proteins, lipids and natural pigments. Previous studies have successfully applied *Chlorella* and/or *Dunaliella* for the production of value-added biomass, which can partially replace aquaculture feed and improve the immunity of aquatic animals (Sirakov et al., 2015; Lu et al., 2019). Last but not least, microalgae with high oxygen-producing capacity can act as a biological pump to aerate aquaculture effluent and regulate microbial communities in water bodies. Therefore, water quality from aquaculture effluent can be properly controlled to avoid blooms or oxygen depletion. Due to the aforementioned benefits, the use of microalgae for remediation of aquaculture wastewater has recently gained more attention. In recent years, the concept of using microalgae in aquaculture has been proposed, and efforts have been made to promote the industrialized implementation of microalgae-assisted aquaculture (Lu et al., 2019; Roy and Pal, 2015). Microalgae play an important role in water quality control, aquaculture feed production and nutrient recovery, and contribute to the sustainable development of aquaculture. At the same time, the related technologies and mechanisms of microalgae-assisted culture are introduced.

Finally, challenges and prospects for the integration of microalgae and aquaculture are discussed. It is expected that in the near future the industrial application of microalgae technology will be a promising approach to overcome traditional aquaculture problems, and upgrade the entire aquaculture industry for global sustainable development.

3. Microalgae-based Aquaculture Wastewater Treatment

3.1 Ammonia removal from wastewater by microalgae

Microalgae can purify different types of wastewaters through bioaccumulation, biodegradation and biosorption. Nitrogen is a key element in the conversion of biomass to such entities as enzymes, peptides, proteins and chlorophyll, as well as energy such as ADP and ATP. Inorganic nitrogen available to microalgae includes nitrite nitrogen (NO_2), ammonia (NH_3) and nitrogen gas (N_2). Nitrogen in wastewater is usually present as the ammonium ion, NH^{4+} (Ross et al., 2018). Microalgae take up NO_3^-, reduce it to NO_2^-, and then convert it to NH_4^+, which is later bound to amino acids, while NH can be taken up and used directly by microalgae. The first step in nitrate assimilation is the use of nitrate reductase (NR), the reduced form of nicotinamide adenine dinucleotide (NADH), which is present within the microalgae. After transferring two electrons in the nitrate to nitrite conversion reaction, the microalgae use nitrite reductase (NADPH) to photosynthesize with ADP, phosphate and NADP. NO is converted back to NH_4^+ (Fig. 11.1) (Emparan et al., 2019).

3.2 Phosphorus removal from wastewater by microalgae

Inorganic phosphorus is present in wastewater in lipids, nucleic acids and proteins, and its transport occurs across the plasma membrane of microalgal cells. The primary forms of phosphorus utilized by microalgae are dihydrogen phosphate ($H_2PO_4^-$) and hydrogen phosphate (HPO_4^-), which convert $H_2PO_4^-$ and HPO_4^- into organic compounds such as ATP through phosphorylation (Fig 11.1) (Cai et al., 2013), while the energy required for phosphorylation can be generated in the oxidation of substrates by respiration, electron transfer in the mitochondria of eukaryotic microalgae, and light utilized during photosynthesis (Cuellar-Bermudez et al., 2015). For photosynthesis, two steps are involved: photochemical and redox reactions. The equation for phosphorylation involving ADP, phosphate (P) and nicotinamide adenine dinucleotide phosphate (NADP) is as follows (Razzak et al., 2013):

$$2H_2O + 2NADP^+ + 3ADP + 3P + Photons\,(light) \rightarrow 2NADPH + 2H^+ + 3ATP + O_2$$

From the equation, it is clear that light energy is used to synthesize energy storage molecules (ATP and NADPH). In addition, at higher pH, phosphate can be removed from water by forming hydroxyapatite and precipitating with Ca^{2+} and Mg^{2+} (Lu et al., 2016). At the end of the two-stage culture, the removal rates of total nitrogen, total phosphorus, ammonia nitrogen, and COD by UMN 280 were 90.60%, 98.48%, 100% and 79.10%, respectively. This also improved the nutrient removal efficiency from the wastewater (Hussain et al., 2017).

3.3 Heavy metals removal from wastewater by microalgae

Current methods for the removal of heavy metals from wastewater include chemical precipitation, membrane separation and adsorption, but these methods all suffer from high costs. Microalgae can be used to remove heavy metals from water bodies by means of biosorption, bioaccumulation or biotransformation for water purification purposes (Zhou et al., 2012). One study used cellulose ester membranes and microalgae to form biofilms for zinc removal, and the results showed that each gram of microalgae could adsorb 15–19 mg of metallic zinc, which had a strong removal capacity.

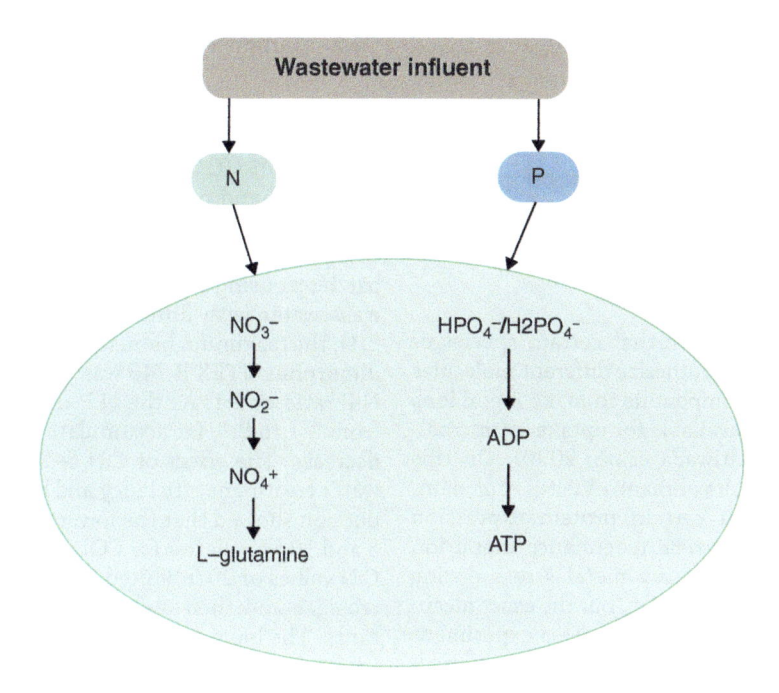

Fig. 11.1. Uptake mechanism of nutrients in intracellular of microalgae.

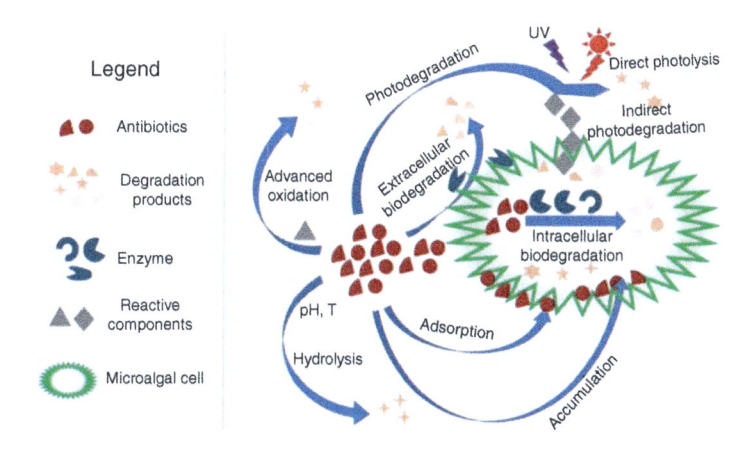

Fig. 11.2. The specific mechanism of antibiotics degradation with microalgae.

Danouche et al. (2021) studied the removal capacity of six microalgae for chromium, and the results showed that brown algae had a strong removal capacity for Cr^{3+} and Cr^{4+}, with different removal capacities for different oxidation states (Murphy et al., 2008). The removal of heavy metals by microalgae is characterized by short treatment time, high sensitivity to heavy metals, low input cost, and environmental friendliness. It has also been shown that heavy metals adsorbed by microalgae can be desorbed under certain conditions, enabling the recovery of heavy metals and improving economic efficiency (Blanco-Vieites et al., 2022).

4. Factors Affecting the Effectiveness of Wastewater Treatment with Microalgae

4.1 Illumination

Studies have shown that certain species of microalgae can synthesize different molecules as chelating compounds to make metal ions more readily available for uptake by microalgal cells (Khatiwada et al., 2020). On this basis, the results of Blanco-Vieites et al. demonstrated that certain protein expression occurs in A. maxima to enhance metal ion sorption under heavy metal stress during heavy metal remediation, but the exact mechanism needs further study. This experiment is the first study of industrial-scale microalgae culture for bioremediation of iron and steel industrial wastewater, illustrating the potential that A. maxima has for heavy metal remediation (Blanco-Vieites et al., 2022).

Microbial communities in wastewater may compete for nutrient resources with exogenous microalgae added to the wastewater, and light intensity and photoperiod are important factors affecting the growth rate of the microalgal biomass, as well as the efficiency of pollutant removal from wastewater (Lee et al., 2013). Researchers have investigated the effect of different light intensities (20 µmol/s·m² -100 µmol/s·m²) on the growth of a mixture of cyanobacteria, Chlorella vulgaris and Chlamydomonas reinhardtii in municipal wastewater (Iasimone et al., 2018). The total biomass of microalgae and the $N-NH_4^+$ removal rate increased significantly with increasing light intensity. At a light intensity of 100 µmol/s·m², the biomass reached 243 mg/L and the $N-NH_4^+$ removal rate increased 40.48%, compared to 20 µmol/s·m². The energy captured by microalgae from light is necessary for CO_2 fixation, and the rate of photosynthesis is proportional to irradiance below the light saturation point, above which the receptor system is impaired, resulting in photo-inhibition (Williams et al., 2010).

4.2 Carbon, nitrogen and phosphorus

The physicochemical characteristics of different types of wastewaters such as carbon-to-nitrogen ratio (C:N), nitrogen-to-phosphorus ratio (N:P) and nutrients have a significant impact on nutrient removal rates and microalgal growth (Faruque et al., 2020). In a batch test using synthetic tertiary municipal wastewater with different N:P ratios (1:1–8:1), the maximum biomass of Scenedesmus dimorphus UTEX B 746 was 733 mg/L at an N:P ratio of 2:1. As the N:P ratio increased from 1:1 to 8:1, fat accumulation tended to decrease. The effect of C:N 5–20 on wastewater treatment efficiency and biomass production showed that the lower C:N values of 5 and 10 were better for COD removal, while C:N values of 20 inhibited the growth of microalgae and their ability to remove pollutants. The biomass tended to decrease with increasing C:N values, while the lipid content showed the opposite trend. At a C:N value of 15, the highest oil and grease yield was achieved, and the removal rates of COD, nitrogen, and phosphorus were 87.2%, 90.5% and 88.6%, respectively (Ferreira et al., 2022).

4.3 Temperature

Ambient temperature also has a significant effect on microalgal biomass and wastewater treatment efficiency. For example, carbon assimilation in photosynthesis (i.e. Calvin cycle) is mediated by enzymes, so the reaction rate is susceptible to temperature (Mohsenpour et al., 2021). One researcher evaluated the removal of NH_4^+-N from wastewater by Scenedesmus sp.; removal rates were 4.3, 6.7, 15.7 and 17 mg/(NL·d) at 15°C, 18°C, 26°C and 34°C, respectively (Ruiz-Martínez et al., 2021). However, the optimum temperature has been shown to depend on the species of microalgae and their adaptation to the particular environment. In general, most microalgae are able to survive at temperatures between 10°C and 30°C, while the optimum temperature range is much narrower, usually between 15°C and 25°C (Ruiz-Martínez et al., 2021). In microalgae treatment of wastewater, it is not feasible to maintain the

Table 11.1. The mechanism of carbon treatment by microalgae.

Nutrient substance	Mechanism	Cell entry mode
CO_2/HCO_3	Calvin cycle	Diffuse or active transport
Organic carbon	Respiratory metabolism	Diffuse or active transport

optimum temperature by heating the reactor, as the reaction equipment is too large and requires significant energy consumption. Therefore, the algal species used to treat wastewater should be selected according to the environmental conditions of the wastewater treatment plant (Williams et al., 2010).

4.4 The specific pollutant antibiotics removal

4.4.1 Adsorption

Adsorption is the process of removing pollutants by passive binding of the pollutants to a solid material (Bai and Acharya, 2016). The removal of antibiotics by adsorbent such as biochar, activated carbon, and nano-materials has been extensively reported (Ahmed et al., 2015). Microalgae can also be an effective adsorbent for antibiotics removal. The adsorption can be accomplished by the functional groups and polymer assemblages (similar to cellulose, hemicelluloses and proteins) on the algal cell walls and it is an extracellular process (Xiong et al., 2018). The effectiveness of the adsorption process may be assessed by using dead algal biomass to adsorb antibiotics. The adsorption performance varies significantly depending on specific antibiotics and microalgae, due to their different hydrophilicity, functionality and structures. In general, it is desirable when the antibiotic is more hydrophobic rather than hydrophilic and when it carries an opposite charge to the microalgae (Leng et al., 2015; Xiong et al., 2018).

Some antibiotics can be removed effectively by adsorption. For example, 82.7% and 71.2% of the cefalexin (at 50 mg/L) in modeling wastewater can be removed by *Chlorella* sp. biomass and the defatted biomass of *Chlorella* sp., and the calculated theoretical adsorption capacities were 129 mg/g and 63 mg/g, respectively (Angulo et al., 2018). Very high adsorption capacity (295 mg/g) was obtained when using lipid-extracted *S. quadricauda* biomass to remove tetracycline (Daneshvar et al., 2018). Adsorption was also found to be one of the main mechanisms for the removal of tetracycline in HRAP (Norvill et al., 2016).

4.4.2 Accumulation

Unlike adsorption, which is extracellular, accumulation is an intracellular process to remove pollutant from water (Bai and Acharya, 2016). Some antibiotics can cross algal cell membranes and then possibly be assimilated by the cells. The intracellular accumulated antibiotics may be extracted by sonication coupled with dichloromethane/methanol mixture (v:v /1:2) extraction (Song et al., 2019). Algae accumulation has been reported to play an important role in the removal of antibiotics such as trimethoprim, sulfamethoxazole and doxycycline (Prata et al., 2018). Some accumulated antibiotics can induce the production of reactive oxygen species, which is essential to control cellular metabolism at normal concentrations, but may result in severe damage to cells, or eventually death, if in excess (Xiong et al., 2018). On the other hand, the algal cell can counteract to deplete the antibiotics by metabolism. In this case, accumulation becomes a pre-step for biodegradation, and the combination of accumulation and biodegradation in the algal cells can contribute considerably to the completion of the assimilation of some antibiotics. The accumulation of sulfamethazine in *C. pyrenoidosa* was observed and then followed by biodegradation (Sun et al., 2017). Levofloxacin was also removed by *C. vulgaris* through accumulation and subsequent intracellular biodegradation (Xiong et al., 2017). Otherwise, antibiotics will be accumulated in the organisms, and the antibiotics can further accumulate and magnify through the food chain and eventually cause antibiotic resistance in the human body.

4.5 Antibiotic

As ecofriendly aquaculture and the recovery of nutrients from wastewater become increasingly important for global sustainable development, the potential benefits that microalgae-assisted aquaculture brings to the natural environment deserves the attention of both academia and industry. With the solution of the above problems, microalgae-assisted farming will move from laboratory research to industrial application (Liu et al., 2014). Generally speaking, the progress brought by the widespread application of microalgal biotechnology to aquaculture can be divided into two aspects: environmental and economic. First, threats to environmental sustainability from aquaculture can be controlled to a certain extent. Due to the performance of microalgae in carbon dioxide fixation and wastewater remediation, pollution caused by aquaculture can be mitigated. In addition, the misuse of antibiotics or drugs in microalgae-assisted farming can be prohibitive, due to the health benefits, or risks, of microalgae feeds to aquatic animals. Therefore, the growing antibiotic resistance in aquaculture can be controlled (Paerl and Otten, 2013). Therefore, with the widespread application of microalgae-assisted jujube farming, its positive impact on environmental safety and sustainability will contribute to global sustainable development. Second, aquaculture costs can be reduced through nutrient recovery and biomass production. In practical applications, the cost of feed accounts for a large part of the total cost of aquaculture, so the financial burden is an obstacle to the development of aquaculture. Assisted cultivation through microalgae may help to reduce operating costs (Lamb et al., 2017).

The nutrients in wastewater are converted into value-added biomass, which can be further developed and utilized to produce aquafeed. Therefore, the recovery of nutrients in aquaculture through microalgal biotechnology can reduce material inputs and overall system costs (Bhatnagar and Devi, 2013). Considering the environmental and economic advancements, with the widespread application of microalgae-assisted aquaculture, it may contribute to the global sustainability of aquaculture.

5. Wastewater Treatment with Microalgae by the Mixotrophic Mode

Microalgae mixed culture is a combination of the photoautotrophic and chemical energy heterogeneous culture modes. Light energy is used to enhance the photo-cooperative photo-reaction stage of microalgae, and exogenously added organic carbon is used to enhance the aerobic respiration reaction of microalgae. It is a culture mode with coordination that is conducive to the growth and development of microalgae (Roostaei et al., 2018). In all kinds of wastewater, such as aquaculture wastewater, agricultural wastewater and industrial wastewater, there are various metal ions and organic matter, including NO_2^-, F^-, CN^-, and compounds containing carbon, nitrogen and phosphorus, most of which are essential substances for the growth of microalgae. The mixed-culture mode of microalgae in wastewater treatment can not only effectively reduce the content of compounds in wastewater but they can also harvest a certain amount of microalgae biomass (Rambabu et al., 2020).

The carbon in wastewater is divided into inorganic carbon (dissolved CO_2 and carbonate etc.) and organic carbon. Microalgae can absorb inorganic carbon through active transport or free diffusion. Some heterotrophic microalgae can utilize organic pollutants in wastewater as a carbon source and energy, for heterotrophic metabolism, thereby reducing the COD content in the wastewater (Carvalho et al., 2013). When microalgae are mixed, they can be divided into two stages. In the first stage, heterotrophic metabolic growth was carried out due to the high initial organic carbon content in the

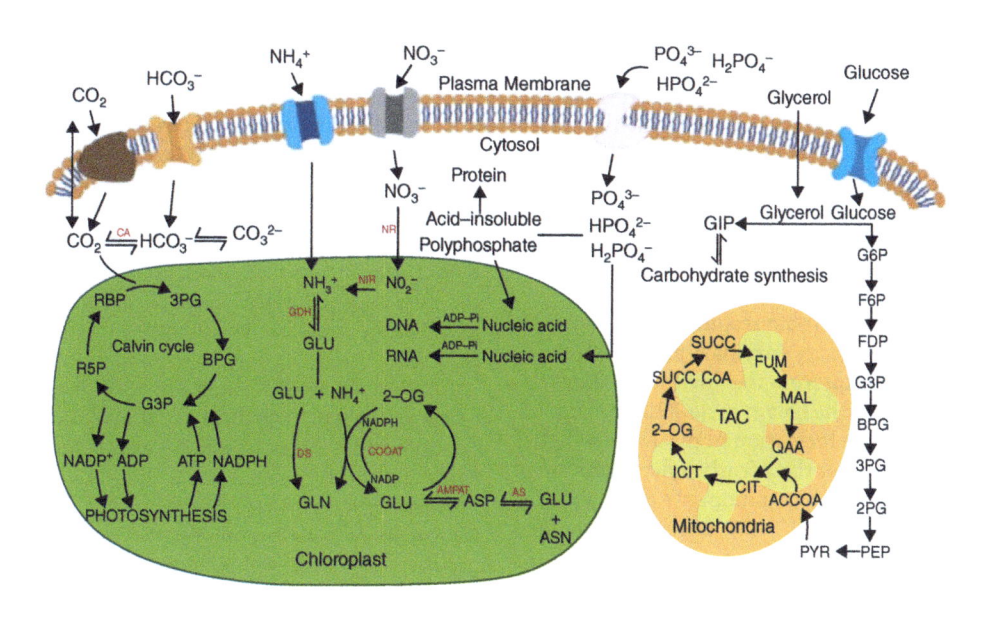

Fig. 11.3. The advantages of microalgae mixed culture in wastewater treatment.

Table 11.2. The mechanism for the nutrients removal with microalgae.

Nutrient substance	Mechanism	Cell entry mode
N_2	Fixed to ammonia by microalgae, and then converted to amino acids	Diffuse or active transport
NO_3-N and NO_2-N	Reduced to ammonia and then converted to amino acids	Active transport
NH_2-N	Directly converted to amino acids	Active transport
PO_4-P	phosphorylation	Active transport

wastewater. When the organic carbon content decreases to a certain level, the second stage of photosynthesis starts to be induced, and the algae absorb CO_2 for photoautotrophic growth. If the light-dark cycle is combined to allow autotrophic and heterotrophic microalgae to proceed under their optimal conditions, the biomass and lipid content of the mixed culture will be more than the sum of the autotrophic and heterotrophic processes alone (Sara et al., 2015).

5.1 Treatment of inorganic carbon in wastewater under mixotrophic mode

Algae can convert inorganic matter in the air into organic matter through photosynthesis to meet their own growth needs (Neilson et al., 1974). Carbon is an important component of cell structure. Algae stem cells contain a large amount of carbon, accounting for 40–50%. Microalgae, on the one hand, can make use of CO_2 from the atmosphere by photosynthesis, and provide a carbon source for itself; on the other hand, it can take advantage of the wastewater that contains negative ion HCO_3^- and give itself a second carbon source, since the diffusion rate of CO_2 in the water is lower than the rate of diffusion in the air (Nirmal et al., 2017). This is why microalgae can treat inorganic carbon in wastewater. In addition to diffusing CO_2 uptake, microalgae have also developed CO_2 concentration mechanisms (CCMs) to promote photosynthesis in response to low CO_2 concentrations. Of all microalgae, CCMs are most capable of actively transporting CO_2 or HCO_3^- into cells through the plasma membrane, or inner membrane, in response to low CO_2 concentration conditions. They can then promote algal photosynthesis, whereas increased external CO_2 concentration generally reduces CCMs activity (Jadwiga and Leo, 2014).

5.2 Treatment of organic carbon in wastewater under mixotrophic mode

Microalgae's aerobic respiratory system utilizes glycolysis to enzymatically break down organic carbon from the environment or that produced by the Calvin cycle. This breakdown results in the formation of pyruvate, which is then utilized in the tricarboxylic acid cycle to meet the growth requirements of microalgae. (Craggs et al., 2013). Domestic sewage, food processing, papermaking and other industrial wastewaters, will contain carbohydrates, proteins, oils, lignin and other organic substances. These substances exist in suspension, or in a dissolved state, in sewage, and most can be absorbed by the microalgae to promote their own growth. At the same time, the absorption of organic carbon by microalgae promotes aerobic respiration, and simultaneously promotes photosynthesis, which is an important role for coordinated growth of the microalgae during its co-culture mode (Anil et al., 2020).

5.3 Treatment of other substances in wastewater with mixotrophic mode

Due to the common occurrence of organic carbon and inorganic carbon in wastewater, microalgae can grow rapidly. This enables the microalgae to absorb a large number of elements, such as N and P in the environment, to maintain their development. Phang et al. (2000) used *Spirulina* to treat starch aquaculture wastewater, and the results showed that it also enhanced the removal rates of ammonia.

5.4 Factors affecting the treatment of wastewater by mixotrophic microalgae

5.4.1 The characteristics of wastewater

Different sources of wastewater have different water-quality characteristics, such as pH value, temperature, ammonia nitrogen, organic matter, heavy metals and other toxic substances, which have a great impact on the growth of microalgae. For example, livestock wastewater has higher ammonia nitrogen content, while beer wastewater has higher organic matter content (Ting et al., 2000). In the process of wastewater treatment, the optimum culture conditions for the microalgae, and the characteristics of the wastewater, sometimes do not completely coincide.

Generally, two methods are used to cultivate microalgae. The first is to screen and cultivate microalgae, suitable for the characteristics of its wastewater, such as the separation of microalgae from wastewater to use in the treatment of wastewater itself; second, the proportion of wastewater components should be optimized to make microalgae more suitable for growing in characteristic wastewater (Katam and Bhattacharyya, 2018).

5.4.2 Light intensity and light photoperiod

Light intensity and light photoperiod are important factors affecting the growth rate and biomass of microalgae, as well as the removal efficiency of pollutants in wastewater. Studies have shown that light intensity is positively correlated with microalgae biomass, and it is found that higher light intensity combined with exogenous CO_2 can not only promote the increase of biomass but also enhance the treatment effect of COD and nitrogen in municipal wastewater (Li et al., 2011). Some researchers have studied the effects of microalgal-bacterial complex on nutrient removal and biomass production under three different light photoperiods (12h/12h, 36h/12h and 60h/12h). The results showed that carbon removal was positively correlated with the length of the dark photoperiod, while nitrogen and phosphorus showed an opposite trend. This indicates that the light-dark photoperiod is the key influencing factor of wastewater treatment by microalgae (Chang et al., 2015). It indicates that the optimization of light intensity and light photoperiod can obtain higher biomass and wastewater treatment effects.

6. High-value Application of Microalgae

Compared to higher plants, microalgae exhibit higher annual photon-to-biomass conversion efficiencies (of about 3%, compared to higher plants at less than 1%) and with no intrinsic sensitivity to seasonality (Perin et al., 2019). This feature, combined with the enormous diversity of algal strains, enables the production of valuable molecules such as proteins, lipids, carbohydrates and pigments, in high yields. They can be grown on wastewater, or non-agricultural land, without the use of pesticides, and without affecting the production of food or other crop products. In addition, they are able to recycle carbon dioxide from the atmosphere, thereby minimizing the associated environmental impact. As a result, microalgal biotechnology has risen steadily, driven by current energy, environmental and food challenges (Hamed, 2016; Milledge, 2011). At present, it can be divided into four research areas, namely: (i) wastewater treatment; (ii) carbon dioxide balance; (iii) biofuel production; and (iv) high value-added molecular production.

The advantage of using microalgae to extract high-value products is that it can be obtained at minimal cost by using only water and atmospheric CO_2 for cultivation. It does not create competition for land and food crops, because microalgae can grow on degraded land. However, microalgal culture requires a medium rich in nutrients and salts (Baicha et al., 2016). In addition, microalgal biomass also possesses high photosynthetic efficiency and, when combined with bioenergy production systems, has the potential to be a sustainable route to renewable energy in the future (Khoo et al., 2011; Cheah et al., 2015).

6.1 Biofuels

Compared with the first generation of edible crops based on sugars, starches and oils, and the non-edible lignocelluloses derived from industrial agricultural residues, the third generation of biofuels based on microalgae can produce cleaner and more sustainable biofuels (Dai et al., 2014). Microalgae can convert CO_2 and some nutrients in wastewater into biomass with more lipids, proteins, polysaccharides and other products, which can be further converted into valuable bioenergy options, such as biodiesel, ethanol or biogas. Biodiesel derived from plants can not only generate the energy required for transportation but it can also reduce the emissions of air pollutants such as polycyclic aromatic hydrocarbons and sulfur (Boelee et al., 2012). Diesel emissions can be improved, and reduced, through the use of biodiesel or oxygenated blending components. As oil-rich plants, microalgae have a short growth cycle and do not occupy arable land. Microalgae can be cultivated and harvested throughout the year without time limit, providing stable, reliable and low-pollution biofuels. Microalgal biofuels have the potential to meet global demand because they do not compete with the production of food products. Microalgae can be used to produce biofuels, such as: biodiesel, biogas, bioethanol and biochar. The realization of biofuels can be achieved through thermochemical conversion and biochemical processing technologies, using microalgae as a feedstock (Ofari-Boateng et al., 2012).

6.2 Protein

In recent years, due to increasing health problems, there is a growing demand for convenient and healthy nutritional products; there is an expanding market for phytonutrients, especially protein sources. Increased demand has prompted industries, such as food production, to look for lucrative sources of proteins (Markou et al., 2012). Important bioactive compounds, such as nutritional proteins, can be isolated from microalgae.

Microalgae contain a variety of biologically active substances, and essential nutrients, required by the human body, and are a promising source of protein. Among the many microalgae, *Chlorella* and *Spirulina* are the most popular in the global microalgae market due to their high protein quantity and quality (50–70% of dry cell weight) and other nutrients (Capelli and Cysewski, 2010). *Spirulina* has high protein quality because it contains all the essential amino acids, accounting for 47% of its total protein (Soni *et al.*, 2017). The quantity, ratio and quality of amino acids, are used to assess the overall quality of the protein (Haoujar *et al.*, 2022). The vegetable protein contained in *Spirulina* is about twice as high as the optimal protein in other vegetables. It is much higher than high-quality protein sources such as soybean (35%), milk powder (35%) and peanut (25%) (AlFadhly *et al.*, 2022). Batista *et al.* found increased protein availability in biscuits by adding *Arthrospira* as a high-protein food source (Batista *et al.*, 2019). His study found that the effect of *Spirulina* on malnourished children is more beneficial than that of milk powder. Milk powder contains lactic acid that can be difficult to digest and absorb, and the protein from *Spirulina* is more effective than that of milk powder (Singh *et al.*, 2021; Seyidoglu *et al.*, 2017).

6.3 Pigment

Three basic classes of natural pigments in microalgae should be highlighted: carotenoids, chlorophyll and phycobiliproteins. These pigments have been used as precursors for vitamins, human food, animal feed additives, cosmetics in the pharmaceutical industry, food colorants, and biomaterials (Krupa *et al.*, 2010).

Carotenoids are fat-soluble pigments that color some plants and are considered auxiliary pigments (Chen *et al.*, 2016). Traditionally, the extraction methods of microalgal carotenoids have used organic solvents and Soxhlet extraction (Cheng *et al.*, 2016), where the choice of solvent is important for determining affinity and promoting carotenoid release. However, low selectivity and excess solvent requirements have led to the use of alternatives, such as using supercritical fluid extraction (SFE) (Nobre *et al.*, 2013).

It is inferred that the addition of a co-solvent can improve the extraction efficiency of carotenoids. In addition to this, a precipitation process has been suggested to increase the carotenoid yield (Liau *et al.*, 2010). Singh *et al.* (2013) performed various cell-disruption methods to maximize the extraction of zeaxanthin, a carotenoid. For chemical cell destruction, microalgal biomass is immersed in dimethyl sulfoxide (DMSO) followed by centrifugation, followed by treatment of freeze-dried cells with strong inorganic acids such as hydrochloric and sulfuric acids or mild organic acids such as citric and acetic acid, and then extract. In chemical cell disruption, lower yields were reported using strong acids, while improved extraction efficiencies were achieved using weak acids, where a 30-fold increase in carnitine yield was observed (Singh *et al.*, 2013). As for mechanical cell disruption, sonication resulted in the highest carotenoid yield, a nearly 40-fold increase over direct extraction.

Phycobiliproteins are the major photosynthetic accessory proteins in microalgae. Major applications of phycobiliproteins include the use of natural dyes in the food industry and pharmaceutical products. It possesses anti-oxidant, anti-viral, anti-cancer, anti-allergic, anti-inflammatory and neuroprotective properties, making it a promising material for health-related applications (Chen *et al.*, 2013). Phycobiliproteins can be obtained by sequential centrifugation, drying, homogenization and repeated freeze-thaw processes (Hemlata and Fareha, 2011). Methods for the isolation and purification of phycobiliproteins have been developed, but many of them are cumbersome, time-consuming, involve numerous steps, and have low yields. A potential alternative to the purification technique is aqueous two-phase extraction (ATPE) (Patil *et al.*, 2008). This application in the extraction of phycobiliproteins was investigated under different process parameters (such as type of salt, connecting line length and volume ratio, final product purity). The integration of membrane processes with ATPE has also shown that separation of phase-forming components is associated with improved product purity (Patil *et al.*, 2008). Pigments such as astaxanthin are potent anti-oxidants, among carotenoids, because of their potent anti-aging, sun protection, anti-inflammatory and immune-system-enhancing effects, useful in the nutrition, food and feed supply industries (Cheng et al., 2016).

Astaxanthin is highly sensitive to heat, light and oxygen and readily degrades when exposed to oxidative stress during conventional organic solvent extraction (Thana et al., 2008). It was found that the use of $SC-CO_2$ with co-solvents such as vegetable oil improved the extraction efficiency of astaxanthin due to the higher solubility of astaxanthin in $SC-CO_2$ and oil mixtures. The enhanced mass transfer rate caused by the expansion of the biomass pores will also facilitate the release of the pigment.

6.4 Other high-value-added products

Other valuable and interesting metabolites can be found in diatom chloroplasts, such as chlorophyll a, c1 and c2, and a complement of accessory pigments, including xanthophylls and carotenes. The light-harvesting pigments chlorophyll a, chlorophyll c and fucoxanthin together constitute the complex fucoxanthin-chlorophyll protein (FCP), which has the function of harvesting the light necessary for diatoms. The pigment fucoxanthin is also responsible for the golden-brown color of these microalgae, being the most abundant pigment. Nowadays, seaweeds represent the main source of fucoxanthin at industrial scale. Nevertheless, marine diatoms accumulate five- to tenfold higher amounts and thereby are regarded as a potential source of this valuable pigment (Pereira et al., 2021). An industrially relevant feature of *P. tricornutum* is that it is a fast-growing species, which makes it suitable to be cultivated at a large scale. To date, in Europe, it is cultivated by at least eight companies reaching a yearly production of four tons of dry biomass (Araújo et al., 2021). Among them, *P. tricornutum* is exploited for fucoxanthin production by Alga Technologies (Israel) and for EPA production by Simris (Sweden). Moreover, *P. tricornutum* extract and EPA-rich oil extract were proposed for authorization as novel food to the European Union, whose evaluations are still ongoing.

Even if less commercially competitive, healthy substances obtained from microalgae have specific advantages over standard synthetic alternatives: from a chemical point of view, synthetic molecules are only present in specific isomers, which are normally much less effective than natural versions for certain applications, such as in artificial milk, dietary supplements or fish pigment enhancers (Enzing et al., 2014) (Fig. 11.4). PUFAs such as EPA and DHA have been shown to improve eye health and brain development, and perform cardiovascular and inflammatory disease prevention, anti-aging function, and treatment of psychiatric disorders. Currently, deep-ocean fish oil represents the primary feedstock for EPA and DHA production; however, unpleasant smell, environmental impacts on the marine ecosystem due to overfishing, heavy metals toxicity, and considerable purification needs demanded sustainable fatty acids production alternatives (Kadalag et al., 2021). Moreover, only 0.2 million tons of EPA/DHA are obtained from fish sources, in the face of 1.3 million tons required for the world population, given the 500 mg/day intake recommended by the World Health Organization. Hence, further studies to increase fatty acids accumulation in microalgae biomass are needed to facilitate their industrial application. Fucoxanthin also exerts multiple pharmacological bioactivities including anti-inflammatory, anti-obesity, anti-diabetic, anti-cancer and anti-oxidant effects (Song et al., 2020). Anti-tumor bioactivities were determined also in chrysolaminarin (Zhang et al., 2018).

Among these, flat-panel configurations have proved to accomplish low shear stress and adequate irradiation which ultimately resulted in higher biomass, EPA and fucoxanthin productivities. Illumination constitutes a primary affecting element in biomass production and biochemical composition, together with light wavelength and light-dark cycle. The increase of light intensity usually involves the rise of the growth rate, until it reaches a saturation value. Nonetheless, above this saturation value, the growth is inhibited by the production of harmful

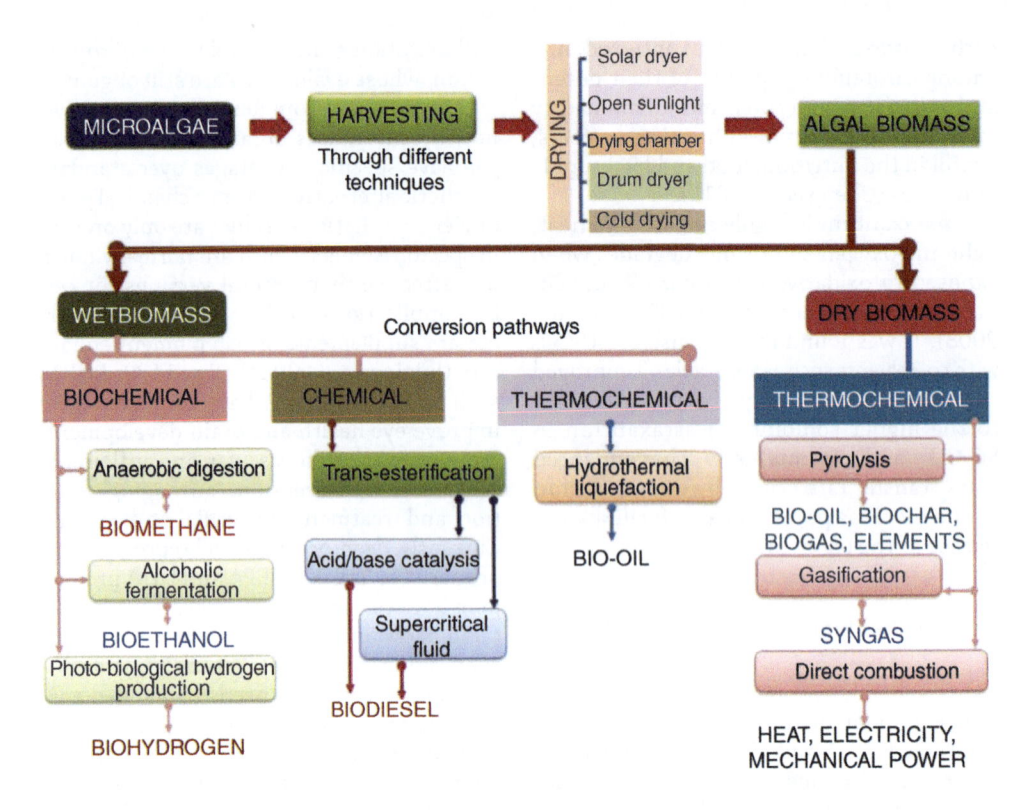

Fig. 11.4. Conversion pathways and processes involved in the recovery of value-added products from microalgae biomass.

substances (i.e. ROS), involved in the photo-inhibition process, which eventually cause a decrease in biomass productivity (Jethani and Hebbar, 2021). Generally, the photosynthetic efficiency in microalgae grown in phototrophic conditions is lower than theoretically achievable values, due to the complex nature of the culture system, in which various parameters influence, differently, the growth and the biochemical composition, resulting in countertrend effects (Butler *et al.*, 2020).

At present, microalgae such as *Spirulina*, *Chlorella* and *Dunaliella salina* have been used in the fields of healthcare products and food. *Spirulina* products sold on the market have the effects of lowering cholesterol and improving the immune system. Since the oxidized form of *Dunaliella salina* carotenoids can be used as anti-cancer agents, the polysaccharides in microalgae have anti-bacterial properties and can kill tumor cells. Future developments in microalgae production technology and market conditions could make microalgae production of high-value products economically viable (Stephens *et al.*, 2010). Microalgal lipids can potentially be converted into biofuels, and microalgal bioactive substances can be used to produce valuable products such as carbohydrates, pigments, polyunsaturated fatty acids and proteins, which can be used in a variety of fields (Lammens *et al.*, 2012). Microalgae are therefore an important source for the production of biochemicals.

Acknowledgments

This research was supported in part by grants from the National Natural Science Foundation of China (32170369), the Natural Science Foundation of Zhejiang Province

(LZJWY22B070001), the Fundamental Research Funds for the Provincial Universities (SJLY2020007), the LiDakSum Marine Biopharmaceutical Development Fund, and the National 111 Project of China, University of Minnesota MnDrive Environment Program MNE12, and University of Minnesota Center for Biorefining.

References

Ahmed, M.B., Zhou, J.L., Ngo, H.H. et al. (2015) Adsorptive removal of antibiotics from water and wastewater: progress and challenges. *Science of the Total Environment* 532, 112–126.
AlFadhly, N.K.Z., Alhelfi, N., Altemimi, A.B. et al. (2022) Trends and technological advancements in the possible food applications of spirulina and their health benefits: a review. *Molecules* 27(17), 5584.
Altmann, J., Rehfeld, D., Traeder, K. et al. (2016) Combination of granular activated carbon adsorption and deep-bed filtration as a single advanced wastewater treatment step for organic micropollutant and phosphorus removal. *Water Research* 92, 131–139.
Angulo, E., Bula, L., Mercado, I. et al. (2018) Bioremediation of Cephalexin with non-living *Chlorella* sp., biomass after lipid extraction. *Bioresource Technology* 257, 17–22.
Anil, K., Yoon, Y. and Sang, J. (2020) Emerging prospects of mixotrophic microalgae: way forward to sustainable bioprocess for environmental remediation and cost-effective biofuels. *Bioresource Technology* 300, 122741.
Ansari, F.A., Singh, P., Guldhe, A. et al. (2017) Microalgal cultivation using aquaculture wastewater: integrated biomass generation and nutrient remediation. *Algal Research* 21, 169–177.
Araújo, R., Sanchez-Lopez, J., Calderon, F.V. et al. (2021) Current status of the algae production industry in Europe: an emerging sector of the blue bioeconomy. *Frontiers in Marine Science* 7, 626389.
Bai, X. and Acharya, K. (2016) Removal of trimethoprim, sulfamethoxazole, and triclosan by the green alga *Nannochloris* sp. *Journal of Hazardous Materials* 315, 70–75.
Baicha, Z., Salar-García, M., Ortiz-Martínez, V. et al. (2016) A critical review on microalgae as an alternative source for bioenergy production: a promising low cost substrate for microbial fuel cells. *Fuel Processing Technology* 154, 104–116.
Batista, A.P., Niccolai, A., Bursic, I. et al. (2019) Microalgae as functional ingredients in savory food products: application to wheat crackers. *Foods* 8(12), 611.
Bhatnagar, A. and Devi, P. (2013) Water quality guidelines for the management of pond fish culture. *International Journal of Environmental Sciences* 3(6), 1980–2009.
Blanco-Vieites, M., Suárez-Montes, D., Delgado, F. et al. (2022) Removal of heavy metals and hydrocarbons by microalgae from wastewater in the steel industry. *Algal Research* 64, 102700.
Boelee, N.C., Temmink, H., Janssen, M., Buisman, C.J.N. and Wijffels, R.H. (2012) Scenario Analysis of Nutrient Removal from Municipal Wastewater by Microalgal Biofilms. *Water* 4(2), 460–473.
Butler, T., Kapoore, R.V. and Vaidyanathan, S. (2020) Phaeodactylum tricornutum: a diatom cell factory. *Trends in Biotechnology* 38, 606–622.
Cai, T., Park, S.Y. and Li, Y. (2013) Nutrient recovery from wastewater streams by microalgae: status and prospects. *Renewable and Sustainable Energy Reviews* 19, 360–369.
Capelli, B. and Cysewski, G.R. (2010) Potential health benefits of spirulina microalgae. *Nutrafoods* 9, 19–26.
Carvalho, P.N., Pirra, A., Basto, M.C.P. and Almeida, C.M.R. (2013) Activated sludge systems removal efficiency of veterinary pharmaceuticals from slaughterhouse wastewater. *Environmental Science and Pollution Research* 20(12), 8790–8800.
Chang, S.L., Sang-Ah, L., Ko, S.R. et al. (2015) Effects of photoperiod on nutrient removal, biomass production, and algal-bacterial population dynamics in lab-scale photobioreactors treating municipal wastewater. *Water Research* 68, 680–691.
Cheah, W.Y., Show, P.L., Chang, J.-S. et al. (2015) Biosequestration of atmospheric CO_2 and flue gas-containing CO_2 by microalgae. *Bioresource Technology* 184, 190–201.
Chen, C.-Y., Kao, P.-C., Tsai, C.-J. et al. (2013) Engineering strategies for simultaneous enhancement of C-phycocyanin production and CO_2 fixation with *Spirulina platensis*. *Bioresource Technology* 145, 307–312.
Chen, L., Zhang, L. and Liu, T. (2016) Concurrent production of carotenoids and lipid by a filamentous microalga *Trentepohlia arborum*. *Bioresource Technology* 214, 567–573.
Cheng, J., Li, K., Yang, Z. et al. (2016) Enhancing the growth rate and astaxanthin yield of *Haematococcus pluvialis* by nuclear irradiation and high concentration of carbon dioxide stress. *Bioresource Technology* 204, 49–54.

Craggs, R.J., Lundquist, T.J. and Benemann, J.R. (2013) Wastewater treatment and algal biofuel production. *Algae for Biofuels and Energy* 5, 153–163.

Cuellar-Bermudez, S.P., Garcia-Perez, J.S., Rittmann, B.E. et al. (2015) Photosynthetic bioenergy utilizing CO_2: an approach on flue gases utilization for third generation biofuels. *Journal of Cleaner Production* 98, 53–65.

Dai, Y.-M., Chen, K.-T. and Chen, C.-C. (2014) Study of the microwave lipid extraction from microalgae for biodiesel production. *Chemical Engineering Journal* 250, 267–273.

Daneshvar, E., Zarrinmehr, M.J., Hashtjin, A.M. et al. (2018) Versatile applications of freshwater and marine water microalgae in dairy wastewater treatment, lipid extraction and tetracycline biosorption. *Bioresource Technology* 268, 523–530.

Danouche, M., Ghachtouli, N.E.I. and El, Arroussi, H. (2021) Phycoremediation mechanisms of heavy metals using living green microalgae: physicochemical and molecular approaches for enhancing selectivity and removal capacity. *Heliyon* 7, e07609.

De-Bashan, L.E., Hernandez, J.-P., Morey, T. et al. (2004) Microalgae growth-promoting bacteria as 'helpers' for microalgae: a novel approach for removing ammonium and phosphorus from municipal wastewater. *Water Research* 38, 466–474.

Emparan, Q., Harun, R. and Danquah, M.K. (2019) Role of phycoremediation for nutrient removal from wastewaters: a review. *Applied Ecology and Environmental Research* 17, 889–915.

Enzing, C., Ploeg, M., Barbosa, M. et al. (2014) Microalgae-based products for the food and feed sector: an outlook for Europe. *JRC Science for Policy Report* 19–37.

Faruque, O.M., Ilyas, M., Hossain, M.M. et al. (2020) Influence of nitrogen to phosphorus ratio and CO_2 concentration on lipids accumulation of *Scenedesmus dimorphus* for bioenergy production and CO_2 biofixation. *Chemistry –An Asian Journal* 15, 4307–4320.

Ferreira, R.C., Dias, D., Fonseca, I. et al. (2022) Multi-component adsorption study by using bone char: modelling and removal mechanisms. *Environmental Technology* 43, 789–804.

Gao, F., Li, C., Yang, Z.-H. et al. (2016) Continuous microalgae cultivation in aquaculture wastewater by a membrane photo bioreactor for biomass production and nutrients removal. *Ecological Engineering* 92, 55–61.

Hamed, I. (2016) The evolution and versatility of microalgal biotechnology: a review. *Comprehensive Reviews in Food Science and Food Safety* 15, 1104–1123.

Haoujar, I., Haoujar, M., Altemimi, A.B. et al. (2022) Nutritional, sustainable source of aqua feed and food from microalgae: a mini review. *International Aquatic Research* 14, 1–9.

Hemlata, P.G. and Fareha, S. (2011) Studies on anabaena sp. NCCU-9 with special reference to phycocyanin. *Journal of Algal Biomass Utilization* 2, 30–51.

Hussain, F., Shah, S.Z., Zhou, W. et al. (2017) Microalgae screening under CO_2 stress: growth and micro-nutrients removal efficiency. *Journal of Photochemistry Photobiology, B* 170, 91–98.

Iasimone, F., Panico, A., De Felice, V. et al. (2018) Effect of light intensity and nutrients supply on microalgae cultivated in urban wastewater: biomass production, lipids accumulation and settleability characteristics. *Journal of Environmental Management* 223, 1078–1085.

Jadwiga, R. and Leo, S. (2014) Recent developments and prospects for algae-based fuels in the US. *Renewable and Sustainable Energy Reviews* 29, 847–853.

Jethani, H. and Hebbar, U.H. (2021) Plant-based biopolymers: emerging bio-flocculants for microalgal biomass recovery. *Reviews in Environmental Science and Bio/Technology* 20(1), 143–165.

Kadalag, N.L., Pawar, P.R. and Prakash, G. (2021) Co-cultivation of Phaeodactylum tricornutum and Aurantiochytrium limacinum for polyunsaturated omega-3 fatty acids production. *Bioresource Technology* 346, 126544.

Katam, K. and Bhattacharyya, D. (2018) Comparative study on treatment of kitchen wastewater using a mixed microalgal culture and an aerobic bacterial culture: kinetic evaluation and FAME analysis. *Environmental Science and Pollution Research* 25, 20732–20742.

Khatiwada, B., Hasan, M.T., Sun, A. et al. (2020) Proteomic response of Euglena gracilis to heavy metal exposure: identification of key proteins involved in heavy metal tolerance and accumulation. *Algal Research* 45, 101764.

Khoo, H., Sharratt, P., Das, P. et al. (2011) Life cycle energy and CO_2 analysis of microalgae-to-biodiesel: preliminary results and comparisons. *Bioresource Technology* 102, 5800–5807.

Krupa, D., Nakkeeran, E., Kumaresan, N. et al. (2010) Extraction, purification and concentration of partially saturated canthaxanthin from Aspergillus carbonarius. *Bioresource Technology* 101(19), 7598–7604.

Lamb, J.B., van de Water, J.A., Bourne, D.G. et al. (2017) Seagrass ecosystems reduce exposure to bacterial pathogens of humans, fishes, and invertebrates. *Science* 355, 731–733.

Lammens, T., Franssen, M.C.R., Scott, E.L. et al. (2012) Availability of protein-derived amino acids as feedstock for the production of bio-based chemicals. *Biomass and Bioenergy* 44, 168–181.

Lee, J., Lee, T.K., Woo, S.G. et al. (2013) In-depth characterization of wastewater bacterial community in response to algal growth using pyrosequencing. *Journal of Microbiology and Biotechnology* 23, 1472–1477.

Leng, L., Yuan, X., Zeng, G. et al. (2015) Surface characterization of rice husk bio-char produced by liquefaction and application for cationic dye (Malachite green) adsorption. *Fuel* 155, 77–85.

Leng, L., Li, J., Wen, Z. et al. (2018) Use of microalgae to recycle nutrients in aqueous phase derived from hydrothermal liquefaction process. *Bioresource Technology* 256, 529–542.

Li, Y., Zhou, W., Hu, B. et al. (2011) Integration of algae cultivation as biodiesel production feedstock with municipal wastewater treatment: strains screening and significance evaluation of environmental factors. *Bioresource Technology* 102, 10861–10867.

Liau, B.C., Shen, C.T., Liang, F.P. et al. (2010) Supercritical fluids extraction and anti-solvent purification of carotenoids from microalgae and associated bioactivity. *Journal of Supercritical Fluids* 55(1), 169–175.

Liu, X., Xu, H., Wang, X. et al. (2014) An ecological engineering pond aquaculture recirculating system for effluent purification and water quality control. *Acta Hydrochimica et Hydrobiologica* 42, 221–228.

Longo, S., d'Antoni, B.M., Bongards, M. et al. (2016) Monitoring and diagnosis of energy consumption in wastewater treatment plants: a state of the art and proposals for improvement. *Applied Energy* 179, 1251–1268.

Lu, H., Wan, J., Li, J. et al. (2016) Periphytic biofilm: a buffer for phosphorus precipitation and release between sediments and water. *Chemosphere* 144, 2058–2064.

Lu, Q., Zhou, W., Min, M. et al. (2015) Growing *Chlorella* sp. on meat processing wastewater for nutrient removal and biomass production. *Bioresource Technology* 198, 189–197.

Lu, Q., Han, P., Xiao, Y. et al. (2019a) The novel approach of using microbial system for sustainable development of aquaponics. *Journal of Cleaner Production* 217, 573–575.

Lu, Q., Ji, C., Yan, Y. et al. (2019b) Application of a novel microalgae-film based air purifier to improve air quality through oxygen production and fine particulates removal. *Journal of Chemical Technology and Biotechnology* 94, 1057–1063.

Markou, G., Angelidaki, I. and Georgakakis, D. (2012) Microalgal carbohydrates: an overview of the factors influencing carbohydrates production, and of main bioconversion technologies for production of biofuels. *Applied Microbiology and Biotechnology* 96(3), 631–645.

Milledge, J.J. (2011) Commercial application of microalgae other than as biofuels: a brief review. *Reviews in Environmental Science and Bio/Technology* 100(1), 31–41.

Mirzoyan, N., Parnes, S., Singer, A. et al. (2008) Quality of brackish aquaculture sludge and its suitability for anaerobic digestion and methane production in an upflow anaerobic sludge blanket (UASB) reactor. *Aquaculture* 279, 35–41.

Mohsenpour, S.F., Hennige, S., Willoughby, N. et al. (2021) Integrating micro-algae into wastewater treatment: a review. *Science of the Total Environment* 752, 142168.

Murphy, V., Hughes, H. and McLoughlin, P. (2008) Comparative study of chromium biosorption by red, green and brown seaweed biomass. *Chemosphere* 70, 1128–1134.

Neilson, A.H. and Lewin, R.A. (1974) The uptake and utilization of organic carbon by algae: an essay in comparative biochemistry. *Phycologia* 13, 227–264.

Nirmal, R., Pranassa, R., Sood, A. et al. (2017) Wastewater grown microalgal biomass as inoculants for improving micronutrient availability in wheat. *Rhizosphere* 3, 150–159.

Nobre, B.P., Villalobos, F., Barragán, B.E. et al. (2013) A biorefinery from *Nannochloropsis* sp. microalga – extraction of oils and pigments. Production of biohydrogen from the leftover biomass. *Bioresource Technology* 135, 128–136.

Norvill, Z.N., Shilton, A. and Guieysse, B. (2016) Emerging contaminant degradation and removal in algal wastewater treatment ponds: identifying the research gaps. *Journal of Hazardous Materials* 313, 291–309.

Ofari-Boateng, C., Lee, K.T. and Lim, J. (2012) Sustainability assessment of microalgal biodiesel production processes: an exergetic analysis approach with Aspen Plus. *International Journal of Energy* 10(4), 400–416.

Paerl, H.W. and Otten, T.G. (2013) Harmful cyanobacterial blooms: causes, consequences, and controls. *Microbial Ecology* 65, 995–1010.

Patil, G., Chethana, S., Madhusudhan, M.C. et al. (2008) Fractionation and purification of the phycobiliproteins from Spirulina platensis. *Bioresource Technology* 99(15), 7393–7396.

Pereira, H., Sa, M., Maia, I. et al. (2021) Fucoxanthin production from *Tisochrysis lutea* and *Phaeodactylum tricornutum* at industrial scale. *Algal Research* 56, 102322.

Perin, G., Bellan, A., Bernardi, A. et al. (2019) The potential of quantitative models to improve microalgae photosynthetic efficiency. *Physiologia Plantarum* 1660(1), 380–391.

Phang, S.M., Miah, M.S., Yeoh, B.G. et al. (2000) *Spirulina* cultivation in digested sago starch factory wastewater. *Journal of Applied Phycology* 12, 395–400.

Prata, J.C., Lavorante, B.R.B.O., Maria, d.M. et al. (2018) Influence of microplastics on the toxicity of the pharmaceuticals procainamide and doxycycline on the marine microalgae *Tetraselmis chuii*. *Aquatic Toxicology* 197, 143–152.

Rambabu, K., Banat, F., Quan M.P. et al. (2020) Biological remediation of acid mine drainage: review of past trends and current outlook. *Environmental Science & Ecotechnology* 2, 100024.

Razzak, S.A., Hossain, M.M., Lucky, R.A. et al. (2013) Integrated CO_2 capture, wastewater treatment and biofuel production by microalgae culturing: a review. *Renewable and Sustainable Energy Reviews* 27, 622–653.

Roostaei, J., Zhang, Y., Gopalakrishnan, K. et al. (2018) Mixotrophic microalgae biofilm: a novel algae cultivation strategy for improved productivity and cost-efficiency of biofuel feedstock production. *Scientific Reports* 8, 12528.

Ross, M.E., Davis, K., McColl, R. et al. (2018) Nitrogen uptake by the macro-algae *cladophora coelothrix* and *cladophora parriaudii*: influence on growth, nitrogen preference and biochemical composition. *Algal Research* 30, 1–10.

Roy, S.S. and Pal, R. (2015) Microalgae in aquaculture: a review with special references to nutritional value and fish dietetics. *Proceedings of the Zoological Society (Calcutta)* 68, 1–8.

Ruiz-Martínez, A., Serralta, J., Seco, A. et al. (2015) Effect of temperature on ammonium removal in *scenedesmus* sp. *Bioresource Technology* 191, 346–349.

Sara, P., Jonathan, S., Bruce, E. et al. (2015) Photosynthetic bioenergy utilizing CO_2: an approach on flue gases utilization for third generation biofuels. *Journal of Cleaner Production* 98, 53–65.

Seyidoglu, N., Inan, S. and Aydin, C. (2017) A prominent superfood: *Spirulina platensis*. *Superfood and Functional Food: The Development of Superfoods and Their Roles as Medicine* 22, 1–27.

Singh, D., Puri, M., Wilkens, S. et al. (2013) Characterization of a new zeaxanthin producing strain of *Chlorella saccharophila* isolated from New Zealand marine waters. *Bioresource Technology* 143, 308–314.

Singh, S., Verma, D.K., Thakur, M. et al. (2021) Supercritical Fluid Extraction (SCFE) as green extraction technology for high-value metabolites of algae, its potential trends in food and human health. *Food Research International* 150, 110746.

Sirakov, I., Velichkova, K., Stoyanova, S. et al. (2015) The importance of microalgae for aquaculture industry. *International Journal of Fisheries and Aquatic Studies* 2, 81–84.

Song, C., Wei, Y., Qiu, Y. et al. (2019) Biodegradability and mechanism of florfenicol via *Chlorella* sp. *UTEX1602* and *L38*: experimental study. *Bioresource Technology* 272, 529–534.

Song, Z., Lye, G.J. and Parker, B.M. (2020) Morphological and biochemical changes in *Phaeodactylum tricornutum* triggered by culture media: implications for industrial exploitation. *Algal Research* 47, 101822.

Soni, R.A., Sudhakar, K. and Rana R.S. (2017) *Spirulina*: from growth to nutritional product: a review. *Trends in Food Science and Technology* 69, 157–171.

Stephens, E., Ross, I.L., King, Z. et al. (2010) An economic and technical evaluation of microalgal biofuels. *Nature Biotechnology* 28, 126–128.

Sun, M., Lin, H., Guo, W. et al. (2017) Bioaccumulation and biodegradation of sulfamethazine in *Chlorella pyrenoidosa*. *Journal of Ocean University of China* 16, 1167–1174.

Sydney, E.B., Novak, A.C., Carvalho, J. et al. (2014) Respirometric balance and carbon fixation of industrially important algae. *Biofuels Algae* 23, 67–84.

Thana, P., Machmudah, S., Goto, M., Sasaki, M., Pavasant, P. and Shotipruk, A. (2008) Response surface methodology to supercritical carbon dioxide extraction of astaxanthin from *Haematococcus pluvialis*. *Bioresource Technology* 99(8), 3110–3115.

Ting, C., Stephen, Y. and Yebo, L. (2000) Nutrient recovery from wastewater streams by microalgae: status and prospects. *Renewable and Sustainable Energy Reviews* 19, 360–369.

Wang, J., Zhou, W., Yang, H. et al. (2015) Trophic mode conversion and nitrogen deprivation of microalgae for high ammonium removal from synthetic wastewater. *Bioresource Technology* 196, 668–676.

Williams, P.J.L.B. and Laurens, L.M.L. (2010) Microalgae as biodiesel & biomass feedstocks: review & analysis of the biochemistry, energetics & economics. *Energy & Environmental Science* 3, 554–590.

Xiong, J.Q., Kurade, M.B. and Jeon, B.H. (2017) Biodegradation of levofloxacin by an acclimated freshwater microalga, Chlorella vulgaris. *Chemical Engineering Journal* 313, 1251–1257.

Xiong, J.-Q., Kim, S.-J., Kurade, M.B. et al. (2018) Combined effects of sulfamethazine and sulfamethoxazole on a freshwater microalga, *Scenedesmus obliquus*: toxicity, biodegradation, and metabolic fate. *Journal of Hazardous Materials* 370, 138–146.

Zhang, W., Wang, F., Gao, B. et al. (2018) An integrated biorefinery process: stepwise extraction of fucoxanthin, eicosapentaenoic acid and chrysolaminarin from the same Phaeodactylum tricornutum biomass. *Algal Research* 32, 193–200.

Zhou, W.G., Min, M., Li, Y.C. et al. (2012) A hetero-photoautotrophic two-stage cultivation process to improve wastewater nutrient removal and enhance algal lipid accumulation. *Bioresource Technology* 110, 448–455.

12 The Role of Microalgae in the Mitigation of the Impact of Chemical Pollution in Freshwater Habitat

Adamu Yunusa Ugya[1,2,3,4] and Qiang Wang[1,2,3]*

[1]*State Key Laboratory of Crop Stress Adaptation and Improvement; Henan University, Kaifeng, China;* [2]*School of Life Sciences, Henan University, Kaifeng, China;* [3]*Academy for Advanced Interdisciplinary Studies, Henan University, Kaifeng, China;* [4]*Department of Environmental Management, Kaduna State University, Kaduna State, Nigeria*

Abstract

Freshwater microalgae have been associated with the production of algae blooms, dangerous toxins and encouragement of bacteria growth, which are detrimental to freshwater biota; although freshwater microalgae can equally be beneficial to freshwater biota due to the production of reactive oxygen species (ROS). This concise review is aimed at accessing the role of freshwater microalgae biofilm in the mitigation of the impact of chemical pollution to freshwater biota. It explains the concept of freshwater chemical pollution and the ecological implications of freshwater chemical pollution, and discusses how freshwater biofilm could be both beneficial and harmful to freshwater biota. It shows that freshwater microalgae play an important role in the mitigation of the impact of pollution resulting from chemical industries, although further research needs to be done to know which environmental factor stimulates the beneficial role of freshwater microalgae.

Keywords: freshwater microalgae; reactive oxygen species; natural resources; chemical pollution; organic pollutants

1. Introduction

The incessant discharge of chemical pollutants into freshwater bodies, including lakes, rivers, streams, ponds, etc., is on the increase due to continuous human anthropogenic activities such as urbanisation, chemical processing, and agricultural activities. (Arenas-Sanchez et al., 2016). These chemical pollutants are created in an effort to improve humans' standard of living but, ironically, the unplanned intrusion of these pollutants reverses that same standard by impacting negatively on the flora and fauna of freshwater habitat (Ugya, 2015; Ugya et al., 2019a, b). The discharged chemical pollutants tend to alter the physicochemical and nutrient status of these water bodies (Ma et al., 2019; Ugya et al., 2019c). This alteration of the physicochemical and nutrient status of freshwater habitat affects the overall quality of freshwater and encourages the growth of algae (Rastogi et al., 2015; Kimambo et al., 2019). The continuous alteration of the nutrient status of the freshwater habitat leads to rapid increase and

*Address for correspondence: wangqiang@henu.edu.cn

accumulation of the toxic algae population leading to the formation of algae blooms (Stauffer et al., 2019). These algae blooms have been associated with the production of dangerous toxins that tend to kill other organisms present in the habitat, particularly fish and aquatic insect (Shilo, 1967; Collins, 1978; Anderson et al., 2012). Algae blooms also encourage the growth of bacteria in freshwater habitat; these bacteria tend to use up the dissolved oxygen present in the habitat, thereby leading to the death of aquatic insects and fish (Yang et al., 2008; Ugya et al., 2019d).

Freshwater microalgae are unicellular and microscopic organisms which range in size from a few μm to a few hundred μm (Lee et al., 2014; Cheng et al., 2017). Many are free-living organisms while few live in colonies with no specific specialization. They occupy the most important level in the food chain and food web of aquatic ecosystems due to their ability to utilise carbon for metabolic activities. (Ndikubwimana et al., 2016). They are also an important ecological tool used in the maintenance of the balance of aquatic ecosystems due to their ability to utilise carbon dioxide and provide oxygen for the ecosystem. They also enhance the quality of water due to the production of reactive oxygen species (ROS) (Ugya et al., 2019c, d). These ROS are a metabolic by-product of molecular oxygen which has been associated with the reduction of risk associated with chemical pollution. Little or no literature exists comparing the beneficial and non-beneficial roles of freshwater microalgae for the effective utilization of them as a solution for the control of water pollution resulting from chemical industries through environmental stimulation (Li et al., 2019). This critical review aims at accessing the role of freshwater microalgae in the mitigation of the impact of chemical pollution on freshwater environment.

2. The Concept of Freshwater Chemical Pollution

The presence of pollutants in freshwater caused by the chemical industry tends to alter the characteristics of the freshwater, thereby making it unsafe for drinking, recreational activities, irrigation, and sometimes habitat, due to the negative effects associated with these chemical pollutants. Water pollution tends to affect the suitability of water for drinking, recreational activities, and irrigation, it also tends to affect the balance of the ecosystem, thereby exerting a negative effect on aquatic organisms. (Malaj et al., 2014; Choudri et al., 2017).

The sources of freshwater chemical pollution are classified into point and non-point sources (Karuppiah and Gupta, 1996; Muller et al., 2002). Point sources are chemical pollutants of specific site of origin that are identifiable as a source of freshwater chemical pollution, e.g. chemical industrial discharge, mines discharge, power plants discharge etc. (van Drecht et al., 2001), while non-point sources of freshwater pollution are chemical pollutants generated over a large area which are not identifiable. Examples include erosion of excess pesticide and fertilizers from agricultural land, residential land and runoff of used chemicals such as oil and grease from urban towns used for energy production (Gao et al., 2015; Kuai et al., 2015).

Chemical pollutants that contaminate freshwater environment can be grouped in two major groups, namely organic and inorganic (Ugya et al., 2019d). Table 12.1 shows various organic and inorganic pollutants that have been shown to be responsible for the alteration of the physicochemical and nutrient status of different freshwater bodies. Organic pollutants are chemical pollutants containing carbon, with the exception of carbonates, cyanides and oxalates (Schweitzer and Noblet, 2018). These pollutants tend to contain other elements such as oxygen, chlorine, bromine, sulfur, nitrogen, phosphorus, hydrogen etc. in addition to carbon atom, which is the principal element. These pollutants are produced as a result of the usage and production of synthetic organic compounds such as pesticides, herbicides, fertilizers, petrochemicals etc. (Carpenter, 2011).

Organic pollutants can be classified into two classes based on polarity, namely hydrophilic and hydrophobic organic pollutants (Schweitzer and Noblet, 2018). Hydrophilic organic pollutants are polar and soluble in

Table 12.1. The ecological fate of chemical pollutants in freshwater habitat.

SN	Type of pollutant	Study area examples	Environmental fate of chemical pollutant	References
1	DDT	Amazon river, Brazil, Sayong River Watershed, Malaysia	Accumulates in sediment, water and organisms of freshwater habitat; is reported to cause reproductive failure in fish and aquatic birds that prey on fish	Torres et al. (2002); Ghani et al. (2017)
2	PCBs	Willamette river basin, Arkansas river basin	Accumulates in sediment, water and organisms of freshwater habitat; is reported to cause reproductive failure in fish and aquatic birds that prey on fish	Belden et al. (2000); Hope (2008)
3	PBDEs	Chaohu Lake China, Asunle Stream, Nigeria	Accumulates in sediment, water and organisms of freshwater habitat; is reported to cause reproductive failure in fish and aquatic birds that prey on fish	Olutona et al. (2017); Liu et al. (2018)
4	PCDDs	Detroit and Rounge, Saginaw river	Accumulates in sediment, water and organisms of freshwater habitat; is reported to cause reproductive failure in fish and aquatic birds that prey on fish	Kannan et al. (2001, 2008)
5	PCDFs	Jukskei and Klip/Vall South Africa, Xiangjiang river	Accumulates in sediment, water and organisms of freshwater habitat; is reported to cause reproductive failure in fish and aquatic birds that prey on fish	Chen et al. (2012); Rimayi et al. (2016)
6	PAHs	Columbia Cauca river, Lower Fox river	PAH is utilized by aquatic organisms including mollusks, crustaceans, polychaetes, fish etc., but at higher concentration PAH tend to be toxic to aquatic organisms leading to reproductive and endocrine failure in fish	Collier et al. (2013); Sarria-Villa et al. (2016); Brewster et al. (2018)
7	BTEX	Tegas river and Marrecas Stream, Buriganga river	BTEX are moderately toxic to aquatic flora and fauna at higher concentration which results from spillage that is unlikely to cause any toxicity at normal environmental concentration because it degrades easily.	Mottaleb et al. (2003); Fernandes et al. (2014)
8	Ammonia	Yellow river, Grand river	Causes eutrophication and algae blooms	Zhang et al. (2007); Sonthiphand et al. (2013)
9	Nitrate	Iowa's rivers	Causes eutrophication and algae blooms	Li et al. (2013)
10	Phosphate	Poyang Lake,	Causes eutrophication and algae blooms	Wang and Liang (2015)
11	Heavy metals	River Pra	Effect depends on the type of heavy metal but toxicity of all heavy metal in freshwater organisms results when the uptake rate exceeds the combined rates of detoxification	May et al. (2008); Duncan et al. (2018)

water while hydrophobic organic pollutants are non-polar and insoluble in water (Nanny and Ratasuk, 2002). Hydrophilic organic pollutants include alcohol and carboxylic acid with relatively small alkyl group (CH_3, C_2H_5, C_3H_7 and C_4H_5). Another important example is methyl tertiary butyl ether (MTBE). These pollutants do not exert pressure on freshwater due to their non-persistent nature and their ability to mix well with water (Gabrielska et al., 1997; Qin et al., 2010). Hydrophobic organic pollutants are organic pollutants including dichlorodiphenyltrichloroethane (DDT), polychlorinated biphenyls (PCBs), polybrominated diphenyl ethers (PBDEs), polychlorinated dibenzodioxins (PCDDs), polychlorinated dibenzofurans (PCDFs), polyaromatic hydrocarbons (PAHs), benzene toluene, ethybenzene and xylene (BTEX) (Jaffe, 1991). These pollutants are problematic due to their persistent nature, which leads to bioaccumulation in freshwater habitat until they reach an elevated concentration that is injurious to freshwater flora and fauna (Zhang and Kelly, 2018).

Inorganic pollutants are pollutants including sulfur and its compounds, ammonia, nitrate, phosphate, heavy metals etc. released into freshwater as a result of industrial and power plants discharge, agricultural and urban runoff, acid mine drainage etc. These pollutants tend to facilitate the depletion of oxygen in freshwater habitat and also lead to the change of the physical chemistry of freshwater bodies, thereby exerting negative impact on the flora and fauna inhabiting freshwater habitat (Camargo and Alonso, 2006; Zhou et al., 2014).

The ecological fates of some chemical pollutants in freshwater habitat are summarized in Table 12.1. The most similar effect attributing to almost all the pollutants is eutrophication. Eutrophication has been defined by Walker et al. (2019) as the incessant accumulation of organic and inorganic material in the aquatic environment. The continuous accumulation of these nutrients stimulates the increase of algae biomass, which is a result of the utilization of the accumulated nutrients by the algae species. These increases in algae biomass lead to toxic algae blooms, which are associated with the release of toxins, which affect the flora and fauna of freshwater habitat (Heisler et al., 2008). The blooms also encourage the growth of bacteria, which use up the available dissolved oxygen present in freshwater habitat, thereby also contributing to the negative effect on the fauna and flora (Ramanan et al., 2016; Yarimizu et al., 2018).

Persistence organic chemical pollutants such as DDT, PCB, PBDE, PCDF, PAH and BTEX tend to accumulate and magnify in sediment and across the aquatic food chain and web, although scant literature exists showing acute toxicity to freshwater fauna resulting from these organic pollutants, but reproductive failure has been reported in fish and some aquatic birds (Opopol, 2008). The presence of these persistent organic compounds in freshwater habitat can stimulate the growth of bacteria, which use up the available dissolved oxygen in the freshwater environment thereby affecting the biodiversity of the habitat in significant ways (Doolotkeldieva et al., 2018).

The environmental fate of heavy metals tends to vary between metals; mercury, for instance, tends to accumulate and magnify across the food chain and food web (Tchounwou et al., 2012). The concentration of mercury accumulated in small fish tends to double the concentration of mercury in the plankton directly exposed to the mercury; so also the concentration of mercury accumulated in larger fish is higher than the concentration of mercury accumulated in the small fish and the concentration of mercury gets higher across the food chain and web (Kim et al., 2006). Mercury tends to accumulate and magnify in bottom sediment of freshwater habitat leading to the formation of methylmercury, which is a chemical highly toxic to the flora and fauna of freshwater (Rice et al., n.d.). Arsenic and cadmium tend to accumulate in water, sediment and tissues of freshwater organisms, thereby resulting in deleterious effects on aquatic fauna (Perera et al., 2016).

Selenium is another heavy metal that accumulates but does not magnify like mercury. Selenium affects the survival and growth rate of juvenile fish and has been linked to ill development of offspring of fish exposed to selenium (May et al., 2008). Other freshwater organisms that prey on fish either experience acute toxicity resulting from the exposure to selenium or produce offspring that are stillborn, possessing skeletal deformity (May et al., 2008). Heavy metals such as lead have been reported to have no marked effect on freshwater habitat due to their low bioavailability, but continuous release into the freshwater environment by point sources can lead to toxicity in freshwater organisms (Paul et al., 2014).

3. Role of Microalgae in Heavy Metals Removal

Increasing industrialization and urbanization has led to the presence of heavy metals in the environment. Heavy metals are elements such as cadmium (Cd), mercury (Hg), lead (Pb), arsenic (As), zinc (Zn), copper (Cu), nickel (Ni) and chromium (Cr). These metals have relatively high densities, atomic weights and atomic numbers (Yan et al., 2020). The exposure of the environment to these heavy metals is due to human activity resulting from the need to enhance standard of living. These activities, which include agriculture, electroplating, mining and smelting, and fossil fuel burning, tend to release these heavy metals into the environment. These metals are persistent and non-degradable by any biological or physical process, thereby endangering the wellbeing of organisms in the environment.

Some heavy metals, such as Cu, Fe, Mn, Ni and Zn, are referred to as essential elements because they are utilized by plants for basic physiological and biochemical processes during the plant's life cycle and due to their role in biological systems. Some heavy metals such as Pb, Cd, As and Hg are referred to as non-essential elements because they are highly toxic and have no particular function in plants. The non-essential heavy metals are also toxic to plants when in excess and, as such, can also contribute to environmental pollution. In aquatic ecosystems, heavy metals tend to enter into the food chain and have a toxic effect on aquatic organisms; the heavy metal can also accumulate and magnify along the food chain thereby posing threat to human health. Hence, it is essential to control the availability of heavy metals in both terrestrial and aquatic environments.

Many technologies, including reverse osmosis, electrodialysis, ultrafiltration, ion exchange and chemical precipitation, are available for the remediation of water contaminated with heavy metals but most of these technologies are not viable due to high cost and inefficiency. It is therefore necessary to utilize an ecofriendly method in the treatment of heavy metal-contaminated water (Priya et al., 2014). Phytoremediation is a natural method which employs plant-based technology in the treatment of polluted water. The technology employs the ability of plants to extract metal from water thereby reducing the bioavailability of the heavy metals in water. A more specialized type of phytoremediation is phycoremediation, which employs the use of algae in the treatment of water polluted with heavy metals. Microalgae is a type of phycoremediation that involves the use of microalgae able to absorb and adsorb heavy metals in polluted water.

Microalgae are able to transform, detoxify and volatilize heavy metals during metabolism without any risk of releasing heavy metals into the environment. Microalgae are able to use the process of biosorption to remove heavy metals from water. Microalgae are able to use the extracellular polymeric substances (EPS) to accumulate heavy metals on their surfaces. The EPS of microalgae are able to successfully absorb heavy metals due to the presence of polysaccharides, proteins or lipids, which make up the EPS and also a basic component of the cell wall. The presence of polysaccharides, proteins and lipid in the cell wall and EPS causes the presence of functional groups such as amino, hydroxyl, carboxyl and sulfate on the surface of the cell walls of the member microalgae cells. These functional groups serve as binding sites for metals, hence the reason why microalgae biofilm is efficient in heavy metals removal from contaminated water.

4. Role of Microalgae Biofilm in Organic Pollutant Removal

One of the problems facing the world today is the emission of organic pollutants into the environment. Aromatic and heteroaromatic compounds such as phenol, pyridine, p-nitrophenol, trichloroethylene and dimethyl phthalate (DMP) are good examples of organic pollutants that contaminate surface water. The contamination of surface water by organic pollutants is largely due to the used of chemical and pharmaceutical products. A larger portion of organic pollutants emissions results from the discharge of industrial wastewater while a smaller portion is due to the use of the finished products in the larger society. Organic pollutants are part of the list of pollutants listed by the US Environmental Protection Agency as priority due to their highly hazardous characteristics. Therefore, there is a need to remediate these pollutants from the environment thereby preventing their negative effect. A number of technologies are available for the remediation of water polluted by organic pollutants. These methods include chlorination, ozonation, adsorption, solvent extraction, membrane process, coagulation, flocculation and biological degradation but these methods are cost-ineffective and difficult to apply under field conditions. The use of microalgae biofilm will solve this problem because the method is more economical and environmentally friendly. Microalgae biofilm formation tends to have high concentration of microalgae in water containing a high concentration of organic pollutants. The constituent microalgae in this association include green algae *Chlamydomonas, Euglena,* diatoms, *Navicula, Synedra* and the blue-green algae *Oscillatoria* and *Phormidium,* which have been shown to tolerate organic pollution (Priya et al., 2014). The association also contains other microorganisms such as bacteria, which are effective in the degradation of these organic pollutants because they utilize organic compounds as a carbon source for vital metabolic activities.

The mechanisms involved in organic pollutant degradation by microalgae biofilm are a natural process resulting from interaction between the microalgae biofilm and the organic pollutants. The thousands of microalgae cells in the biofilm association tend to biotransform these organic pollutants. The ability of microalgae biofilm to biotranform organic pollutants is due to the enzymatic activities of the enzymes produced by the individual microalgae cells present in the biofilm. This enzymatic biotransformation also causes degradation (Ugya et al., 2021a). Another reason why microalgae biofilm is able to degrade or transform organic pollutants in contaminated water is the increased level of ROS produced as a result of the high amount of microalgae cells forming the biofilm (Ugya et al., 2021b). A number of previous studies show that microalgae biofilm produces a high amount of ROS, which can oxidize complex organic compounds such as trichlorobenzene, tetrachlorobenzene, alkyl halides, and some PCBs. The process of oxidation leads to the degradation of these complex compounds into CO_2, CO_3^{2-} and H_2O. The products of the oxidation reaction are utilized by the individual microalgae forming the biofilm for biomass production. Ugya et al. (2020) show that microalgae biofilm is able to produce ROS of high concentration, which are capable of degrading organic compounds and sulfur-containing pollutants. The study shows that ROS produced by microalgae biofilm are capable of desulfurizing thiocarbonyl compounds such as thiouracils, thioamides and thioureas to carbonyl compounds (Ugya et al., 2020).

5. The Harmful Effect of Microalgae to Freshwater Habitat

Freshwater microalgae can be very dangerous to freshwater habitat due to the formation of algae blooms and production and release of ROS (Mohamed, 2018). Freshwater microalgae blooms vary in color from green, yellow or brown color depending on the species of microalgae involved (Wells et al., 2015). This bloom is caused as a result of excessive utilization of nutrients by freshwater microalgae particularly in slow-moving water bodies such as lakes and ponds (Mohamed, 2018). These blooms produced by freshwater microalgae are extremely toxic to freshwater biota because they tend to block the gills of fish and other freshwater microinvertebrates and also deplete the available dissolved oxygen in freshwater habitat leading to the death of these freshwater organisms (Dorantes-Aranda et al., 2015). These blooms also cause an ecosystem to be unsustainable because they tend to be poisonous to the smaller fish and other filter feeder organisms that feed on them.

This toxicity is passed across the food web leading to the death of macroorganisms in the freshwater environment (Paerl et al., 2016; Paerl, 2018; Schmale et al., 2019). Algae bloom can also affect aquatic productivity because they prevent sunlight energy from penetrating deep into other freshwater zones leading to the reduction in the diversity of organisms in the freshwater habitat (Wu et al., 2015). The prevention of sunlight by algae bloom is due to the decoloration of freshwater habitat and the formation of thick film on the surface of freshwater by microalgae (Wang et al., 2017). Algae bloom blooms also affect the mobility of freshwater fauna such as coral and can also affect other freshwater flora from getting sunlight for photosynthesis because they submerge this freshwater flora and also serve as a shade, thereby preventing sunlight energy reaching these organisms (Havens, 2008). Microalgae are able to produce microtoxins that have detrimental effects on freshwater biota, particularly fish and other micro- and macroinvertebrates (Lowenstine, 2008; Ferrão-Filho and Kozlowsky-Suzuki, 2011; Fox, 2012).

6. Freshwater Microalgae as Important Natural Resources

Freshwater microalgae are abundant species of organisms which have been shown to produce important substances such as carotenoids, antioxidants, fatty acid, enzymes, peptides, steroids, polymers etc. (Chaudhuri et al., 2014; Sathasivam et al., 2019). These chemical products are dependent on the growth conditions of the freshwater microalgae, hence the reason why change in environmental conditions like temperature, light intensity, PH, nutrients, carbon sources etc. in the freshwater microalgae environment tend to affect the amount of important resources harnessed for the benefit of humans (Yu et al., 2015; Panahi et al., 2019). Microalgae are readily utilized in aquaculture due to the fact that they form an important part of the freshwater food web (Allen et al., 2019). They are utilized in aquaculture to increase the nutritional value of fish (Wells et al., 2017). The ability of freshwater microalgae to utilize sunlight and carbon to produce organic substances such as lipid is the reason why fish oil consists of omega 3 fats (Tocher et al., 2019). Omega 3 fat is the class of polyunsaturated fatty acid that has been of great importance to man because of its role in preventing diseases such as cancer, rheumatoid arthritis, eczema, heart disease, stroke etc. (Gammone et al., 2018). Freshwater microalgae also provide bivalves with polyunsaturated fatty acid, which is necessary for their growth and development due to the inability of bivalves to produce polyunsaturated fatty acids (Shanab et al., 2018). Freshwater microalgae are also used in the treatment of wastewater from chemical industries; the utilization of freshwater microalgae for wastewater treatment is cost-effective, easily accessible and has the full potential of eradicating the environmental implications associated with the use of chemicals for wastewater treatment (Ray et al., 2019). The method of nutrient removal from wastewater by freshwater microalgae is through assimilation and utilization leading to the production and release of oxygen, which, in turn, stimulates the growth of bacteria that bring about the degradation of organic pollutants present in the wastewater. The heavy metal present in wastewater is removed by freshwater microalgae either by the production of ROS or by the process of adsorption. The ROS produced by freshwater microalgae are able to reduce heavy metals in wastewater by the process of auto-reduction, which also increases the ability of the freshwater microalgae cells to serve as a biosorbent to precipitate heavy metals from wastewater.

7. Comparative Analysis of Beneficial and Harmful Roles of Microalgae to Freshwater Habitat

The presence of freshwater microalgae in freshwater habitats is of great significance. Literature has shown that the presence of freshwater microalgae in the freshwater environment tends to stimulate the presence of bacteria, which use up available dissolved oxygen meant to be used by freshwater organisms, but studies have also shown that freshwater microalgae constantly release oxygen into the freshwater habitat to counter the oxygen used by bacteria. Freshwater microalgae also release ROS along with oxygen and these ROS are capable of destroying bacteria, thereby preventing the total depletion of dissolved oxygen in the freshwater environment. Many studies have shown that the continuous utilization of phosphorus, ammonia and nitrogen compounds present in freshwater habitats leads to algae blooms, which are toxic to flora and fauna organisms inhabiting freshwater, but recent studies have shown that the presence of higher nutrient concentration and optimum other growth parameters of freshwater microalgae lead to the production and release of higher ROS by freshwater microalgae which, in turn, prevent the occurrence of algae blooms. Since increased nutrients in freshwater habitats leads to increased ROS production, which avoids algae blooms production in the freshwater habitat, the negative effect associated with the production of algae blooms by freshwater microalgae will be completely avoided. There is no doubt that freshwater microalgae are an important tool for the natural clean-up of freshwater habitats but scant literature exists on the extent of how ROS affect the survival of bacteria in freshwater environments; this is important because bacteria also play a natural role in the clean-up of the freshwater environment. There is a need to do further research to know how freshwater microalgae are able to avoid the formation of algae blooms in order to know the role of abiotic factors in ROS formation.

8. Conclusion

These studies show that freshwater microalgae biofilm has a more beneficial than harmful effect on freshwater biota. Environmental factors such as nutrients and light are important tools that, if well stimulated, will help freshwater microalgae prevent the occurrence of algae blooms, which are the greater threat to freshwater microalgae. The stimulation of these factors can also make freshwater microalgae release higher amounts of ROS that can regulate the growth of bacteria, thereby minimizing dissolved oxygen loss. Further research is recommended to know how these factors can stimulate the production and release of ROS by freshwater microalgae and how ROS are able to regulate the growth of bacteria.

Acknowledgement

The author of this chapter is grateful to the head of the department of the Department of Environmental Management at Kaduna State University for his support during the drafting and publication of this chapter.

References

Allen, K.M. *et al.* (2019) Freshwater microalgae (Schizochytrium sp.) as a substitute to fish oil for shrimp feed. *Scientific Reports* 9(1), 6178.

Anderson, D.M., Cembella, A.D. and Hallegraeff, G.M. (2012) Progress in understanding harmful algal blooms: paradigm shifts and new technologies for research, monitoring, and management. *Annual Review of Marine Science* 4, 143–176.

Arenas-Sanchez, A., Rico, A. and Vighi, M. (2016) Effects of water scarcity and chemical pollution in aquatic ecosystems: state of the art. *Science of the Total Environment* 572, 390–403.

Belden, J.B., Eaton, H.J. and Lydy, M.J. (2000) Occurrence of polychlorinated biphenyls in fish and sediment of the Arkansas River Basin, Kansas. *Transactions of the Kansas Academy of Science (1903)* 103(3/4), 177–184.

Brewster, C.S. *et al.* (2018) Occurrence, distribution and composition of aliphatic and polycyclic aromatic hydrocarbons in sediment cores from the Lower Fox River, Wisconsin, US. *Environmental Science and Pollution Research* 25(5), 4974–4988.

Camargo, J.A. and Alonso, A. (2006) Ecological and toxicological effects of inorganic nitrogen pollution in aquatic ecosystems: a global assessment. *Environment International* 32(6), 831–849.

Carpenter, D.O. (2011) Health effects of persistent organic pollutants: the challenge for the Pacific Basin and for the world. *Reviews on Environmental Health* 26(1), 61–69.

Chaudhuri, D. et al. (2014) Assessment of the phytochemical constituents and antioxidant activity of a bloom forming microalgae Euglena tuba. *Biological Research* 47(1), 24.

Chen, Z. et al. (2012) Distribution patterns of polychlorinated dibenzo-p-dioxins and polychlorinated dibenzofurans in sediments of the Xiangjiang River, China. *Environmental Monitoring and Assessment* 184(12), 7083–7092.

Cheng, P. et al. (2017) Biofilm attached cultivation of chlorella pyrenoidosa is a developed system for swine wastewater treatment and lipid production. *Frontiers in Plant Science* 8, 1594–1594.

Choudri, B.S. et al. (2017) Effects of pollution on freshwater organisms. *Water Environment Research* 89(10), 1676–1703.

Collier, T.K. et al. (2013) Effects on fish of polycyclic aromatic hydrocarbons (PAHS) and naphthenic acid exposures. In: Tierney, K.B., Farrell, A.P. and Brauner, C.J. (eds) *Fish Physiology*. Academic Press, pp. 195–255.

Collins, M. (1978) Algal toxins. *Microbiological Reviews* 42(4), 725–746.

Doolotkeldieva, T., Konurbaeva, M. and Bobusheva, S. (2018) Microbial communities in pesticide-contaminated soils in Kyrgyzstan and bioremediation possibilities. *Environmental Science and Pollution Research International* 25(32), 31848–31862.

Dorantes-Aranda, J.J. et al. (2015) Progress in understanding algal bloom-mediated fish kills: the role of superoxide radicals, phycotoxins and fatty acids. *PLOS ONE* 10(7), e0133549.

Duncan, A.E., de Vries, N. and Nyarko, K.B. (2018) Assessment of heavy metal pollution in the sediments of the River Pra and its tributaries. *Water, Air, and Soil Pollution* 229(8), 272.

Fernandes, A.N. et al. (2014) Determination of monoaromatic hydrocarbons (BTEX) in surface waters from a Brazilian subtropical hydrographic basin. *Bulletin of Environmental Contamination and Toxicology* 92(4), 455–459.

Ferrão-Filho, A.d.S. and Kozlowsky-Suzuki, B. (2011) Cyanotoxins: bioaccumulation and effects on aquatic animals. *Marine Drugs* 9(12), 2729–2772.

Fox, J.W. (2012) Venoms and poisons from marine organisms. In: Goldman, L. and Schafer, A.I. (eds) *Goldman's Cecil Medicine*, 24th edn. W.B. Saunders, Philadelphia. Pennsylvania, pp. 697–700.

Gabrielska, J., Sarapuk, J. and Przestalski, S. (1997) Role of hydrophobic and hydrophilic interactions of organotin and organolead compounds with model lipid membranes. *Zeitschrift für Naturforschung C* 52(3–4), 209–216.

Gammone, M.A. et al. (2018) Omega-3 polyunsaturated fatty acids: benefits and endpoints in sport. *Nutrients* 11(1), 46.

Gao, X.P. et al. (2015) Modeling the effects of point and non-point source pollution on a diversion channel from Yellow River to an artificial lake in China. *Water Science and Technology* 71(12), 1806–1814.

Ghani, Z.A. et al. (2017) Multimedia Environmental Fate and Transport Model of dichlorodiphenyltrichloroethane (DDT): case study Sayong River Watershed, Johor, Malaysia. *Sains Malaysiana* 41(12), 2383–2392.

Havens, K.E. (2008) Cyanobacteria blooms: effects on aquatic ecosystems. *Advances in Experimental Medicine and Biology* 619, 733–747.

Heisler, J. et al. (2008) Eutrophication and harmful algal blooms: a scientific consensus. *Harmful Algae* 8(1), 3–13.

Hope, B.K. (2008) A model for the presence of polychlorinated biphenyls (PCBs) in the Willamette River Basin (Oregon). *Environmental Science & Technology* 42(16), 5998–6006.

Jaffe, R. (1991) Fate of hydrophobic organic pollutants in the aquatic environment: a review. *Environmental Pollution* 69(2–3), 237–257.

Kannan, K. et al. (2001) Polychlorinated naphthalenes, biphenyls, dibenzo-p-dioxins, and dibenzofurans as well as polycyclic aromatic hydrocarbons and alkylphenols in sediment from the Detroit and Rouge Rivers, Michigan, USA. *Environmental Toxicology and Chemistry* 20(9), 1878–1889.

Kannan, K. et al. (2008) Dioxin-like toxicity in the Saginaw River Watershed: polychlorinated dibenzo-p-dioxins, dibenzofurans, and biphenyls in sediments and floodplain soils from the Saginaw and Shiawassee Rivers and Saginaw Bay, Michigan, USA. *Archives of Environmental Contamination and Toxicology* 54(1), 9–19.

Karuppiah, M. and Gupta, G. (1996) Impact of point and nonpoint source pollution on pore waters of two Chesapeake Bay tributaries. *Ecotoxicology and Environmental Safety* 35(1), 81–85.

Kim, E.H. *et al.* (2006) The effect of fish consumption on blood mercury levels of pregnant women. *Yonsei Medical Journal* 47(5), 626–633.

Kimambo, O.N., Gumbo, J.R. and Chikoore, H. (2019) The occurrence of cyanobacteria blooms in freshwater ecosystems and their link with hydro-meteorological and environmental variations in Tanzania. *Heliyon* 5(3), e01312.

Kuai, P., Li, W. and Liu, N. (2015) Evaluating the effects of land use planning for non-point source pollution based on a system dynamics approach in China. *PLOS ONE* 10(8), e0135572.

Lee, K. *et al.* (2014) Isolation and screening of microalgae from natural habitats in the midwestern United States of America for biomass and biodiesel sources. *Journal of Natural Science, Biology, and Medicine* 5(2), 333–339.

Li, D., Chan, K.S. and Schilling, K.E. (2013) Nitrate concentration trends in Iowa's rivers, 1998 to 2012: What challenges await nutrient reduction initiatives? *Journal of Environmental Quality* 42(6), 1822–1828.

Li, L. *et al.* (2019) The effect and mechanism of organic pollutants oxidation and chemical energy conversion for neutral wastewater via strengthening reactive oxygen species. *Science of the Total Environment* 651(Pt 1), 1226–1235.

Liu, J. *et al.* (2018) Polybrominated diphenyl ethers (PBDEs) in a large, highly polluted freshwater lake, China: occurrence, fate, and risk assessment. *International Journal of Environmental Research and Public Health* 15(7), 1529.

Lowenstine, L.J. (2008) Algal bloom toxicity in marine animals. In: Fowler, M.E. and Miller, R.E. (eds) *Zoo and Wild Animal Medicine*, 6th edn. W.B. Saunders, Saint Louis, Missouri, pp. 341–348.

Ma, J. *et al.* (2019) Evaluation of Pistia stratiotes fractions as effective larvicide against Anopheles mosquitoes. *Artificial Cells, Nanomedicine, and Biotechnology* 47(1), 945–950.

Malaj, E. *et al.* (2014) Organic chemicals jeopardize the health of freshwater ecosystems on the continental scale. *Proceedings of the National Academy of Sciences of the United States of America* 111(26), 9549–9554.

May, T.W. *et al.* (2008) An evaluation of selenium concentrations in water, sediment, invertebrates, and fish from the Solomon River Basin. *Environmental Monitoring and Assessment* 137(1–3), 213–232.

Mohamed, Z.A. (2018) Potentially harmful microalgae and algal blooms in the Red Sea: current knowledge and research needs. *Marine Environmental Research* 140, 234–242.

Mottaleb, M.A., Abedin, M.Z. and Islam, M.S. (2003) Determination of benzene, toluene, ethylbenzene and xylene in river water by solid-phase extraction and gas chromatography. *Analytical Sciences* 19(10), 1365–1369.

Muller, K. *et al.* (2002) Point- and nonpoint-source pesticide contamination in the Zwester Ohm catchment, Germany. *Journal of Environmental Quality* 31(1), 309–318.

Nanny, M.A. and Ratasuk, N. (2002) Characterization and comparison of hydrophobic neutral and hydrophobic acid dissolved organic carbon isolated from three municipal landfill leachates. *Water Research* 36(6), 1572–1584.

Ndikubwimana, T. *et al.* (2016) Harvesting of freshwater microalgae with microbial bioflocculant: a pilot-scale study. *Biotechnology for Biofuels* 9, 47.

Olutona, G.O. *et al.* (2017) Concentrations of Polybrominated Diphenyl Ethers (PBDEs) in water from Asunle Stream, Ile-Ife, Nigeria. *Toxics* 5(2), 13.

Opopol, N. (2008) Impact of pops on the Republic of Moldova environment and public health. In: Mehmetli, E. and Koumanova, B. (eds) *The Fate of Persistent Organic Pollutants in the Environment*. Springer, Dordrecht, The Netherlands, pp. 405–424.

Paerl, H.W. (2018) Mitigating toxic planktonic cyanobacterial blooms in aquatic ecosystems facing increasing anthropogenic and climatic pressures. *Toxins* 10(2), 76.

Paerl, H.W. *et al.* (2016) Mitigating cyanobacterial harmful algal blooms in aquatic ecosystems impacted by climate change and anthropogenic nutrients. *Harmful Algae* 54, 213–222.

Panahi, Y. *et al.* (2019) Impact of cultivation condition and media content on chlorella vulgaris composition. *Advanced Pharmaceutical Bulletin* 9(2), 182–194.

Paul, N., Chakraborty, S. and Sengupta, M. (2014) Lead toxicity on non-specific immune mechanisms of freshwater fish Channa punctatus. *Aquatic Toxicology* 152, 105–112.

Perera, P.A.C.T. *et al.* (2016) Arsenic and cadmium contamination in water, sediments and fish is a consequence of paddy cultivation: evidence of river pollution in Sri Lanka. *Achievements in the Life Sciences* 10(2), 144–160.

Priya, M. et al. (2014) Microalgae in removal of heavy metal and organic pollutants from soil. In: Das, S. (ed.) *Microbial Biodegradation and Bioremediation.* Elsevier, Oxford, UK, pp. 519–537.

Qin, W.C. et al. (2010) Toxicity of organic pollutants to seven aquatic organisms: effect of polarity and ionization. *SAR and QSAR in Environmental Research* 21(5–6), 389–401.

Ramanan, R. et al. (2016) Algae–bacteria interactions: evolution, ecology and emerging applications. *Biotechnology Advances* 34(1), 14–29.

Rastogi, R.P., Madamwar, D. and Incharoensakdi, A. (2015) Bloom dynamics of cyanobacteria and their toxins: environmental health impacts and mitigation strategies. *Frontiers in Microbiology* 6, 1254.

Ray, M. et al. (2019) Microalgae: a way forward approach towards wastewater treatment and bio-fuel production. In: Shukla, P. (ed.) *Applied Microbiology and Bioengineering.* Academic Press, pp. 229–243.

Rice, K.M. et al. Environmental mercury and its toxic effects. *Journal of Preventive Medicine and Public Health = Yebang Uihakhoe chi* 47(2), 74–83.

Rimayi, C. et al. (2016) Distribution of 2,3,7,8-substituted polychlorinated dibenzo-p-dioxin and polychlorinated dibenzofurans in the Jukskei and Klip/Vaal catchment areas in South Africa. *Chemosphere* 145, 314–321.

Sarria-Villa, R. et al. (2016) Presence of PAHs in water and sediments of the Colombian Cauca River during heavy rain episodes, and implications for risk assessment. *Science of the Total Environmental* 540, 455–465.

Sathasivam, R. et al. (2019) Microalgae metabolites: a rich source for food and medicine. *Saudi Journal of Biological Sciences* 26(4), 709–722.

Schmale, D.G. et al. (2019) Perspectives on harmful algal blooms (HABs) and the cyberbiosecurity of freshwater systems. *Frontiers in Bioengineering and Biotechnology* 7, 128.

Schweitzer, L. and Noblet, J. (2018) Water contamination and pollution. In: Török, B. and Dransfield, T. (eds) *Green Chemistry.* Elsevier, pp. 261–290.

Shanab, S.M.M., Hafez, R.M. and Fouad, A.S. (2018) A review on algae and plants as potential source of arachidonic acid. *Journal of Advanced Research* 11, 3–13.

Shilo, M. (1967) Formation and mode of action of algal toxins. *Bacteriological Reviews* 31(3), 180–193.

Sonthiphand, P. et al. (2013) Wastewater effluent impacts ammonia-oxidizing prokaryotes of the Grand River, Canada. *Applied and Environmental Microbiology* 79(23), 7454–7465.

Stauffer, B.A. et al. (2019) Considerations in harmful algal bloom research and monitoring: perspectives from a consensus-building workshop and technology testing. *Frontiers in Marine Science* 6, 399.

Tchounwou, P.B. et al. (2012) Heavy metal toxicity and the environment. *Experientia Supplementum* 101, 133–164.

Tocher, D.R. et al. (2019) Omega-3 long-chain polyunsaturated fatty acids, EPA and DHA: bridging the gap between supply and demand. *Nutrients* 11(1), 89.

Torres, J.P. et al. (2002) Dichlorodiphenyltrichloroethane in soil, river sediment, and fish in the Amazon in Brazil. *Environmental Research* 88(2), 134–139.

Ugya, A.Y. (2015) The efficiency of Lemna minor L. in the phytoremediation of Romi stream: a case study of Kaduna refinery and petrochemical company polluted stream. *Journal of Applied Biology and Biotechnology* 3, 11–14.

Ugya, A.Y., Hua, X. and Ma, J. (2019a) Phytoremediation as a tool for the remediation of wastewater resulting from dyeing activities. *Applied Ecology and Environmental Research* 17(2), 3723–3735.

Ugya, A.Y. et al. (2019b) Efficacy of Eicchornia Crassipes, Pistia stratiotes and nymphaea lotus in the biosorption of nickel from refinery wastewater. *Applied Ecology and Environmental Research* 17(6), 13075.

Ugya, A.Y., Hua, X. and Ma, J. (2019c) Biosorption of Cr3+ and Pb2+ from tannery wastewater using combined fruit waste. *Applied Ecology and Environmental Research* 17(2), 1773–1787.

Ugya, A.Y. et al. (2019d) Molecular approach to uncover the function of bacteria in petrochemical refining wastewater: a mini review. *Applied Ecology and Environmental Research* 17(2), 3645–3665.

Ugya, A.Y. et al. (2020) Antioxidant response mechanism of freshwater microalgae species to reactive oxygen species production: a mini review. *Chemistry and Ecology* 36(2), 174–193.

Ugya, Y.A. et al. (2021a) Microalgae biofilm cultured in nutrient-rich water as a tool for the phycoremediation of petroleum-contaminated water. *International Journal of Phytoremediation* 23(11), 1175–1183.

Ugya, A.Y., Ajibade, F.O. and Hua, X. (2021b) The efficiency of microalgae biofilm in the phycoremediation of water from River Kaduna. *Journal of Environmental Management* 295, 113109.

van Drecht, G. et al. (2001) Global pollution of surface waters from point and nonpoint sources of nitrogen. *Scientific World Journal* 1(Suppl 2), 632–641.

Walker, D.B. et al. (2019) Surface water pollution. In: Brusseau, M.L., Pepper, I.L. and Gerba, C.P. (eds) *Environmental and Pollution Science*, 3rd edn. Academic Press, pp. 261–292.

Wang, L. and Liang, T. (2015) Distribution characteristics of phosphorus in the sediments and overlying water of Poyang lake. *PLOS ONE* 10(5), e0125859.

Wang, L. et al. (2017) Analysis of algae growth mechanism and water bloom prediction under the effect of multi-affecting factor. *Saudi Journal of Biological Sciences* 24(3), 556–562.

Wells, M.L. et al. (2015) Harmful algal blooms and climate change: learning from the past and present to forecast the future. *Harmful Algae* 49, 68–93.

Wells, M.L. et al. (2017) Algae as nutritional and functional food sources: revisiting our understanding. *Journal of Applied Phycology* 29(2), 949–982.

Wu, T.T. et al. (2015) Impacts of algal blooms accumulation on physiological ecology of water hyacinth. *Huan Jing Ke Xue* 36(1), 114–120.

Yan, A. et al. (2020) Phytoremediation: a promising approach for revegetation of heavy metal-polluted land. *Frontiers in Plant Science* 11.

Yang, X.-E. et al. (2008) Mechanisms and assessment of water eutrophication. *Journal of Zhejiang University. Science B* 9(3), 197–209.

Yarimizu, K., Cruz-López, R. and Carrano, C.J. (2018) Iron and harmful algae blooms: potential algal-bacterial mutualism between lingulodinium polyedrum and marinobacter algicola. *Frontiers in Marine Science* 5, 180.

Yu, X., Chen, L. and Zhang, W. (2015) Chemicals to enhance microalgal growth and accumulation of high-value bioproducts. *Frontiers in Microbiology* 6, 56–56.

Zhang, H. and Kelly, B.C. (2018) Sorption and bioaccumulation behavior of multi-class hydrophobic organic contaminants in a tropical marine food web. *Chemosphere* 199, 44–53.

Zhang, X.Q., Xia, X.H. and Yang, Z.F. (2007) Reasons of high concentration ammonium in Yellow River, China. *Huan Jing Ke Xue* 28(7), 1435–1441.

Zhou, G.J. et al. (2014) Simultaneous removal of inorganic and organic compounds in wastewater by freshwater green microalgae. *Environmental Science: Processes and Impacts* 16(8), 2018–2027.

13 Microalgae-based Bioremediation of Heavy Metals and Emerging Contaminants

Vishal Rajput[1], Vinod Kumar[2,3]*, Krishna Kumar Jaiswal[4], Sanjay Gupta[5], Anna I. Kurbatova[3] and Mikhail S. Vlaskin[5]

[1]Department of Biosciences, Swami Rama Himalayan University, Doiwala, Uttarakhand, India; [2]Algal Research and Bioenergy Laboratory, Department of Life Sciences, Graphic Era, Dehradun, Uttarakhand, India; [3]Department of Environmental Safety and Product Quality Management, Institute of Environmental Engineering, Peoples' Friendship University of Russia (RUDN University), Moscow, Russia; [4]Department of Green Energy Technology, Pondicherry University, Puducherry, India; [5]Joint Institute for High Temperatures of the Russian Academy of Sciences, Moscow, Russia

Abstract
During the last decades, special attention has been paid to the management of environmental pollution caused by hazardous materials. Several techniques have been implemented to eliminate toxic elements; however, these techniques have several shortcomings. The present attempt highlights the alternative biological agent abundantly present in nature, i.e. microalgae. Microalgae exist in freshwater and marine ecosystems and have been considered remarkably vital in the global ecology due to their photosynthetic efficiency and taxonomically diverse nature. Microalgae are considered a potential sink for removing various toxic substances from the ecosystem. Microalgae exhibit the ability to metabolize or accumulate or adsorb the toxic substance in the ecosystem to a substantial level. Also, the growing demand for pesticides is causing serious equilibrium problems in the soil and the aquatic environment. Nowadays, a large proportion of contamination comes from micro-pollutants, which have become a severe concern to human health and the natural ecosystem. Recent research attempts have shown that bioremediation through microalgae has become a novel and attractive tool. Bioremediation with microalgae has shown promising prospects since it is an ecological, efficient and economical method to prevent the accumulation of toxic substances in the ecosystem. In the current scenario, microalgae bioremediation supports the biological remediation of environmental toxins in the ecosystem. This chapter offers a comprehensive review of the available techniques for bioremediation and deliberates the potential, impact and challenges of microalgae-based bioremediations.

Keywords: microalgae; bioremediation; heavy metals; pesticides; emerging contaminants

*Address for correspondence: vinodkdhatwalia@gmail.com

1. Introduction

In the current scenario, different chemical and commercial forms of (toxic) pesticides have been reported, such as herbicides, fungicides, insecticides and plant growth regulators. It is evident that the agricultural fields have indisputable strata of different toxins and are used worldwide in various sectors, including the livestock sector and the industrial domain, to raise the annual yield and financial prosperity. Simultaneously, the uncontrolled use of pesticides can manifest itself in the form of increased use and accumulation in different spheres of the ecosystem, resulting in severe environmental toxicity (Norvill et al., 2016; Donmez et al., 2020). The globally uncontrolled use of pesticides in all aspects of agriculture results in severe toxicity in every stratum of the ecosystem and the environment. In addition, agricultural runoff also adds and increases toxicity in freshwater streams and reservoirs (Gentili and Fick, 2017; Osundeko et al., 2020). The ecosystem receives the toxicity of pesticides from various sources, such as effluent from untreated pesticide manufacturing industries, non-uniform spraying of pesticides, any previously possible agricultural toxicity, improper or defective management of pesticides during transport, any possible runoff from agriculture, insufficient or damaged pesticide storage facilities, etc.

The discharge of effluents from the agricultural industry causes toxicity in freshwater streams as well as in reservoirs worldwide. This toxicity results in severe pollution and is manifested by the entry of pesticides and heavy metals. Different pesticides are used near water bodies (streams and reservoirs) to eliminate or control the population of harmful insects in various agricultural crops. Moreover, these toxins and pesticides cause different harmful effects such as hormonal disorders, hepatotoxicity, and genetic disorders in organisms and humans. Recently, heavy metals and pesticides have become the primary source of toxicity among freshwater streams and reservoirs (Liu et al., 2015). As several researchers have investigated, bioaccumulation can occur in the form of pesticide accretion. It can also lead to the alteration of the food chain and risks to human health. Pesticides can accumulate in the water, making them capable of increasing toxicity to habituating organisms. As reported in several studies, the bioaccumulation of heavy metals and pesticides can result in various severe biochemical and histological alterations in the form of deterioration or even mortality in freshwater and marine organisms (Pan et al., 2018).

Several techniques are available to treat wastewater: physical, chemical, precipitation, microfiltration, carbon adsorption, and activated sludge. These techniques are expensive and ineffectual in all conditions. Bioremediation shows promising possibilities to treat toxic substances and heavy metals from the ecosystem. In addition, several microorganisms, such as photoautotrophs and green algae, have shown promising prospects in the wastewater remediation process due to the harmful accumulation capacity of ecosystem pesticides (Tiwari et al., 2017; Dutta et al., 2021). As it is evident that green algae exhibit the ability to use organic carbon and sunlight simultaneously, green algae also have the ability to act in a synergistic pattern to use organic pesticides as well as toxic heavy metals during metabolic activity very effectively (Villar-Navarro et al., 2018). Microalgae and photosynthetic algae have shown the efficient ability to transform sunlight into a chemical form of energy. Microalgae exhibit simpler organization of cell structure and different factors available in surrounding help better absorption of water and nutrition (Jaiswal and Pandey, 2014). As reported by researchers, microalgae show effective adaptations and have the ability to grow in various modes, such as heterotrophic, mixotrophic or autotrophic. These capabilities provide algae with the strength to grow in an environment of emerging pollutants and heavy metals (Kumar et al., 2021). Microalgae bioremediation shows promising aspects to treating wastewater by working on TDS reduction, COD and BOD balance, and, most importantly, pH amendment. In addition, microalgae have a promising bioremediation capacity and can be used to produce different products with high added value (Xiong et al., 2017; Dutta et al., 2020).

2. Ecological Impact of Heavy Metals and Emerging Pollutants

Heavy metals have been concerned with biological toxicity, including cadmium (Cd), mercury (Hg), chromium (Cr), arsenic (As) etc. Heavy-metal contamination occurs when water retains upper limits of heavy metals and related compounds (Fatima et al., 2020; Nanda et al., 2021). Moreover, these compounds are not biodegradable and lead to bio-accumulation and bio-amplification. Besides, when these heavy metals enter the organism's body through different channels, they cause symptoms such as liver abnormalities, kidney failure, cancer etc. Many innovations and technological advances have substantially amplified the contact between heavy metals and organisms in various ecosystems (Keryanti and Mulyono 2022; Yan et al., 2022). Heavy metals in various concentrations and chemical forms are present in food sources, ecosystems, and the organism's body, generating hazardous conditions for human health. Also, industries are causing a large influx of heavy metals into the ecosystem. Industrial facilities can also introduce heavy metals through the air and soil. Industries release a large amount of waste gas, which contains a large reserve of heavy metals, and finally forms toxic aerogel, and this aerogel enters the aquatic ecosystem by sedimentation.

Wastewater from industries accumulates in various ecosystems (aquatic zones) and deposits a large amount of heavy metals. This significant accumulation of heavy metals in water and soil passes into groundwater through leaching and results in excessive accumulation of heavy metals in water (Sutherland et al., 2018; Keryanti and Mulyono, 2022; Yan et al., 2022). Heavy metals have been reported as highly toxic, even at low concentrations; therefore, heavy-metal contamination needs a global remediation approach (Fig. 13.1).

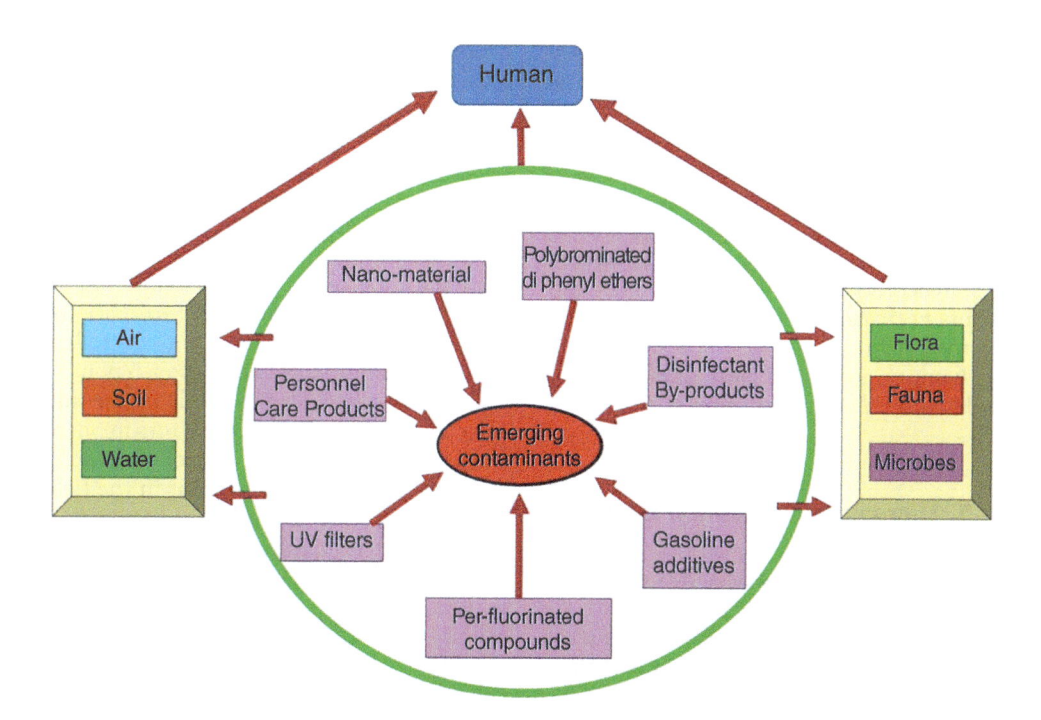

Fig. 13.1. Schematic representation of the movement of emerging pollutants in the ecosystem.

The water quality conservation process focuses mainly on heavy metals, pathogens, sediments and nutrients. However, in previous years an alarming situation has arisen in the form of emerging contaminants in the aquatic ecosystem. Emerging pollutants are synthetic organic compounds present in the environment and recently detected. Emerging contaminants can be described as those compounds that do not have current or emerging standards but can still cause severe damage to the aquatic ecosystem and human health. Most emerging pollutants enter the human body through direct discharge, groundwater, industrial waste and landfill leachate. Preliminary investigations have shown that municipal wastewater treatment plants contribute a large proportion of emerging pollutants to the aquatic ecosystem, while a large number (30–190 compounds) of emerging pollutants are released through industrial wastewater (Ali *et al.*, 2018; Wolfaardt *et al.*, 2018).

Emerging contaminants fall into several categories, including artificial sweeteners, retardants, personal care products, drugs and pharmaceuticals. However, due to their increased use and high solubility, pesticides, personal care products and pharmaceuticals have been reported to be the main contributors to emerging contaminants. The exact number and concentration of emerging pollutants reported in the aquatic ecosystem are highly dependent on the socioeconomic framework of the community. Additionally, several researchers have noted that emerging contaminants in the aquatic ecosystem still lack the ecotoxic data set to determine their ability to hamper ecosystem and human health (Aderemi *et al.*, 2018; Escapa *et al.*, 2019).

3. Microalgae Cultivation System

The microalgae cultivation process seems complex but can be easily achieved with the precise applications. Several approaches can be followed for the cultivation of microalgae, e.g. photoautotrophic, photoheterotrophic and mixoautotrophic techniques. To implement the photoautotrophic strategy, three central components must be considered, such as light, carbon dioxide, carbon and nutrients. This technique uses sunlight to convert carbon dioxide into chemical energy, while, in the case of the heterotrophic algae cultivation process, a source of carbon, nutrients and water is required (Escapa *et al.*, 2017; Jaiswal *et al.*, 2021). The second method has been considered better due to not requiring sunlight or any light source. In addition, microalgae can also be cultivated by mixotrophic cultivation technique; this method is based on the mixed approach of photoautotrophic and heterotrophic techniques (Chen *et al.*, 2015; Keryanti and Mulyono, 2022; Yan *et al.*, 2022).

3.1. Open-pond algal cultivation system

The open-pond cultivation method consists of non-shallow raceway ponds (depth of 0.20–0.50 m) built with clay, concrete or plastic lines to cultivate microalgae. Due to the lower cost of the feedstocks, in this pond, the construction cost can remain affordable. For effective suspension of the microalgae culture, a paddle wheel can be provided for regular mixing. However, since the mixing energy requirement for the blades remains low, resulting in lower gas transfer efficiency, artificial aerators can be employed to offset CO_2 input (Choi and Lee, 2015; Gentili and Fick 2017). The pond's temperature is not controlled and still depends on the environmental conditions, while the light input is also not controlled; therefore, the efficiency of the open pond is highly dependent on the local climate and solar irradiations. Different factors in open-pond culture may not be controlled, making open-pond culture less efficient in the field than algal monoculture due to native algal grazers. Different ways out have been proposed to overcome that hurdle, such as maintaining higher temperature, pH and salinity to control the growth of contaminating microalgae strains. *Spirulina* sp. monoculture has been commercially maintained in an open-pond system with a higher pH level (9.5–11.5) (Yan *et al.*, 2022).

3.2. Closed cultivation system

The closed-culture system has been carried out through a closed tubular reactor and has been reported to be very effective in reducing the chances of contamination. The closed system exhibits higher productivity than the open-pond system and achieves better efficiency of fixation of introduced CO_2. The closed-culture method is capable of maintaining the desired conditions for the growth of the selected strains, as well as controlling the chances of the strains being invasive. The closed photobioreactor can maintain high cell density and achieve higher productivity (Ding et al., 2017; Benemann, 2008). Different closed systems have been proposed to design the microalgae culture configuration, e.g. bag reactor, flat tube and tubular bioreactor (Fig. 13.2). As reported by researchers, closed systems have been proved to be a better alternative in microalgae cultivation for high-value products. In addition, high biomass productivity followed by the least possible contamination has been recorded, but closed-culture systems are not the most suitable for wastewater treatment due to their higher operating cost and lack of technical expertise. The fundamental principle of bioreactors used for wastewater treatment is based on closed photobioreactor and is considered a complex treatment at the industrial level. In recent efforts, a tubular bubble column bioreactor has been used for *Chlorella zofiniensis* in a temperature range of 21–31°C, and a yield of 18.1 g/day was reported (Ebele et al., 2019; Donmez et al., 2020).

3.3. Open-raceway pond cultivation system

The open-raceway pond culture system is a very simple technique with a closed path and a motorized paddle wheel (Fig. 13.3). The desired microalgae strain can be inoculated and cultivated in the raceway pond, and the motorized paddle wheel maintains the regular circulation of the wastewater. As reported by researchers, open-raceway pond cultivation system needs lower wastewater treatment cost and biomass production can

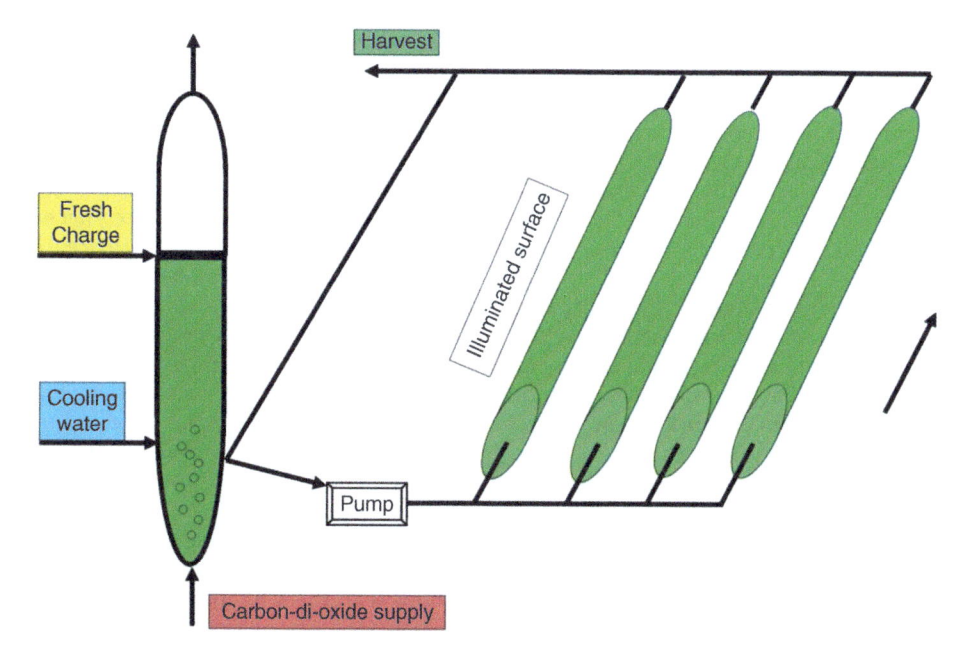

Fig. 13.2. Horizontal tube photobioreactor for algal cultivation.

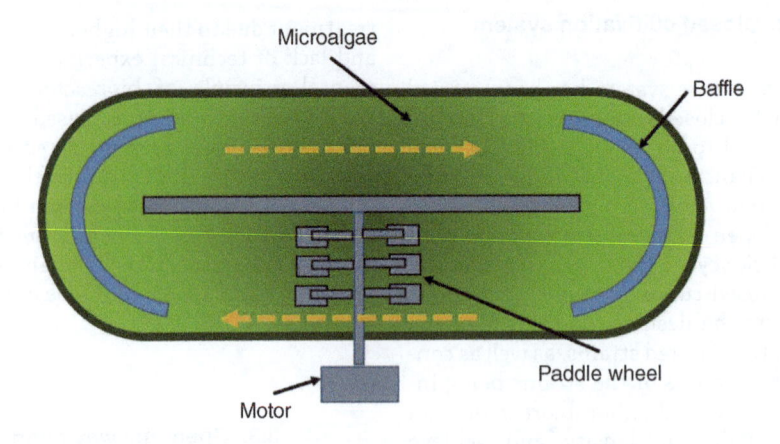

Fig. 13.3. Open-raceway pond for microalgae cultivation.

be achieved at a lower running cost. However, researchers have reported flaws of this set-up in the form of bacterial contamination and a challenging environment for microalgal growth (Kelly and Brooks, 2018; Ibuot et al., 2019; Keryanti and Mulyono, 2022).

To obtain the desired result of the open-raceway microalgae culture system, certain vital factors must be carefully considered, such as the depth of the open-raceway pond configuration, which must be maintained in accordance with adequate light penetration. The vitality of this characteristic is directly related to the concentration of suspended solids present in the environment since suspended solids can alter the water's turbidity. If these conditions can be managed efficiently, they can result in higher yields due to high biomass accumulation through mixotrophic metabolism. As reported by various researchers, disturbed turbidity can cause the sustainability of bacterial reproduction and yield to be drastically hampered. In addition, the low ambient temperature in winter can inhibit or hinder microalgae growth in wastewater treatment. The open-raceway pond system has an advantage in the form of lower investment and operating costs (Lee et al., 2019). A balanced relationship between environmental factors and microalgae cultivation has been considered a key component in wastewater treatment and biomass production; thus, the open-raceway pond cultivation system has shown remarkable suitability to fulfill the requirement with better results.

4. Bioremediation of Emerging Contaminants

Emerging contaminants, also called 'chemicals of emerging contaminants', have attracted considerable interest from environmentalists due to global awareness of freshwater and marine ecosystems and human health (Llad et al., 2016). The most abundant emerging contaminants are related to personal care products, gasoline additives, disinfectant by-products, algae toxins, various surfactants and endocrine disrupting compounds etc. Furthermore, until now, approximately 11,000 tons per year of personal care products and pharmaceutical compounds are used worldwide (Li et al., 2015; Kumar et al., 2016). This huge input of emerging contaminants can lead to high polarity, persistent biodegradation and bioaccumulation, which can cause serious harm to human health and the aquatic ecosystem. Based on the findings reported by several researchers, it can be stated that pesticides and pharmaceutical steroids can act as endocrine disruptors and cause reproductive disorders or feminization in the human and aquatic ecosystems (Fig. 13.4). Almost all emerging contaminants lack available regulatory standards, resulting in a massive non-transparent environment for any control measures. In the current

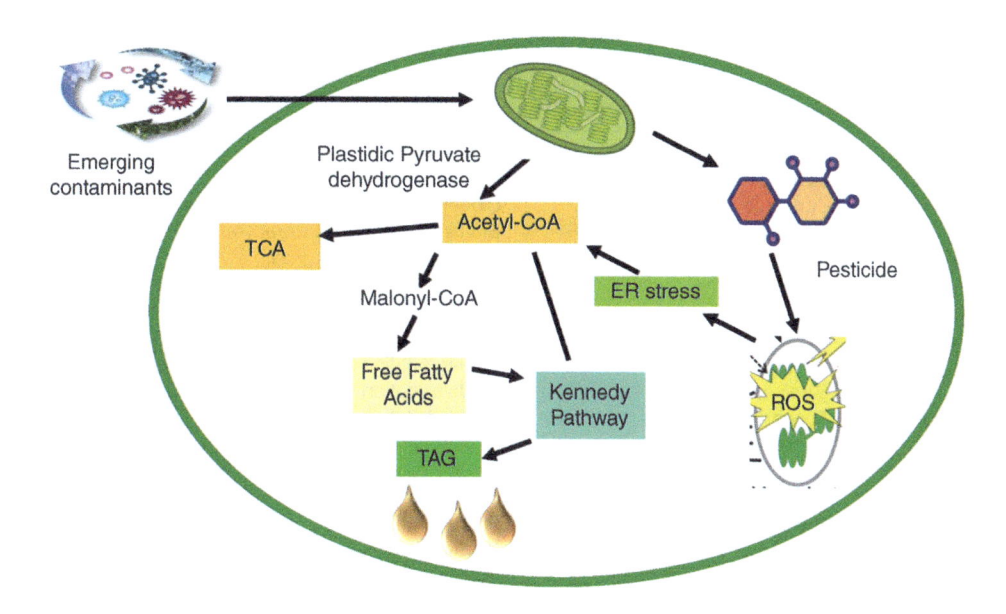

Fig. 13.4. Illustration of the TAG pathway under stress from emerging contaminants.

scenario, contemporary wastewater treatment plants are not efficient enough to bioremediate or degrade emerging contaminants. Various other methodologies have been employed to treat or remove emerging contaminants, e.g. constructed wetlands, solvent extraction, anaerobic bed reactors, activated sludge, and electrocoagulation. However, this field still has a long way to go (Choi and Lee, 2015).

The algae-based methodology has been reported to be very effective in removing heavy metals, organic and inorganic toxic substances. Furthermore, microalgae have gained remarkable attention in nutrient removal from the ecosystem. Several researchers have started advocating for an efficient role of microalgae in removing emerging contaminants, as multiple findings have reported the remarkable efficiency of microalgae in dealing with nutrients and emerging contaminants. In light of recent advances, it can be mentioned that microalgae have been growing as an innovative and cost-effective alternative for removing and treating nutrients and emerging contaminants (Ding *et al.*, 2017). As illustrated in Table 13.1, different microalgae have been employed for the bioremediation of pharmaceutical compounds. It can be seen that *Chlamydomonas* sp., *Scenedesmus* sp. and *Chlorella* sp. have shown remarkable outcomes. Such remedial capacity of microalgae may have been achieved due to their robustness and adaptability to adverse environmental conditions. However, despite having a rich diversity of microalgae, many species have been examined for the treatment and bioremediation of emerging contaminants (Bai and Acharya, 2020). In the current situation, extensive studies and evaluations are required to explore and validate the specific removal of emerging contaminants in a wide range of microalgae species. Furthermore, a detailed diversity of microalgae offers excellent research prospects in the bioremediation of organic/inorganic toxic substances and emerging contaminants (Guo *et al.*, 2016; Keryanti and Mulyono, 2022).

5. Microalgae-based Bioremediation of Organic Pollutants and Heavy Metals

5.1. Microalgae-based bioremediation of organic pollutants

Microalgae have the ability to accumulate organic substances in the cytoplasm. Therefore,

Table 13.1. Tabular representation of microalgae efficiency on emerging contaminants.

Emerging contaminants	Microalgae	Removal efficiency (%)	Pathways	References
Biperiden	C. vulgaris	88	Bioadsorption	Rykowska and Wasiak (2020)
Hydroxyzine	Consortia	79	Biodegradation	Donmez et al. (2020)
Lincomycin	Consortia	67	Biodegradation	Ali et al. (2018)
Orphenadrine	C. sorokiniana	86	Bio-uptake	Zeraatkar et al. (2016)
Codeine	S. dimorphus	72	Biodegradation	Ebele et al. (2019)
Estrone	Consortia	88	Biodegradation	Ahmed et al. (2017)
Fluconazol	Desmodesmus sp.	38	Bioadsorption	Rykowska and Wasiak (2020)
Ibuprofen	C. sorokiniana	97	Bio-uptake	Rykowska and Wasiak (2020)
Levofloxacin	C. vulgaris	70–95	Biodegradation	Bai and Acharya (2021)
Metoprolol	C. reinhardtii	92	Biodegradation	Escapa et al. (2019)
Norfloxaxin	Consortia	45–67	Biodegradation	Bai and Acharya (2021)
Ofloxacin	Consortia	44–69	Bio-uptake	Rykowska and Wasiak (2020)
Paracetamol	Consortia	87–94	Photo degradation	Escapa et al. (2019)
Progesterone	C. pyrenoidosa	94	Biodegradation	Bai and Acharya (2021)
Tetracycline	C. vulgaris	70	Photo degradation	Escapa et al. (2019)
Lincomycin	Consortia	82	Biodegradation	Zhou et al. (2020)
Naproxen	Consortia	65	Biodegradation	Xiong et al. (2017)
Lorazepam	Consortia	55	Photo degradation	Ali et al. (2018)
Paroxetine	Consortia	98	Biodegradation	Cho et al. (2016)
Hydroxyzine	C. sorokiniana	74	Biodegradation	Gentili and Fick (2017)
Fluxonazole	Consortia	35	Biodegradation	Guerra et al. (2018)
Diltiazen	Consortia	73–78	Photo degradation	Ibuot et al. (2019)
Codeine	C. sorokiniana	55	Biodegradation	Lee et al. (2019)
Clofibric	Consortia	10–35	Biodegradation	Huang et al. (2016)
Cefradine	C. pyrenoidosa	82	Biodegradation	Sutherland et al. (2018)
Bupropion	C. sorokiniana	68	Bioadsorption	Tran and Gin (2017)
Atenolol	Consortia	88–92	Biodegradation	Stravs et al. (2017)
Amoxicillin	C. pyrenoidosa	72	Biodegradation	Bwapwa et al. (2017)
Enrofloxacin	Consortia	71–74	Biodegradation	Ebele et al. (2019)
Carbamazepine	C. maxicana	42	Biodegradation	Hansda and Kumar (2016)
Mitrazapine	C. sorokiniana	68	Biodegradation	Knillmann et al. (2018)
Tramadol	Desmodesmus sp.	22–39	Bio-uptake	Rykowska and Wasiak (2020)
Tylosin	Consortia	81	Biodegradation	Bai and Acharya (2021)
Sulfapyridine	Consortia	94	Biodegradation	Donmez et al. (2020)
Sulfamethazine	S. obliquus	22	Biodegradation	Xiong et al. (2017)
Salinomycin	Consortia	68–73	Bio-uptake	Sutherland et al. (2018)
Triclosan	C. pyrenoidosa	79	Biodegradation	Ahmed et al. (2017)
Testosterone	Consortia	100	Biodegradation	Huang et al. (2016)
Salicylic acid	C. sorokiniana	70	Biodegradation	Bwapwa et al. (2017)
Norfloxacin	Consortia	45–60	Biodegradation	Cho et al. (2016)
Flecainide	C. saccharophila	100	Biodegradation	Tran and Gin (2017)
Ciprofloxacin	Nannochloris	96	Biodegradation	Ebele et al. (2019)

microalgae can play a vital role in wastewater treatment. Furthermore, microalgae are the pioneer producer in the aquatic ecosystem and can thrive in the presence of different aquatic organic pollutants due to adopted mechanisms such as biosorption, bioaccumulation, biodegradation and phycoremediation (Liu *et al.*, 2018; Pan *et al.*, 2018; Osundeko *et al.*, 2020). Microalgae carry out a biosorption process that is based on several aspects, such as absorption, ion exchange and precipitation. Biosorption is considered a very efficient method in the field of toxic substance removal as it is a cost-effective and environmentally friendly approach. The biosorption method is considered an environmentally friendly approach as it adsorbs and degrades organic contaminants and converts them into simpler non-hazardous compounds with the help of specific enzymes. Biodegradation can occur at the intracellular or extracellular level; however, preliminary degradation occurs at the extracellular level. Biodegradation shows a trend of high selectivity and high yield at lower operating cost (Llad et al., 2016; Keryanti and Mulyono, 2022; Yan et al., 2022). On the other hand, phycoremediation is also considered an effective and efficient methodology in the field of organic pollutant removal. In phycoremediation, microalgae convert sunlight into valuable biomass by using various nutrients present in the ecosystem. Several studies have identified microalgae's effectiveness in removing organic contaminants present in the aquatic ecosystem.

5.2. Microalgae-based bioremediation of heavy metals

Microalgae use heavy metals such as cobalt (Co), iron (Fe), zinc (Zn) etc. in the form of trace metals for cell metabolism and enzymatic processes. Also, certain heavy metals (such as arsenic, mercury and cadmium) have been found to be toxic to microalgae (Fig. 13.5). According to the reports of several investigations, it can be stated that the low concentration of toxic heavy metals can promote the growth of microalgae due to the phenomenon of hormesis (Saavedra et al., 2018). Microalgae have developed various methodologies against heavy-metal toxicity, such as gene regulation, exclusion, immobilization and chelation process. In addition, microalgae also carry out redox reactions to degrade heavy metals with the help of reducing enzymes. Microalgae can create a heavy metals-cellular protein complex. These organometallic complexes can be separated into cellular vacuoles to help maintain a lower concentration of heavy-metal ions, further reducing heavy-metal toxicity.

Additionally, heavy-metal accumulations initiate the synthesis of phytochelatins to minimize stress caused by heavy metals. Also, bioaccumulation of heavy metals occurs in the cytoplasm of the cell at a very gradual rate. Heavy metals are actively transported across the cell membrane and eventually

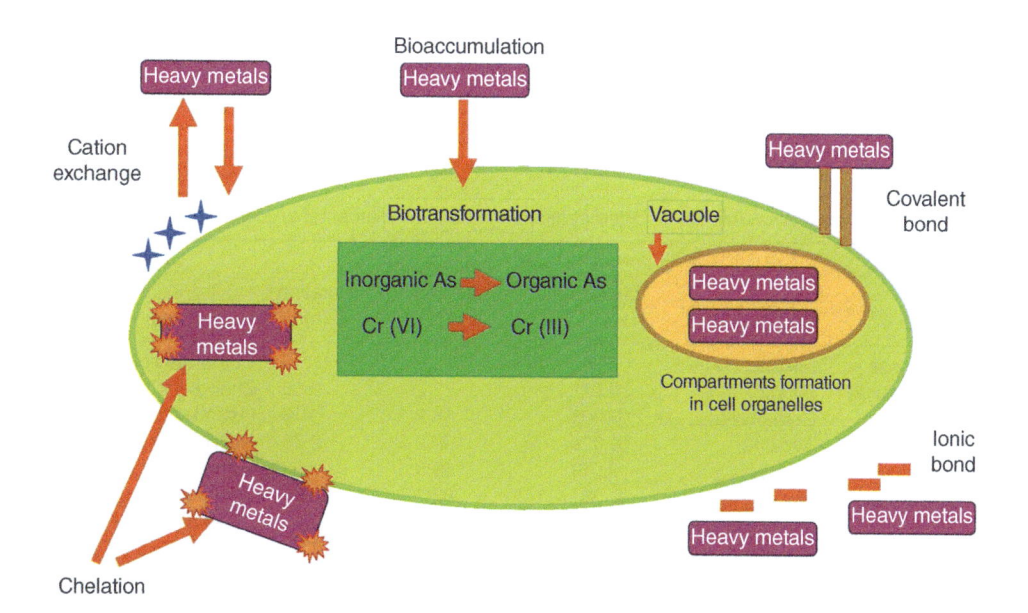

Fig. 13.5. Bioremediation of heavy metals by microalgae.

enter the cytoplasm through diffusion and then bind at the internal binding sites of peptides and proteins (Rühmland et al., 2015; Tran et al., 2018).

Heavy metals can be degraded by microalgae administration through a process that requires an energy source from an external medium. According to the various investigations, several microalgae strains need different environmental attributes to sustain themselves. Furthermore, various processes in microalgae strains are highly dependent on certain environmental factors, such as pH, temperature and TDS, while the metabolic rate of activities exhibits elevated focus with increasing atmospheric temperature (Petrie et al., 2015; Stravs et al., 2017). Bioremediation of organic pollutants and heavy metals can be achieved through microalgae cultivation through direct employment, which requires the energy of some external source. It also includes rapid physicochemical interaction toward the cell wall (Manamsa et al., 2016; Zhang et al., 2020). Several bioremediation attempts of organic pollutants and heavy metals through microalgae have been administered, summarized in Tables 13.2 and 13.3. As several researchers have mentioned, different methods are being applied to remove pesticides using microalgae via the processes of bioadsorption, bioaccumulation and biodegradation.

5.3. Role of bioadsorption in biodegradation

Evidently, some microalgae strains have been reported to exhibit the ability to adsorb toxic substances and heavy metals. As noted, 72–91% of organic toxins and heavy metals have been degraded with the help of bioadsorption process accomplished by microalgae (Xiong et al., 2017; Kumar et al., 2021). The bioadsorption process has multiple pathways or mechanisms, e.g. ion exchange, complex surface process, precipitation and adsorption. Besides, the adsorption of organic toxins and heavy metals largely depends on various active groups present on the surface and in microalgae (Aderemi et al., 2018; Wolfaardt et al., 2018). Moreover, the adsorption of organic toxic substances and heavy metals can also be supported by the presence of polysaccharides, carbohydrates and intercellular space in the cell wall. In addition, it can be mentioned that microalgae can remove organic toxins and heavy metals (Ebele et al., 2019; Zhou et al., 2020). It can be stated that the degradation process of organic pollutants and heavy metals mainly depends on two vital factors, i.e. the sustainable capacity of the biome as well as the supportive environment and the source and organization of organic pesticides and heavy metals and the factors involved for microalgae such as light, temperature, pH, and also the presence of carbon substrates.

Table 13.2. Biodegradation efficacy of different microalgae on environmental toxicants.

Microalgae	Toxicants	Concentration of substrate (mg/L)	Degradation efficiency (%)	Duration (h)	References
C. sorokiniana	Diazinon	18	83.3	245	Sutherland et al. (2018)
C. saccharophila	Pyriproxin	0.011	73.2	2.56	Zhang et al. (2020)
C. vulgaris	Carbafuran	0.003	88.2	1.16	Osundeko et al. (2020)
C. pyrenoidosa	Atrazine	0.029	69.2	2.76	Norvill et al. (2016)
C. maxicana	Molinate	0.019	59.5	1.83	Manamsa et al. (2016)
C. astroideum	Propanil	0.031	82.7	2.57	Escapa et al. (2017)
S. platensis	Malathion	0.03–90	49.3	389	Liu et al. (2018)
N. muscorum	Malathion	0.02–89	55.2	371	Tran and Gin (2017)
S. obliquus	Diazinon	24	44.7	221	Tran et al. (2018)
C. reinhardtii	Prometryne	7800	69.6	177	Zeraatkar et al. (2016)
C. reinhardtii	Fluroxypyr	0.7	51.4	160	Zeraatkar et al. (2016)

Table 13.3. Capacity of microalgae to bioremediate heavy-metal contaminants.

Heavy-metal contaminants	Microalgae	pH	Uptake (mg/g)	References
Hg^{+2}	C. reinhardtii	5	42.5	Rafiee et al. (2020)
	C. vulgaris	6	22	
	S. acutus	7	21	
Fe^{+3}	C. vulgaris	3	28.2	Agüera et al. (2020)
Cu^{+2}	C. vulgaris	5	78	Sakarika et al. (2020)
	Synechocystis sp.	5	28	
Zn^{+2}	C. vulgaris	5.5	48	Papazi et al. (2019)
	Cyclotella cryptica	6	100	
	D. pleiomorphus	6	92	
	P. lanceolatum	6.5	85	
	S. subspicatus	5	78	
Ni^{+2}	C. vulgaris	5	62	Gojkovic et al. (2019)
Pb^{+2}	C. reinhardtii	6	88.7	Serejo et al. (2020)
	S. acutus	6.5	95	
Cr^{+3}	C. miniata	5	32	Kiki et al. (2020)
	C. reinhardtii	3	37	
	Dunaliella sp.	3	64	
Co	C. reinhardtii	6	5.6	Loftus and Johnson (2019)
Cd^{+2}	C. calcitrans	9	146	García et al. (2020)
	C. reinhardtii	5	89	
	C. sorokiniana	6	114	
	S. abundans	8.2	167	
	S. acutus	8	73	

5.4. Role of bioaccumulation in biodegradation

In the active form, bioaccumulation is believed to be a process that can be denoted with the help of the bio-concentration factor. This factor focuses on expressing the concentration ratio of a specific organic toxicant or heavy metal in the organism's metabolism related to the ecosystem (Zhang et al., 2020; Bai and Acharya, 2021). The analyzable value of the bio-concentration factor can be vitally affected by the change in concentration, organic matter, ionization of metabolic components, and any physical interference (Abo et al., 2016; Bai and Acharya, 2020). Several researchers have reported that, in case of interaction of microalgae with toxic organic substances and heavy metals, a higher yield of ROS can be generated from any living cell. As a result of this scenario, the amount of ROS obtained exhibits a greater oxidative tendency. In addition, the atoms responsible for DNA oxidation can cause various algae disorders and cell death (Cho et al., 2016; Guerra et al., 2018). Moreover, any possibility of toxicity can trigger the gene expression process within the microalgae; thus, antioxidant enzymes can be produced. Previous investigations have explained that microalgae can start the bioaccumulation process and, as a result, the degradation of organic toxic substances and heavy metals can take place (Kelly and Brooks, 2018; Ebele et al., 2019). This phenomenon can be better understood with an example since *Scenedesmus obliquus* showed a remarkable accumulation of triadimefon. In contrast, the elimination of organic toxic substances and heavy metals showed a remarkable degradation by microalgae. Several experts strongly recommended combining these two processes to remove organic toxins and heavy metals using microalgae (Hom-Diaz et al., 2015, 2017).

5.5. Biodegradation as a vital tool in the elimination of toxic substances

Efficient and effective removal of toxic organic substances and heavy metals can be achieved through biodegradation. However, it has been recognized that microalgae have the ability to degrade organic toxins and heavy metals found in wastewater by breaking down organic toxins and heavy metals into small molecules, as these small molecules serve as a good source of nutrition in the growth of microalgae (Ibuot et al., 2019). In the case of the biodegradation of malathion (inorganic toxicant) and cadmium (heavy metal), some microalgae strains turned out to be very effective. *S. platensis* has demonstrated the breakdown of 58% of the malathion, while *N. muscorum* has shown 69% heavy-metal toxicity remediation. Additionally, several other microalgae strains (*Chlorella* sp.) exhibited remarkable progress, degrading 89% of organic toxicants and heavy metals. Microalgae possess several enzymes *viz.*, transferase, oxygenase, oxyreductase and phosphotriesterase, and these enzymes play a vital role in the biodegradation and detoxification process (Kumar et al., 2016). However, we have a long way to go to explore the entire biodegradation mechanism of organic toxicants and heavy metals through enzymes (Guerra et al., 2018).

6. Mechanisms Adapted by Algae for the Bioremediation of Emerging Contaminants

Efficient bioremediation of emerging contaminants, including pharmaceutical compounds using microalgae, has become necessary in the current situation. Microalgae such as *Chlamydomans* sp., *Chlorella* sp., and some other species have been widely used and extensively studied to validate the concept. Such a remarkable result may have been achieved due to the efficacy and robustness of microalgae in the polluted and adverse environment. Accurate exposure set-up is another essential aspect to consider during microalgae selection against environmental stress. However, the uncertain aspects related to the operation of the reactors during the hydraulic retention time have their own challenges (Gojkovic et al., 2019; Lee et al., 2019; Jaiswal et al., 2020). Bioreactors operating on an industrial scale work in continuous or semi-continuous mode with hydraulic retention time. Also, optimization of microalgae removal efficiency is still reported under laboratory conditions, and the limited amount of literature also makes the situation challenging. The toxicity of wastewater directly depends on the type and source of contamination, several factors such as high O_2 concentration and heavy metals, and high ammonium content in municipal wastewater can create hindrances in water treatment with contemporary methods. The application of microalgae in such wastewater can produce some noteworthy and encouraging results.

The studies revealed that some types of genetic adaptations could improve the tolerance capacity of microalgae against emerging contaminants at the laboratory level. Also, genetic adaptations equip microalgae against heavy metals, intense sunlight and fluctuating salinity. The interaction of microalgae with emerging contaminants releases degrading enzymes to counteract them (Nakayama et al., 2019; Osundeko et al., 2020). Traces of emerging contaminants such as pesticides, personal care products and pharmaceuticals become another complication confronted by microalgae in treating emerging contaminants in wastewater. In the case of *Chlorella vulgaris*, it was found that the insecticide (diazinon) at 18% concentration retarded microalgae growth (Villar-Navarro et al., 2018; Rykowska and Wasiak, 2020). *Chlorella luteoviridis* was reported to be well acclimatized to the secondary treated wastewater supply after the six weeks of exposure time. Moreover, this acclimatization and tolerance was recorded due to high ascorbate peroxidase activities and carotenoid pigment assimilation. Another, the microalga *P. kessleri* isolated from municipal wastewater, exhibited notable abilities against environmental stresses such as oxidation, high salinity, high temperature and altered pH (Zhang et al., 2017; Zhou et al., 2020).

6.1. Presence of multiple emerging contaminants in microalgae cells

In wastewater treatment, different emerging contaminants can compete for binding sites and result in instability due to the interaction between emerging contaminants and microalgae. Several investigations have mentioned the emerging contaminants and the interaction of microalgae for the presence of an additive effect, synergistic effect and antagonistic effect due to multiple contaminants in microalgal cells. In a recent attempt, the EC_{50} of *Chlorella vulgaris* range for enrofloxacin, erythromycin and enrofloxacin-erythromycin complex recorded to be 119.3 mg/L, 93.4 mg/L and 52.1 mg/L, respectively. The lowest EC_{50} value recorded for the complex indicated the presence of a synergistic impact of related antibiotics (Bwapwa et al., 2017). In another attempt, a combination of sulfamethazine and sulfamethoxazole was used and reported a fourfold increase in sulfamethazine (19.6%). As suggested by researchers, sulfamethoxazole causes the induction of several catalytic enzymes (Ebele et al., 2019). Co-metabolism has attracted considerable attention in the field of removal of emerging contaminants by microalgae. The transformation of a non-growth substrate can explain co-metabolism during the appearance of the growth substrate. However, it can be stated that co-metabolism supports the degradation of specific compounds through the common biochemical impact of the organism. The specified perception was supported by the elevation in the elimination efficiency (15.7–42%) of ciprofloxacin using *Chlamydomonas mexicana* due to the presence of sodium acetate (Gentili and Fick, 2017; Escapa et al., 2019). The bioremediation of emerging pollutants by microalgae can be carried out by several routes such as bioadsorption, bioaccumulation and biodegradation. Biosorption includes the adsorption of emerging contaminants to the extracellular polymeric substance (EPS) of the microalgae. Extracellular polymeric substances are released from the cell wall and further synthesize and release forms of lipids, proteins and nucleic acids. Moreover, in the process of bioaccumulation, emerging contaminants passed through the cell wall and bound to intracellular peptides (Cho et al., 2016). Various environmental factors such as redox, temperature, salinity and pH can alter the rate of bioadsorption and bioaccumulation. Many pilot-scale experiments are essentially required to validate the concept of removal of emerging contaminants using microalgae through the bioadsorption and bioaccumulation process (Hom-Diaz et al., 2017).

6.2. Fate of contaminants within the microalgae cells

The bioaccumulation process evidently exhibits efficiency in removing emerging contaminants, but some contrary theories related to the safe removal of emerging contaminants after the bioaccumulation process have still been reported. In a reported study, the removal rate of triexypenidyl was 92% by *Coelastrella* sp. and similar findings were reported in *Chlorella saccharophila*, where a removal efficiency of over 90% was recorded (Guerra et al., 2018). On the basis of comparison, it can be stated that bioadsorption and bioaccumulation have tremendous potential to remove emerging contaminants within the microalgae cells or even in the bulk medium. Emerging contaminants can function as a rich carbon source for microalgae during the metabolic degradation process. Catalytic enzymes of the substrate control the degradation of emerging contaminants. In addition, hormones such as estrogens can play a vital role in removing toxic emerging contaminants by microalgae. The enzymes responsible for biodegradation and enzyme activation have been monitored and determined by the concentration of emerging contaminants. The maximum concentration of emerging contaminants plays a vital role in activating enzymes for biodegradation. Many speculations about catalytic enzymes' role in the removal process of emerging contaminants by microalgae have still been raised (Gojkovik et al., 2019).

7. Challenges and Future Prospects in the Bioremediation of Emerging Contaminants

Certainly, microalgae possess prominent characteristics, such as high treatment capacity, fast growth cycle, and strong tolerance to emerging contaminants, making them a noticeable and efficient biosorption approach supported with broad application prospects (Kelly and Brooks 2018; Knillmann et al., 2018; Lee et al., 2019). In addition, microalgae can be used in various ways, such as fertilizers, medicinal feed supplements and biofuels (Chowdhury et al., 2021). Despite several efficient advantages of microalgae in the bioremediation of emerging contaminants, some aspects of this phenomenon face particular challenges, such as immature processing, harvest complications, the need for specific nutrients, and contamination by non-target strains. The biodegradation of emerging contaminants by microalgae is a complex method and its operating cost depends on several indeterminate factors (such as temperature, pH, sunlight etc.). Also, microalgae are very small in size and the surface of microalgae is negatively charged, making them challenging to harvest (Liu et al., 2018; Nakayama et al., 2019). The possibilities of contamination comprise great concern in the field of microalgae cultivation in open reactors. *Cladocerans* and rotifers can drastically discourage algae accumulation in a very short period of time. To overcome this barrier, a few studies have suggested that mixed cultures of microalgae can be used in raceway ponds to enhance the removal of emerging contaminants. Several researchers have defended the efficacy of photobioreactors, but the productivity and cost of management of this approach on a large scale is still being investigated. Extremophile algae can play a revolutionary role as they can tolerate harsh climatic conditions such as intense light, high or low temperature, and fluctuating pH. Also, the use of extremophile algae in the biodegradation of emerging contaminants still requires extensive research (Pan et al., 2018; Osundeko et al., 2020; Yan et al., 2022). Various methods, *viz.*, granulation, cell immobilization and microalgae biofilm, exhibited significant potential for industrial application. Chemical and surface modification of microalgal biomass and complexation with other degradation techniques can improve the removal capacity (Tiwari et al., 2017; Tran et al., 2017; Banerjee et al., 2020). Future research attempts should consider the biodegradation of emerging contaminants by microalgae as an efficient technique compared to conventional methodologies. The selection and isolation of robust microalgae strains and the selection of cost-effective pretreatment can play a decisive role in improving efficiency and reducing operating costs at the industrial level.

8. Conclusion

Microalgae exist in freshwater and marine ecosystems and have been considered vital in the global ecology due to their efficiency and taxonomically diverse nature. Microalgae are considered a potential sink for the removal of toxic substances from the ecosystem. A large proportion of pollution comes from micro-pollutants, which have become a severe concern to human health and the ecosystem. The ecosystem receives the toxicity of pesticides due to some important sources, such as effluent from untreated pesticide-manufacturing industries, non-uniform spraying of pesticides, any previously possible agricultural toxicity, and improper or defective management of pesticides during transport. Various techniques are available to treat wastewater involving physical and chemical precipitation, microfiltration, carbon adsorption and activated sludge. However, these techniques are expensive and less effective in all conditions. Microalgae have the ability to accumulate organic substances in the cytoplasm. Therefore, microalgae play a vital role in wastewater treatment. Moreover, microalgae are the pioneer producer in the aquatic ecosystem and microalgae can thrive in the presence of different aquatic organic pollutants due to the adopted mechanisms such as biosorption, bioaccumulation, biodegradation and phytoremediation. Emerging contaminants or chemicals of emerging contaminants have attracted considerable interest from environmentalists due to global awareness of freshwater and marine ecosystems as well as human health. The algae-based treatment practice has proved very effective in removing heavy metals and emerging toxic substances. Also, microalgae have gained significant attention in the field of nutrient removal from the ecosystem. It has been advocated for an efficient role of microalgae in removing emerging contaminants, as multiple findings have been reported for the remarkable efficiency of microalgae in treating emerging contaminants.

References

Abo, R., Kummer, N.A. and Merkel, B.J. (2016) Optimized photo degradation of Bisphenol A in water using ZnO, TiO$_2$ and SnO$_2$ photocatalysts under UV radiation as a decontamination procedure. *Drinking Water Engineering and Science* 9(2), 27–35.

Aderemi, A.O., Novais, S.C., Lemos, M.F., Alves, L.M., Hunter, C. et al. (2018) Oxidative stress responses and cellular energy allocation changes in micro algae following exposure to widely used human antibiotics. *Aquatic Toxicology* 203, 130–139.

Agüera, A., Plaza, P. and Fernandez, F.A. (2020) Removal of contaminants of emerging concern by microalgae-based wastewater treatments and related analytical techniques. *Current Developments in Biotechnology and Bioengineering*, 503–525.

Ali, M.E., El-Aty, A.M.A., Badawy, M.I. and Ali, R.K. (2018) Removal of pharmaceutical pollutants from synthetic wastewater using chemically modified biomass of green alga Scenedesmus obliquus. *Ecotoxicology and Environmental Safety* 151, 144–152.

Bai, X. and Acharya, K. (2020) Removal of trimethoprim, sulfamethoxazole, and triclosan by the green alga Nannochloris sp. *Journal of Hazardous Materials* 315, 70–75.

Bai, X. and Acharya, K. (2021) Removal of seven endocrine disrupting chemicals (EDCs) from municipal wastewater effluents by a freshwater green alga. *Environmental Pollution* 247, 534–540.

Banerjee, I., Dutta, S., Pohrmen, C.B., Verma, R. and Singh, D. (2020) Microalgae-based carbon sequestration to mitigate climate change and application of nanomaterials in algal biorefinery. *Octa Journal of Biosciences* 8, 129–136.

Benemann, J.R. (2008) *Opportunities and Challenges in Algae Biofuels Production*. Algae World, Singapore.

Bwapwa, J.K., Jaiyeola, A.T. and Chetty, R. (2017) Bioremediation of acid mine drainage using algae strains: a review. *South African Journal of Chemical Engineering* 24, 62–70.

Chen, J., Zheng, F. and Guo, R. (2015) Algal feedback and removal efficiency in a sequencing batch reactor algae process (SBAR) to treat the antibiotic cefradine. *PLOS ONE* 10(7), 273.

Cho, K., Lee, C.H., Ko, K., Lee, Y.J., Kim, K.N. et al. (2016) Use of phenol-induced oxidative stress acclimation to stimulate cell growth and biodiesel production by the oceanic microalga Dunaliella salina. *Algal Research* 17, 61–66.

Choi, H.J. and Lee, S.M. (2015) Heavy metal removal from acid mine drainage by calcined egg shell and microalgae hybrid system. *Environmental Science and Pollution Research* 22(17), 13404–13411.

Chowdhury, C.R., Banerjee, I. and Trivedi, A. (2021) Microalgae for the production of biofuels and other value-added products. *Octa Journal of Biosciences* 9(2), 50–54.

Ding, T., Lin, K., Yang, B., Yang, M., Li, J. et al. (2017) Biodegradation of naproxen by freshwater algae Cymbella sp. and Scenedesmus quadricauda and the comparative toxicity. *Bioresource Technology* 238, 164–173.

Donmez, G.Ç., Aksu, Z., Oztürk, A. and Kutsal, T. (2020) A comparative study on heavy metal biosorption characteristics of some algae. *Process Biochemistry* 34(9), 885–892.

Dutta, S., Banerjee, I. and Jaiswal, K.K. (2020) Graphene and graphene-based nanomaterials for biological and environmental applications for sustainability. *Octa Journal of Biosciences* 8(2), 106–112.

Dutta, S., Pohrmen, C.B., Banerjee, I., Trivedi, A., Verma, R. et al. (2021) Bio-derived metal and metal oxide incorporated biopolymer nanocomposites for dye degradation applications: a review. *Octa Journal of Biosciences* 9(1), 30–36.

Ebele, A.J., Abdallah, M.A.E. and Harrad, S. (2019) Pharmaceuticals and personal care products (PPCPs) in the freshwater aquatic environment. *Emerging Contaminants* 3(1), 1–16.

Escapa, C., Coimbra, R.N., Paniagua, S., García, A.I. and Otero, M. (2017) Comparison of the culture and harvesting of *Chlorella vulgaris* and *Tetradesmus obliquus* for the removal of pharmaceuticals from water. *Journal of Applied Phycology* 29(3), 1179–1193.

Escapa, C., Coimbra, R.N., Paniagua, S., García, A.I. and Otero, M. (2019) Nutrients and pharmaceuticals removal from wastewater by culture and harvesting of Chlorella sorokiniana. *Bioresource Technology* 185, 276–284.

Fatima, N., Kumar, V., Jaiswal, K.K., Vlaskin, M.S., Gururani, P. et al. (2020) Toxicity of cadmium (Cd) on microalgal growth (IC50 value) and its exertions in biofuel production. *Biointerface Research in Applied Chemistry* 10(4), 5828–5833.

García, M.J., Monllor-Alcaraz, L.S., Postigo, C., Uggetti, E., Alda, M.L. et al. (2020) Microalgae-based bioremediation of water contaminated by pesticides in peri-urban agricultural areas. *Environmental Pollution*, 114579.

Gentili, F.G. and Fick, J. (2017) Algal cultivation in urban wastewater: an efficient way to reduce pharmaceutical pollutants. *Journal of Applied Phycology* 29(1), 255–262.

Gojkovic, Z., Lindberg, R.H., Tysklind, M. and Funk, C. (2019) Northern green algae have the capacity to remove active pharmaceutical ingredients. *Ecotoxicology and Environmental Safety* 170, 644–656.

Guerra, P., Kim, M., Shah, A., Alaee, M. and Smyth, S.A. (2018) Occurrence and fate of antibiotic, analgesic/anti-inflammatory, and antifungal compounds in five wastewater treatment processes. *Science of the Total Environment* 473, 235–243.

Guo, W.Q., Zheng, H.S., Li, S., Du, J.S., Feng, X.C. et al. (2016) Removal of cephalosporin antibiotics 7-ACA from wastewater during the cultivation of lipid-accumulating microalgae. *Bioresource Technology* 221, 284–290.

Hansda, A. and Kumar, V. (2016) A comparative review towards potential of microbial cells for heavy metal removal with emphasis on biosorption and bioaccumulation. *World Journal of Microbiology and Biotechnology* 32(10), 170.

Hom-Diaz, A., Llorca, M., Rodríguez-Mozaz, S., Vicent, T., Barcel O.D. et al. (2015) Microalgae cultivation on wastewater digestate: β-estradiol and 17α-ethynylestradiol degradation and transformation products identification. *Journal of Environmental Management* 155, 106–113.

Hom-Diaz, A., Norvill, Z.N., Blánquez, P., Vicent, T. and Guieysse, B. (2017) Ciprofloxacin removal during secondary domestic wastewater treatment in high rate algalponds. *Chemosphere* 180, 33–41.

Huang, X., Tu, Y., Song, C., Li, T., Lin, J. et al. (2016) Interactions between the antimicrobial agent triclosan and the bloom-forming cyanobacteria Microcystis aeruginosa. *Aquatic Toxicology* 172, 103–110.

Ibuot, A.A., Gupta, S.K., Ansolia, P. and Bajhaiya, A.K. (2019) Heavy metal bioremediationby microalgae. In: Chang, Y.-C. (ed.) *Microbial Biodegradation of Xenobiotic Compounds*. CRC Press, Boca Raton, Florida.

Jaiswal, K.K. and Pandey, H. (2014) Next generation renewable and sustainable micro-fuels from Chlorella pyrenoidosa. *International Journal of Recent Scientific Research* 5(4), 767–769.

Jaiswal, K.K., Banerjee, I., Singh, D., Sajwan, P. and Chhetri, V. (2020) Ecological stress stimulus to improve microalgae biofuel generation: a review. *Octa Journal of Biosciences* 8, 48–54.

Jaiswal, K.K., Dutta, S., Banerjee, I., Pohrmen, C.B. and Kumar, V. (2021) Photosynthetic microalgae-based carbon sequestration and generation of biomass in biorefinery approach for renewable biofuels for a cleaner environment. *Biomass Conversion and Biorefinery* 13, 7403–7421.

Kelly, K.R. and Brooks, B.W. (2018) Global aquatic hazard assessment of ciprofloxacin: exceedances of antibiotic resistance development and ecotoxicological thresholds. *Progress in Molecular Biology and Translational Science* 159, 59–77.

Keryanti, K. and Mulyono, E.W.S. (2022) Determination of optimum condition of lead (Pb) biosorption using dried biomass microalgae Aphanothece sp. *Periodica Polytechnica Chemical Engineering* 65(1), 116–123.

Kiki, C., Rashid, A., Wang, A., Li, Y., Zeng, Q. et al. (2020) Dissipation of antibiotics by microalgae: kinetics, identification of transformation products and pathways. *Journal of Hazardous Materials* 387, 121985.

Knillmann, S., Orlinskiy, P., Kaske, O., Foit, K. and Liess, M. (2018) Indication of pesticide effects and recolonization in streams. *Science of the Total Environment* 630, 1619–1627.

Kumar, D., Pandey, L.K. and Gaur, J.P. (2016) Metal sorption by algal biomass: from batch to continuous system. *Algal Research* 18, 95–109.

Kumar, V., Jaiswal, K.K., Tomar, M.S., Rajput, V., Upadhyay, S. et al. (2021) Production of high value added biomolecules by microalgae cultivation in wastewater from anaerobic digestates of food waste: a review. *Biomass Conversion and Biorefinery*. DOI: 10.1007/s13399-021-01906-y

Lee, K.Y., Lee, S.H., Lee, J.E. and Lee, S.Y. (2019) Biosorption of radioactive cesium from contaminated water by microalgae Haematococcus pluvialis and Chlorella vulgaris. *Journal of Environmental Management* 233, 83–88.

Li, H., Pan, Y., Wang, Z., Chen, S., Guo, R. et al. (2015) An algal process treatment combined with the Fenton reaction for high concentrations of amoxicillin and cefradine. *RSC Advances* 5(122), 100775–100782.

Liu, Y., Wang, F., Chen, X., Zhang, J. and Gao, B. (2015) Cellular responses and biodegradation of amoxicillin in Microcystis aeruginosa at different nitrogen levels. *Ecotoxicology and Environmental Safety* 111, 138–145.

Liu, W., Li, J., Gao, L., Zhang, Z., Zhao, J. et al. (2018) Bioaccumulation and effects of novel chlorinated polyfluorinated ether sulfonate in freshwater alga Scenedesmus obliquus. *Environmental Pollution* 233, 8–15.

Llad, O.J., Sol, S.M., Lao-Luque, C., Fuente, E. and Ruiz, B. (2016) Removal of pharmaceutical industry pollutants by coal-based activated carbons. *Process Safety and Environmental Protection* 104, 294–303.

Loftus, S.E. and Johnson, Z.I. (2019) Reused cultivation water accumulates dissolved organic carbon and uniquely influences different marine microalgae. *Frontiers in Bioengineering and Biotechnology* 7, 101.

Manamsa, K., Crane, E., Stuart, M., Talbot, J., Lapworth, D. et al. (2016) A national scale assessment of micro-organic contaminants in groundwater of England and Wales. *Science of the Total Environment* 568, 712–726.

Nakayama, S.F., Yoshikane, M., Onoda, Y., Nishihama, Y., Iwai-Shimada, M. et al. (2019) Worldwide trends in tracing poly- and perfluoroalkyl substances (PFAS) in the environment. *TrAC Trends Analytical Chemistry* 21(3), 56–68.

Nanda, M., Jaiswal, K.K., Kumar, V., Vlaskin, M.S., Gautam, P. et al. (2021) Micro-pollutant Pb (II) mitigation and lipid induction in oleaginous microalgae Chlorella sorokiniana UUIND6. *Environmental Technology & Innovation* 23, 101613.

Norvill, Z.N., Shilton, A. and Guieysse, B. (2016) Emerging contaminant degradation and removal in algal wastewater treatment ponds: identifying the research gaps. *Journal of Hazardous Materials* 313, 291–309.

Osundeko, O., Dean, A.P., Davies, H. and Pittman, J.K. (2020) Acclimation of microalgae to wastewater environments involves increased oxidative stress tolerance activity. *Plant and Cell Physiology* 55(10), 1848–1857.

Pan, C.G., Peng, F.J. and Ying, G.G. (2018) Removal, biotransformation and toxicity variations of climbazole by freshwater algae Scenedesmus obliquus. *Environmental Pollution* 240, 534–540.

Papazi, A., Karamanli, M. and Kotzabasis, K. (2019) Comparative biodegradation of all chlorinated phenols by the microalga Scenedesmus obliquus: the biodegradation strategy of microalgae. *Journal of Biotechnology* 296, 61–68.

Petrie, B., Barden, R. and Kasprzyk-Hordern, B. (2015) A review on emerging contaminants in wastewaters and the environment: current knowledge, understudied areas and recommendations for future monitoring. *Water Research* 72, 3–27.

Rafiee, P., Ebrahimi, S., Hosseini, M. and Tong, Y.W. (2020) Characterization of Soluble Algal Products (SAPs) after electrocoagulation of a mixed algal culture. *Biotechnology Reports* 25, e00433.

Rühmland, S., Wick, A., Ternes, T.A. and Barjenbruch, M. (2015) Fate of pharmaceuticals in a subsurface flow constructed wetland and two ponds. *Ecological Engineering* 80, 125–139.

Rykowska, I. and Wasiak, W. (2020) Research trends on emerging environment pollutants: a review. *Open Chemistry* 13, 1353–1370.

Saavedra, R., Munoz, R., Taboada, M.E., Vega, M. and Bolado, S. (2018) Comparative uptake study of arsenic, boron, copper, manganese and zinc from water by different green microalgae. *Bioresource Technology* 263, 49–57.

Sakarika, M., Koutra, E., Tsafrakidou, P., Terpou, A. and Kornaros, M. (2020) Microalgae-based remediation of wastewaters. *Microalgae Cultivation for Biofuels Production*, 317–335.

Serejo, M.L., Farias, S.L., Ruas, G., Paulo, P.L. and Boncz, M.A. (2020) Surfactant removal and biomass production in a microalgal-bacterial process: effect of feeding regime. *Water Science Technology*. DOI: 10.2166/ wst.2020.22

Stravs, M.A., Pomati, F. and Hollender, J. (2017) Exploring micropollutant biotransformation in three freshwater phytoplankton species. *Environmental Science: Processes and Impacts* 19(6), 822–832.

Sutherland, D.L., Heubeck, S., Park, J., Turnbull, M.H. and Craggs, R.J. (2018) Seasonal performance of a full-scale wastewater treatment enhanced pond system. *Water Research* 136, 150–159.

Tiwari, B., Sellamuthu, B., Ouarda, Y., Drogui, P., Tyagi, R.D. and Buelna, G. (2017) Review on fate and mechanism of removal of pharmaceutical pollutants from wastewater using biological approach. *Bioresource Technology* 224, 1–12.

Tran, N.H. and Gin, K.Y.H. (2017) Occurrence and removal of pharmaceuticals, hormones, personal care products, and endocrine disrupters in a full-scale water reclamation plant. *Science of the Total Environment* 599, 1503–1516.

Tran, N.H., Reinhard, M. and Gin, K.Y.H. (2018) Occurrence and fate of emerging contaminants in municipal wastewater treatment plants from different geographical regions: a review. *Water Research* 133, 182–207.

Villacorte, L.O., Ekowati, Y., Neu, T.R., Kleijn, J.M., Winters, H. *et al.* (2015) Characterisation of algal organic matter produced by bloom-forming marine and freshwater algae. *Water Research* 73, 216–230.

Villar-Navarro, E., Baena-Nogueras, R.M., Paniw, M., Perales, J.A. and Lara-Martín, P.A. (2018) Removal of pharmaceuticals in urban wastewater: high rate algae pond (HRAP) based technologies as an alternative to activated sludge based processes. *Water Research* 139, 19–29.

Wolfaardt, G.M., Lawrence, J.R., Robarts, R.D. and Caldwell, D.E. (2018) The role of interactions, sessile growth, and nutrient amendments on the degradative efficiency of a microbial consortium. *Canadian Journal of Microbiology* 40(5), 331–340.

Xiong, J.Q., Kurade, M.B. and Jeon, B.H. (2017) Biodegradation of levofloxacin by an acclimated freshwater microalga, Chlorella vulgaris. *Chemical Engineering Journal* 313, 1251–1257.

Yan, C., Qu, Z., Wang, J., Cao, L. and Han, Q. (2022) Microalgal bioremediation of heavy metal pollution in water: Recent advances, challenges, and prospects. *Chemosphere* 286, 131870.

Zeraatkar, A.K., Ahmadzadeh, H., Talebi, A.F., Moheimani, N.R. and McHenry, M.P. (2016) Potential use of algae for heavy metal bioremediation: a critical review. *Journal of Environmental Management* 181, 817–831.

Zhang, D., Gersberg, R.M., Ng, W.J. and Tan, S.K. (2020) Removal of pharmaceuticals and personal care products in aquatic plant-based systems: a review. *Environmental Pollution* 184, 620–639.

Zhang, S., Deng, R., Lin, D. and Wu, F. (2017) Distinct toxic interactions of TiO_2 nanoparticles with four coexisting organochlorine contaminants on algae. *Nanotoxicology* 11(9–10), 1115–1126.

Zhou, G.J., Ying, G.G., Liu, S., Zhou, L.J., Chen, Z.F. *et al.* (2020) Simultaneous removal of inorganic and organic compounds in wastewater by freshwater green microalgae. *Environmental Science: Processes and Impacts* 16(8), 2018–2027.

14 Construction of Microalgal Chloroplast Organelle Factory for the Carbon-neutralized Future

Zhu Zhen[1,2], Tian Jing[2] and Cao Xupeng[1,*]

[1]State Key Laboratory of Catalysis of China, Dalian Institute of Chemical Physics, Chinese Academy of Sciences, Dalian National Laboratory for Clean Energy, Dalian, China; [2]School of Bioengineering, Dalian Polytechnic University, Dalian, China

Abstract

Differing to the multiple copies of chloroplasts in single higher plant cells, some microalgae own a solo large chloroplast in their cells, which contributes to about 70% of cell biomass, and will help obtain pure construction after limited screening. Amongst several editable species, *Chlamydomonas reinhardtii* is the most studied one, with huge potential in carbon neutralization. Here, some progress on the chloroplast organelle factory construction are introduced and summarized to help the development in this field.

1. Introduction

Microalgae is a kind of photosynthetic organism, which is widely distributed in various waters. It has a much higher photosynthetic carbon fixation efficiency than terrestrial plants and a growth rate close to that of microorganisms. Microalgae can synthesize a variety of bioactive substances through metabolism, such as protein, fatty acid, vitamins, pigments, sterols, polyphenols, flavonoids, terpenoids, polysaccharides etc. (Kselikova *et al.*, 2022; Munoz *et al.*, 2021; Wichmann *et al.*, 2018; Tran and Kaldenhoff, 2020). The history of microalgae as a nutritional supplement and animal feed is hundreds of years old (Kusmayadi *et al.*, 2021). With the development and rise of bioenergy, microalgae are considered to be the most important next-generation biomass provider due to the advantages of high photosynthetic carbon-fixation efficiency and easy industrialization. Microalgae is a natural 'cell factory' for the production of high-value-added chemicals and bulk biomass (Gerotto *et al.*, 2020). In addition, excess carbon dioxide from industry and the continued use of fossil fuels are the feedstock for microalgal cell factories that drive green production (Mutale-Joan *et al.*, 2022). This is one of the sustainable and economical strategies for achieving carbon neutrality, which is essential for regulating the global carbon cycle (Behrenfeld *et al.*, 2001). Chloroplast is the important place for photosynthesis and synthesis of various biomass. Therefore, the construction of a chloroplast organelle factory is one of the ways to improve the biochemicals' production by microalgae.

*Address for correspondence: c_x_p@dicp.ac.cn

© CAB International 2023. *Algal Biotechnology* (Qiang Wang)
DOI: 10.1079/9781800621954.0014

Chlamydomonas reinhardtii is an important model for chloroplast transformation of microalgae (Dreesen *et al.*, 2010). Chloroplast genome of *C. reinhardtii* only encodes about 100 key proteins, while most of the proteins required for chloroplast function, about 2900, are encoded by the nucleus and are introduced into the chloroplasts to play a role through transport peptides (Xing *et al.*, 2022). Chloroplasts have a high tolerance to the accumulation of foreign proteins, so there are obvious advantages to using chloroplasts as bioreactors, and it is meaningful to build a chloroplast organelle factory.

The development of a microalgae biotechnology platform has made continuous progress. From the perspective of genetic engineering, microalgae have the potential to produce compounds and proteins at an economic level like a cell factory (Shi *et al.*, 2021; Russell *et al.*, 2022). So far, more than 40 different microalgae, including *C. reinhardtii*, *Dunaliella salina*, *Chlorella vulgaris* and *Haematococcus pluvialis*, have been successfully genetically modified (Shi *et al.*, 2021). Common eukaryotic microalgae chloroplast expression systems can be broadly divided into two categories. First, use the chloroplast expression system to homologously recombine foreign genes into chloroplast DNA for expression (Noordally *et al.*, 2013; Gutensohn *et al.*, 2006). In 1988, *C. reinhardtii* chloroplasts were first transformed by particle bombardment (Boynton *et al.*, 1988). Now, microalgae chloroplasts have been successfully used for light-driven production of recombinant proteins, vaccines, antigens, commercial enzymes and metabolites. *C. reinhardtii* is the most fully developed microalgae in chloroplast transformation. At present, the method of introducing and expressing transgenes into the chloroplast genome of *C. reinhardtii* has been established, and more than 100 different proteins have been successfully produced (Dyo and Purton, 2018). Second, there is the strategy of constructing a chloroplast organelle factory by using chloroplast transit peptide sequence as a target to achieve chimeric nuclear transformation expression system. For example, the transit peptide sequence of Rubisco small subunit (rbcS, Genbank Access Number: AB058647) of *Chlorella vulgaris* is fused with enhanced green fluorescent protein (EGFP) and expressed in the chloroplast of *C. vulgaris*. Using this strategy, the functional role of cyanobacterial fructose 1,6-bisphosphate aldolase in *C. vulgaris* chloroplast was studied (Yang *et al.*, 2017a). In addition, the transport peptide chimeric fluorescent protein was localized in chloroplasts and expressed in *C. reinhardtii* (Zedler *et al.*, 2016) and tobacco (Kobayashi *et al.*, 2017). By fusing the N-terminus of *C. reinhardtii* β-carotene ketolase (CrBKT) sequence with the chloroplast transit peptide from photosystem I subunit D (PsaD), the enzyme can be localized and expressed in chloroplasts. The optimized CrBKT overexpression extended native carotenoid biosynthesis to generate ketocarotenoids, making the green algal reddish brown (Perozeni *et al.*, 2020).

One of the advantages of chloroplast transformation is that foreign genes can be easily integrated through homologous recombination, but their insertion position and length are limited. However, the nuclear transformation of microalgae is mostly random integration, resulting in great differences between different transformants. Therefore, extensive screening is required to isolate transformants that effectively express transgenes (Sproles *et al.*, 2022). At present, the development of the CRISPR-CAS9 system in microalgae may provide a solution for transgenic targeting to specific sites in the nuclear genome (Lu *et al.*, 2021; Zhang *et al.*, 2019). The improvement of transformation tools is also a dynamic field. Taking advantage of the transformation of microalgae nucleus, foreign proteins can be targeted and expressed in chloroplasts, which can carry out glycosylation of proteins and/or additional post-translational modification and secretion. These advantages are particularly important for the industrial production of microalgae for developing recombinant proteins.

Nuclear-encoded chloroplast proteins generally use N-terminal sequence characteristics to locate proteins in the chloroplast cavity. The N-terminal transport peptide is recognized by the Toc complex, and the Tic complex finally delivers proteins to the chloroplast matrix. In the process of transport, TP is cut, so a new N-terminal is generated, which will be the terminal of mature protein or a new TP (Lee and Hwang, 2018; Radhamony and Theg, 2006). However, these transit peptides or recognition sequences have large differences in length and lack the conservatism of the primary structure. At the same time, people have not been able to find reliable identification features based on the secondary structure and have not provided clear prediction rules. They can only be predicted through experimental verification or data mining based on experiments.

Since different nuclear-encoded chloroplast proteins have great functional differences, our team considered combining RbcS2 with cTP with different characteristics to construct expression vectors for research. First, chlorophyllous glyceraldehyde 3-phosphate dehydrogenase (GAPDH, Cre01.g010900.t1.2) (Zhu et al., 2019, 2020) was transferred into C. reinhardtii. Later, we further tried to transfer Saccharomyces cerevisiae PDAT (ScPDAT) (Zhu et al., 2018), which fused the cTP sequence of C. reinhardtii chloroplast phospholipid: diacylglycerol acyltransferase (CrPDAT, GenBank: AFB73928.1), and found the same growth coupled expression trend. This is not consistent with the stable RbcS protein content and the constitutive expression driven by the conventional RbcS2 promoter during normal culture. Based on the previous reports on the stability of RbcS2 system, we speculate that there are two possible mechanisms. One is the exogenous RbcS2 system, especially after the introduction of cTP, which is affected by the regulatory mechanism inconsistent with the endogenous system, or the internal/external RbcS2 promoters interact in the regulatory process. For example, Yamasaki et al. found that after the exogenous RbcS2 promoter was integrated into the nuclear DNA, it would lose the stability of the endogenous RbcS2 promoter due to the influence of histone modification changes, and its activity could be characterized by the level of histone H3 acetylation and H3K9 monomethylation (Yamasaki et al., 2008). Another possibility is that after the expressed protein enters the chloroplast, it will be regulated as a whole by the chloroplast. For the speculation of the latter, using the characteristic fatty acid technology established by the research team, it is found that the transformed ScPDAT mainly affects the fatty acid metabolism and lipid recombination in chloroplasts, which proves that the fused ScPDAT enters and plays a role in chloroplasts, so the above possibility exists.

2. Recent Progress

2.1 Functional study on heterologous expression of ScPDAT in chloroplasts of *C. reinhardtii*

Using Hsp70A-RbcS2 promoter, the chloroplast signal peptide (from CrPDAT) was fused with PDAT from *Saccharomyces cerevisiae* to construct the chloroplast expression system, which was expressed in *C. reinhardtii* CC-137. The results of PCR, sequencing and Western Blot experiments showed that the expression of ScPDAT in the transformant was synchronized with the growth in the exponential phase. Compared with wild type, the expression of ScPDAT in the transformant promoted the total fatty acid content to increase by 22%, and the triacylglycerol (TAG) content to increase by 32%. In addition, the fluctuations of C16 series fatty acids in monogalactosyldiacylglycerol (MGDG), diacylglyceryltrimethylhomoserine (DGTS) and TAG indicate the enhancement of TAG accumulation pathway.

Scpdat and wild type (WT) were incubated in TAP of a shaker with 24-hour light photoperiod and 50-µmol photons m^{-2} s^{-1} irradiance. The initial growth of the *Scpdat* was slower, but it finally reached the same level as the WT, i.e. the dry weight was 0.80 g/L (Fig. 14.1a). For the Φ_{PSII}, there is no significant difference between the *Scpdat* and WT. Except that the F_v/F_m of the *Scpdat* decreases significantly and temporarily on the second day, which was at the same level as WT (Fig. 14.1b). The function of PDAT is to promote the synthesis of TAG. Quantitative determination of total fatty acids showed that *Scpadt* was 23% higher than WT. The

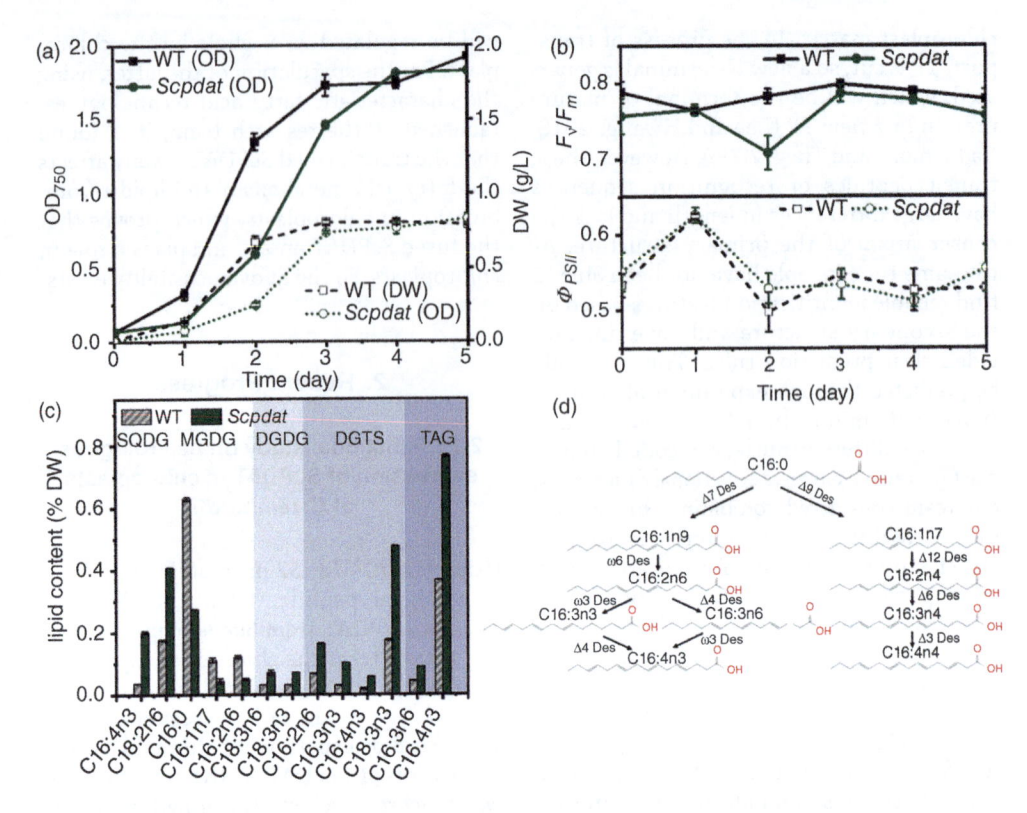

Fig. 14.1. Growth status and lipid analysis of *Scpdat* and WT.

fatty acid (FA) on the third day were analyzed in detail by TLC separation and GC quantitative method, including five types of lipids, i.e. sulfoquinovosyldiacylglycerol (SQDG), digalactosyldiacylglycerol digalactosyldiacylglycerol (DGDG), MGDG, DGTS, and TAG. The content of TAG increased by 32% (Fig. 1c). In addition, C16:4n3 and C18:3n3 are two kinds of chloroplast biomarkers FA, which mainly exist in chloroplasts (Yang et al., 2017b) and are also the main reason for the increase of lipid content.

2.2 Overexpression of chloroplast GAPDH in *C. reinhardtii* enhances carbon fixation

The chloroplast expression system was constructed by using Hsp70A-RbcS2 promoter, the chloroplast GAPDH of *C. reinhardtii* and 3'UTR sequences, and expressed in *C. reinhardtii* CC-137. The transformants were selected and cultured in shake flask and reactor successively. From the analysis of growth state to biomass content, the transformants showed the advantages of high carbon fixation capacity.

2.2.1 Transformants showed high carbon fixation ability in shake flask culture

The difference between P1-GAPDH and P3-GAPDH was whether there was 3'UTR sequence in the expression system. P3-GAPDH represents the transformant containing 3 'UTR sequence. The three algal strains were cultured under Tris-Acetate-Phosphate medium (TAP) in a shaker with 24-hour light photoperiod and 50-μmol photons m^{-2} s^{-1} irradiance. By measuring the growth state of microalgae, it was found that there was no significant difference in their growth and photosynthetic activity. The cell density and dry weight were 1.9 ~ 2.0 and 0.5 ~ 0.6 g/L (Fig. 14.2a, b), respectively. There was no

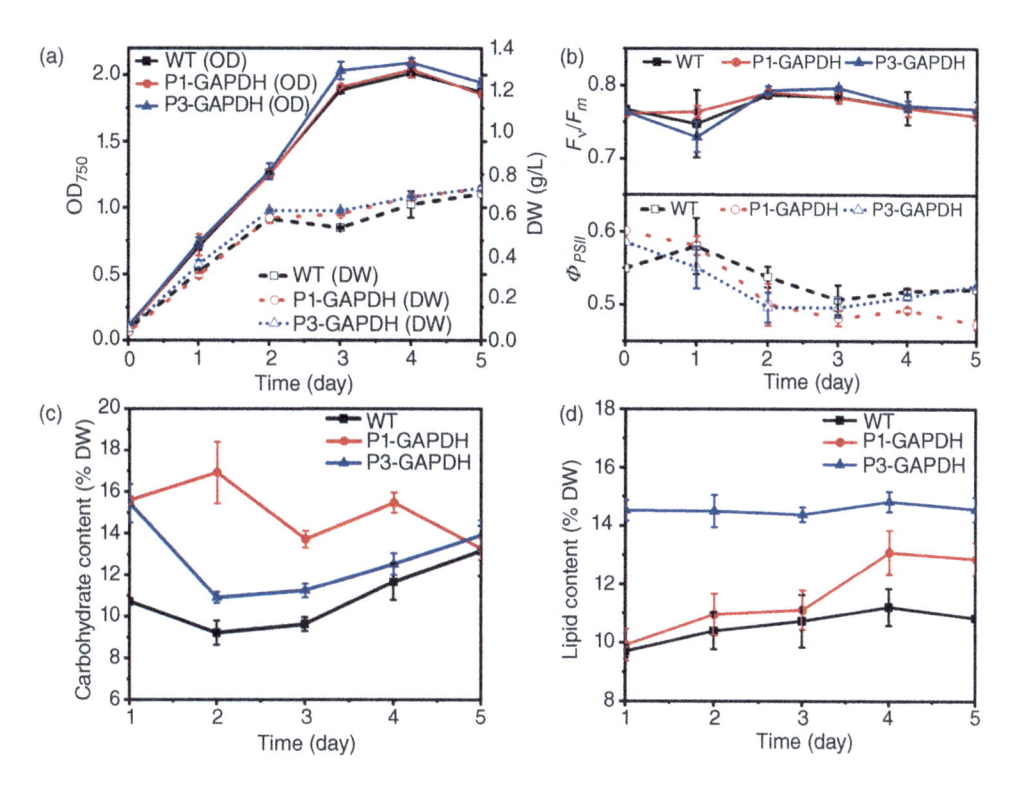

Fig. 14.2. Growth status, carbohydrate and lipid content of microalgae under shake flask culture.

significant difference in F_v/F_m and $\Phi_{PS\,II}$ between WT and transformants. However, there are obvious differences in biomass content. The carbohydrate and lipid contents of the two transformants were higher than those of the WT (Fig. 14.2c, d). It is worth noting that the fatty acid content of P3-GAPDH has been stable and at a high level, and its fatty acid content is about 15%. Compared with the WT, the production of fatty acids increased by 55%, while the production of energy storage compounds increased by 75%. Based on the above results, the overexpression of the chloroplast GAPDH significantly improved the carbon fixation of photosynthesis of C. reinhardtii.

2.2.2 The transformants showed the ability of simultaneous enhancement of growth and carbon fixation

By analyzing the results obtained under shaking flask culture, the outstanding advantage of P3-GAPDH was found. To further explore its carbon fixation level, the cultivation experiment was carried out in 1.5 L flatplate poly (polymethyl methacrylate) (PMMA) photobioreactorplate on Algal Station (AS) system. P3-GAPDH and WT were incubated in TAP with 300 mmol m^{-2} s^{-1} light and 14/10 light/dark cycle, respectively. There were significant differences in the growth of P3-GAPDH and WT, and the dry weight content of cell growth was 1.74 ± 0.09 g/L and 1.23 ± 0.13 (Fig. 14.3a), respectively. For F_v/F_m, the overall level of P3-GAPDH was slightly higher than that of WT. For $\Phi_{PS\,II}$, P3-GAPDH and WT, there was significant difference in trend. WT showed a correlation with photoperiodic changes, while P3-GAPDH remained stable and higher than WT (Fig. 14.3b). The carbon fixation ability can be explained by measuring the content of carbohydrate and lipid. The results showed that the maximum biomass productivity of P3-GAPDH and WT was 28.54 ± 1.43 mg L^{-1} h^{-1} and 24.30 ± 1.65 mg L^{-1} h^{-1}, respectively. In terms of energy storage composition, the

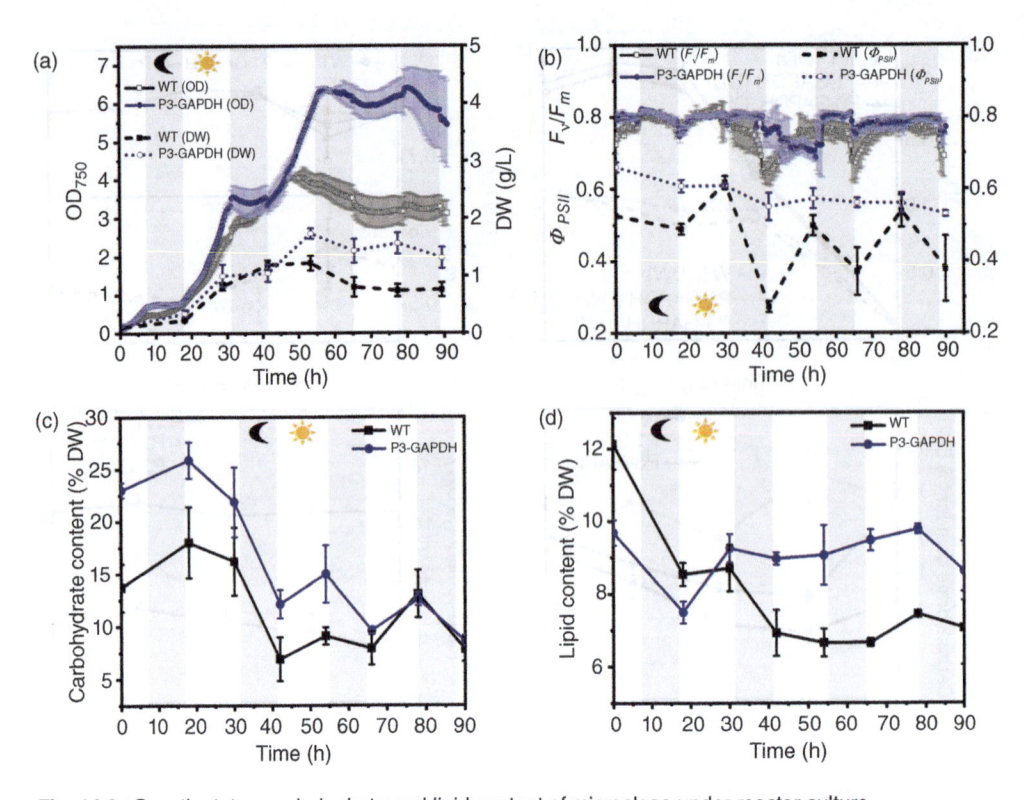

Fig. 14.3. Growth status, carbohydrate and lipid content of microalgae under reactor culture.

content of carbohydrate and fatty acid in P3-GAPDH doubled, with carbohydrate reaching 0.26 ± 0.04 g/L and fatty acid reaching 0.16 ± 0.01 g/L (Fig. 14.3 c, d).

3. Discussion

Using the Hsp70A-RbcS2 promoter, the chloroplast signal peptide was fused with exogenous or endogenous proteins to construct a chloroplast expression system, which was expressed in *C. reinhardtii* CC-137. By analyzing the culture results of *Scpdat*, combining the changes of different lipid components and FAs spectrum (Fig. 14.3 d), especially the changes of C16 series FAs, it can be assumed that exogenous ScPDAT mainly affects FA and lipid metabolism in chloroplasts with the help of cTP of CrPDAT. However, the chloroplast localization of cTP fused protein needs to be proved more directly and will be carried out by fluorescent protein labeling (Yang et al., 2017a; Mori et al., 2016). According to the culture results of P1-GAPDH and P3-GAPDH, the chloroplast GAPDH can enhance the carbon fixation capacity of *C. reinhardtii* under this chloroplast expression system. The results of P3-GAPDH also proved the importance of adding 3'UTR sequence, and chlorophyllous GAPDH can enhance the Calvin cycle, promote carbon conversion and subsequent accumulation of energy storage compounds.

As an important photosynthetic carbon fixation organelle, the chloroplast of microalgae has contributed significant amounts of biomass to support the biosphere. Together with quickly developed artificial low-carbon conversion technologies, a chloroplast factory will be a key solution to the future's environment, foods, drugs, biodegradable materials and energy problems.

References

Behrenfeld, M.J. et al. (2001) Biospheric primary production during an ENSO transition. *Science* 291, 2594–2597.

Boynton, J.E. et al. (1988) Chloroplast transformation in Chlamydomonas with high velocity microprojectiles. *Science* 240, 1534–1538.

Dreesen, I.A.J., Charpin-El Hamri, G. and Fussenegger, M. (2010) Heat-stable oral alga-based vaccine protects mice from Staphylococcus aureus infection. *Journal of Biotechnology* 145, 273–280.

Dyo, Y.M. and Purton, S. (2018) The algal chloroplast as a synthetic biology platform for production of therapeutic proteins. *Microbiology* 164, 113–121.

Gerotto, C., Norici, A. and Giordano, M. (2020) Toward enhanced fixation of CO_2 in aquatic biomass: focus on microalgae. *Frontiers in Energy Research* 8.

Gutensohn, M. et al. (2006) Toc, Tic, Tat et al.: structure and function of protein transport machineries in chloroplasts. *Journal of Plant Physiology* 163, 333–347.

Kobayashi, Y. et al. (2017) Holliday junction resolvases mediate chloroplast nucleoid segregation. *Science* 356, 631–634.

Kselikova, V., Singh, A., Bialevich, V., Cizkova, M. and Bisova, K. (2022) Improving microalgae for biotechnology: from genetics to synthetic biology – moving forward but not there yet. *Biotechnology Advances* 58, 107885.

Kusmayadi, A., Leong, Y.K., Yen, H.W., Huang, C.Y. and Chang, J.S. (2021) Microalgae as sustainable food and feed sources for animals and humans: biotechnological and environmental aspects. *Chemosphere* 271, 129800.

Lee, D.W. and Hwang, I. (2018) Evolution and design principles of the diverse chloroplast transit peptides. *Molecullar Cells* 41, 161–167.

Lu, Y., Zhang, X., Gu, X., Lin, H. and Melis, A. (2021) Engineering microalgae: transition from empirical design to programmable cells. *Critical Review of Biotechnology* 41, 1233–1256.

Mori, N., Moriyama, T., Toyoshima, M. and Sato, N. (2016) Construction of global acyl lipid metabolic map by comparative genomics and subcellular localization analysis in the red alga Cyanidioschyzon merolae. *Frontiers in Plant Science* 7, 958.

Munoz, C.F., Sudfeld, C., Naduthodi, M.I.S., Weusthuis, R.A., Barbosa, M.J., Wijffels, R.H. and D'Adamo, S. (2021) Genetic engineering of microalgae for enhanced lipid production. *Biotechnology Advances* 52, 107836.

Mutale-Joan, C., Sbabu, L. and Hicham, El A. (2022) Microalgae and cyanobacteria: how exploiting these microbial resources can address the underlying challenges related to food sources and sustainable agriculture: a review. *Journal of Plant Growth Regulation* 42, 1–20.

Noordally, Z.B. et al. (2013) Circadian control of chloroplast transcription by a nuclear-encoded timing signal. *Science* 339, 1316–1319.

Perozeni, F. et al. (2020) Turning a green alga red: engineering astaxanthin biosynthesis by intragenic pseudogene revival in Chlamydomonas reinhardtii. *Plant Biotechnology Journal* 18, 2053–2067.

Radharmony, R.N. and Theg, S.M. (2006) Evidence for an ER to Golgi to chloroplast protein transport pathway. *Trends in Cell Biology* 16, 385–387.

Russell, C., Rodriguez, C. and Yaseen, M. (2022) High-value biochemical products & applications of freshwater eukaryotic microalgae. *Science of the Total Environment* 809, 151111.

Shi, Q. et al. (2021) Transgenic eukaryotic microalgae as green factories: providing new ideas for the production of biologically active substances. *Journal of Applied Phycology* 33, 705–728.

Sproles, A.E., Berndt, A., Fields, F.J. and Mayfield, S.P. (2022) Improved high-throughput screening technique to rapidly isolate Chlamydomonas transformants expressing recombinant proteins. *Applied Microbiology and Biotechnology* 106, 1677–1689.

Tran, N.T. and Kaldenhoff, R. (2020) Metabolic engineering of ketocarotenoids biosynthetic pathway in Chlamydomonas reinhardtii strain CC-4102. *Scientific Reports* 10, 10688.

Wichmann, J. et al. (2018) Tailored carbon partitioning for phototrophic production of (E)-alpha-bisabolene from the green microalga Chlamydomonas reinhardtii. *Metabolic Engineering* 45, 211–222.

Xing, J. et al. (2022) The plastid-encoded protein Orf2971 is required for protein translocation and chloroplast quality control. *Plant Cell* 34, 3383–3399.

Yamasaki, T., Miyasaka, H. and Ohama, T. (2008) Unstable RNAi effects through epigenetic silencing of an inverted repeat transgene in Chlamydomonas reinhardtii. *Genetics* 180, 1927–1944.

Yang, B. et al. (2017a) Genetic engineering of the Calvin cycle toward enhanced photosynthetic CO(2) fixation in microalgae. *Biotechnol Biofuels* 10, 229.

Yang, M. et al. (2017b) Chloroplasts isolation from Chlamydomonas reinhardtii under nitrogen stress. *Frontiers in Plant Science* 8, 1503.

Zedler, J.A.Z., Mullineaux, C.W. and Robinson, C. (2016) Efficient targeting of recombinant proteins to the thylakoid lumen in Chlamydomonas reinhardtii using a bacterial Tat signal peptide. *Algal Research* 19, 57–62.

Zhang, Y.T. et al. (2019) Application of the CRISPR/Cas system for genome editing in microalgae. *Applied Microbiology and Biotechnology* 103, 3239–3248.

Zhu, Z. et al. (2018) The synchronous TAG production with the growth by the expression of chloroplast transit peptide-fused ScPDAT in Chlamydomonas reinhardtii. *Biotechnology for Biofuels and Bioproducts* 11, 156.

Zhu, Z. et al. (2019) Studies on the effect of overexpressed chloroplast glyceraldehyde-3-phosphate dehydrogenase on carbohydrate and fatty acid contents of Chlamydomonas reinhardtii. *Periodical of Ocean University of China* 49, 50–58.

Zhu, Z. et al. (2020) A carbon fixation enhanced Chlamydomonas reinhardtii strain for achieving the double-win between growth and biofuel production under non-stressed conditions. *Frontiers in Bioengineering and Biotechnology* 8, 603513.

Index

Note: The page numbers in italics and bold represents figures and tables respectively.

Agrobacterium tumefaciens 66
algae-based synthetic biology 69, 71
anoxygenic phototrophic purple bacteria 78
aquaculture
    harvested microalgae 154
    high-value application, microalgae
        biofuels 163
        conversion pathways and processes *166*
        microalgal lipids 166
        pigment 164–165
        proteins 163–164
        valuable molecules 163
    wastewater 154
    wastewater treatment 155–156
    water quality control 154–155
aqueous twophase extraction (ATPE) 164
aqueous two-phase system (ATPS) 123
artificial coculture systems 34, 37, 39, 43
astaxanthin 108, 165
atomic force microscopy (AFM)
    artificial photosynthetic systems 85
    cyanobacterial thylakoid membrane (*see* thylakoid membrane architecture)
    photosynthetic membrane 77–79
    thylakoid membranes 79–80

bioactive compounds
    carbohydrate 118
    cardiovascular disease 115
    ionic liquids
        cell wall crush, extraction 119, *120*, **121**, 122–123
        conventional organic solvents 119
        multiple products extraction 125
        two-phase systems, extraction 123–124, *124*
    microalgae an activities **116**
biobutanol 105, 106
bioremediations
    algae-based treatment practice 196
    emerging contaminants 188–189, *189*, **190**, 194–195, 196
    heavy metals 184, 197
    heavy metals/emerging pollutants 185–186
    microalgae cultivation process
        open-pond cultivation method 186
        open-raceway pond 187–188
        photoautotrophic strategy 186
    organic pollutants/heavy metals
        bioaccumulation 193
        bioadsorption 192
        heavy metals *191*, 191–192, **193**
        organic pollutants 189–191
        toxic substances 193–194
    (toxic) pesticides 184

capillary electrophoresis coupled to mass spectrometry (CE-MS) 132
capsular polysaccharides (CPS) 147
carbon neutralization
    chloroplast transformation 202
    GAPDH in *C. reinhardtii*, carbon fixation 204–206
    heterologous expression of ScPDAT, *C. reinhardtii* 203–204
    microalgae 201
    microalgae chloroplasts 202

carbon neutralization (*Continued*)
    next-generation biomass 201
    nuclear-encoded chloroplast proteins 202, 203
    RbcS2 promoter 203
    reactor culture 206
    Scpdat and WT, growth status/lipid analysis 204
carotenoids 108, 118, 164
cauliflower mosaic virus (CaMV35S) 8
Chlamydomonas reinhardtii 118, 202
chloroplast organelle factory *see* carbon neutralization
clustered regularly interspaced palindromic sequences (CRISPR/Cas9) technology 19
clustered regularly interspaced short palindromic repeats (CRISPR) system 54
coculture systems 33
control-knob gene-based approach 22
C-phycocyanin 108, 177
C. reinhardtii β-carotene ketolase (CrBKT) sequence 202
CRISPR RNA (crRNA) 55
cyanobacterial FBP/SBPase 67
cyanobacterial species 33
cyanobacterial thylakoid membranes 78
CyanoMetDB 110

diacylglyceryltrimethylhomoserine (DGTS) 203
dichlorodiphenyltrichloroethane (DDT) 173
dimethyl phthalate (DMP) 175
docosahexaenoic acid (DHA) 106
double-strand breaks (DSBs) 54

editing methods
    gene overexpression 16–17
    GM to non-GM engineering 20–21
    nucleus to organelle engineering 19–20
    random mutagenesis 18
    RNAi-based gene silencing 18
    targeted genome editing 18–19
eicosapentaenoic acid (EPA) 106
electron microscopy (EM) 79
electrospray ionization (ESI) 131, 132
energy metabolism 44, 135
engineering microalgae
    conventional gene engineering 21–22
    cyanobacteria 2
    digital photosynthetic cells 22–23, 23–24
    DNA delivery methods, modern breeding strategies **6–7**
    empirical to quantitative designs, transition 3
        empirical designs 8, 15
        empirical to quantitative designs, transition tools 8
        engineering vectors 16
        nanoparticle-mediated DNA delivery 5
        transformation methods 5
        transforming cassette design, quantitative methods 15–16
    engineering strategies 17 (*see also* editing methods)
    genetic engineering, examples **9–14**
    life cycle 2
    metabolic pathways 2
    mutagenesis strategies 24
    photosynthetic cell factories
        heterologous compounds 3
        natural compounds 2–3
        targeted compounds **4**
    representative expression systems **5**
    support ecosystems 2
    synthetic biology 22
    transcriptional engineering 22
    vector design/engineering, methods 2
enhanced green fluorescent protein (EGFP) 202
Euglena-based synthetic biology
    chromosomes 62
    molecular phylogenetic studies 61
Euglena gracilis (*E. gracilis*)
    biological characterization 62–64
    dark anaerobic conditions 62
    definition 62
    delivery methods, exogenous materials
        Agrobacterium-mediated transformation 66
        biolistic bombardment 64–65
        electroporation 65–66
        glass beads 66
        single-cell microinjection 66–67
    gene function, RNAi technique **70**
    genetic transformation 71
    genetic transformation/editing tools
        chloroplast genome 68
        CRISPR-Cas9 application 68–69
        nuclear genome 67
        RNAi technology 69
    industrial species 62
    macromolecular exosomes 70
    paramylon 62
eukaryotic microalgae 2
eukaryotic microalgae chloroplast expression systems 202
exocellular polymeric matrix (EPM) 148
exopolysaccharide (EPS) 118, 147
extremophile algae 196

far-red light photoacclimation (FaRLiP) 83–84
fourier transform infrared spectroscopy (FT-IR) 132
freeze-drying 123
freshwater microalgae
    beneficial/harmful roles 177–178
    chemical pollutants 171, 172–174, **173**
    harmful effect 176–177
    heavy metals removal 175

natural resources 177
organic pollutant removal, biofilm 175–176
ROS 172, 178
fucoxanthin-chlorophyll binding proteins (FCP) 15

gas chromatography (GC) 132
genes of interest (GOIs) 16
genome editing
    CRISPR/Cas systems 55
    CRISPR system 54
    definition 54
    diatom cells 55–57
    HDR 55
    screening 57
    TALEN and CRISPR systems 55, 58
    tools for diatom 57–58
glass-bead method 5
glyceraldehyde 3-phosphate dehydrogenase (GAPDH) 203
GM organisms (GMOs) 20

high-cell-density cultivation, microalgae
    combined culture strategies
        heterotrophic-autotrophic cultivation mode 97
        inducers/stress factors 97–98
    culture strategies
        continuous cultivation 97
        fed-batch cultivation 95–96
        perfusion culture systems 97
        repeated-batch cultivation 96
    heterotrophic cultivation 91, 98
    heterotrophic growth 91, **92**
        environmental factors 94–95
        nutritional factors 91–94
    open culture systems 90
    PBRs 91
    technoeconomic (TE) analysis 98
high-resolution melting analysis (HRMA) 57
high-throughput omics techniques 22
homology-directed recombination (HDR) 55
Hsp70A-RbcS2 promoter 206
hydrophobic organic pollutants 173

intracytoplasmic membranes (ICMs) 78

large supramolecular antenna system 78
light-driven symbiotic systems 33
lipid droplet surface protein (LDSP) promoter 8
liquid chromatography (LC) 132

metabolomics, algae
    analytical tools 135

analytic methods
    GC-MS 133
    LC-MS 132–133
    NMR 132
applications
    economical high-value-added products 135
    ecotoxicology research 135
    stress-resistance research 134–135
data analysis
    annotation 134
    preprocessing 133
    statistical analysis 133–134
metabolites 130
sample preparation/extraction 131–132
metabolites, microalgae
    AACT/Acetyl-coa acetyltransferase *109*
    AGPP/ADP-glucose pyrophosphorylase *107*
    amino acids/peptides/proteins 104–105
    carbohydrates 105–106
    high-value bioactive compounds 103
    large-scale new germplasm screening 110
    lipids 106–107
    PEP *104*
    pigments 107–109
    primary/secondary metabolites 102
    TCA *104*
    types, active metabolites **103**
metabolomics 43
metagenomics 43, 45
methyl tertiary butyl ether (MTBE) 173
microalgae-based aquaculture wastewater treatment
    ammonia removal 156
    antibiotics degradation *157*
    factors affecting effectiveness
        antibiotic 160
        illumination 158
        pollutant antibiotics removal 159
        temperature 158–159
    heavy metals removal 156–158
    mixotrophic mode
        carbon 160
        factors affecting 162
        inorganic carbon 161–162
        microalgae mixed culture *161*
        nutrients removal **161**
        organic carbon 162
    phosphorus removal 156
    uptake mechanism, nutrients *157*
microalgae-based bioremediations *see* bioremediations
microbial cell factories 33
microbial coculture systems, bioproduction
    artificial consortium system *40*
    leguminous plants 33
    natural photosynthetic 34 (*see also* natural photosynthetic microbial coculture systems)
    photoheterotrophic coculture systems, applications **35**

microbial coculture systems, bioproduction *(Continued)*
    photosynthetic cell factories 33
    photosynthetic/heterotrophic species 47
    symbiosis 33
    synthetic biology
        cyanobacteria chassis, modifying 38–39
        heterotrophy chassis 39–40
        natural biological systems 37
        site-directed elimination, targeted genes 38
monogalactosyldiacylglycerol (MGDG) 203
mycosporine-like amino acids (MAAs) 104, 118

natural photosynthetic microbial coculture systems
    applications **35**
    biodegradation 37
    environmental engineering 34
    soil remediation 36
    surface effluent treatment 35–36
natural symbiosis systems 33, 34, 46
nitrate reductase (NR) promoters 56
non-homologous end joining (NHEJ) 55
N-terminal transport peptide 202
nuclear magnetic resonance (NMR) 132
nucleoside diphosphate (NDP) sugars 105
nucleoside monophosphate (NMP) sugars 105

open-pond cultivation method 186
open-raceway pond cultivation system 187, *188*
organic pollutants 172
orthogonal partial least squares discriminant analysis (OPLS-DA) 134

perfusion culture systems 97
Phormidium treleasei Gomont
    biofilms/microbial mats, hot springs
        anaerobic cyanobacteria 146
        cyanobacteria 146, *148*
        EPSs 148
        gas bubbles *147*
        Leptolyngbya biotechnology applications 148–151
        microbes 147
        micro-ecosystems 147
        noble spring channel *149*
        tufa formations *149*
    display springs 145–146
    polyphasic approach 143–145
    taxonomic classifications, Cyanobacteria
        botanical code 140
        history, *P. treleasei* 141–143, **142**
        Leptolyngbya, history 141
        microbiologists 141
photoautotrophic species 33
photobioreactor (PBR) systems 91

photoheterotrophic coculture systems
    omics analysis
        cell metabolism, aspects 44–46
        tools 42–44
    value-added chemicals
        biological hydrogen production 41–42
        lipid production 40–41
        polysaccharide production 41
phycobiliprotein 117, 164
pilot-scale photobioreactors (PBRs) 24
plasmid expression systems 55
poly(3-hydroxybutyrate) (PHB) 150
polyhydroxyalkanoates (PHAs) 40, 150
prokaryotic or prokaryotic-derived genomes 16
proteomics 43
protospacer adjacent motif (PAM) 55
pulsed feeding strategy 95–96

quadrupole-time-of-flight mass spectrometry (Q-TOF) 132

reactive oxygen species (ROS) 34, 172
repeat variable di-residue (RVD) 54
RNA interference (RNAi) 69

Saccharomyces cerevisiae PDAT (ScPDAT) 203
signal transduction 44–45
Simian virus 40 (SV40) 8
single-cell microinjection 66–67
single guide RNA (sgRNA) 55
*Streptomyces hygroscopicus aph*7 gene 8
succinate dehydrogenase (SDH) 78
synthetic coculture systems 33

thylakoid membrane architecture
    far-red light photoacclimation 83–84, *84*
    high light (HL) acclimation 83
    marine ecological niches, *Prochlorococcus* 82–83
    photosynthetic apparatus *81*
    photosynthetic complexes 80
    PSI–NDH-1 supercomplex, structural model *82*
    red algae 84–85, *85*
    Synechococcus/Synechocystis/Thermosynechococcus 80–82
trans-activating crRNA (tracrRNA) 55
transcription activator-like effector (TALE) 54
transcription activator-like effector nucleases (TALEN) 19, 54
transcriptomics 43
transfer DNA (T-DNA) 18
triacylglycerol (TAG) 203
type-I NADH dehydrogenase-like complex (NDH-1) 78

zinc-finger nuclease (ZFN) 18